Plasmonics and Its Applications

Plasmonics and Its Applications

Special Issue Editor
Grégory Barbillon

MDPI • Basel • Beijing • Wuhan • Barcelona • Belgrade

MDPI

Special Issue Editor
Grégory Barbillon
EPF-Ecole d'Ingénieurs
Sceaux, France

Editorial Office
MDPI
St. Alban-Anlage 66
4052 Basel, Switzerland

This is a reprint of articles from the Special Issue published online in the open access journal *Materials* (ISSN 1996-1944) from 2018 to 2019 (available at: https://www.mdpi.com/journal/materials/ special_issues/plasmonics_applications).

For citation purposes, cite each article independently as indicated on the article page online and as indicated below:

LastName, A.A.; LastName, B.B.; LastName, C.C. Article Title. *Journal Name* **Year**, *Article Number*, Page Range.

ISBN 978-3-03897-914-2 (Pbk)
ISBN 978-3-03897-915-9 (PDF)

Contents

About the Special Issue Editor

Grégory Barbillon completed his PhD in Physics (2007) with greatest distinction at the University of Technology of Troyes (France). He then obtained his Habilitation (HDR) in Physics (2013) at the University of Paris Sud (Orsay, France). He has been a Professor of Physics at the Faculty of Engineering "EPF-Ecole d'Ingénieurs" (Sceaux, France) since his appointment in September 2017. His research interests are focused on plasmonics, nano-optics, nonlinear optics, biosensing, optical sensing, condensed matter physics, nanophotonics, nanotechnology, surface-enhanced spectroscopies, sum frequency generation spectroscopy, materials chemistry, physical chemistry, and fluorescence.

materials

MDPI

Editorial
Plasmonics and its Applications

Grégory Barbillon(ORCID)

EPF-Ecole d'Ingénieurs, 3 bis rue Lakanal, 92330 Sceaux, France; gregory.barbillon@epf.fr

Received: 1 May 2019; Accepted: 4 May 2019; Published: 8 May 2019

Abstract: Plasmonics is a quickly developing subject that combines fundamental research and applications ranging from areas such as physics to engineering, chemistry, biology, medicine, food sciences, and the environmental sciences. Plasmonics appeared in the 1950s with the discovery of surface plasmon polaritons. Then, plasmonics went through a novel impulse in mid-1970s when the surface-enhanced Raman scattering was discovered. Nevertheless, it is in this last decade that a very significant explosion of plasmonics and its applications has occurred. Thus, this special issue reports a snapshot of current advances in these various areas of plasmonics and its applications presented in the format of several articles and reviews written by worldwide researchers of this topic.

Keywords: plasmonics; sensing; surface-enhanced Raman scattering; sum-frequency generation; third harmonic generation; surface-enhanced fluorescence; metasurfaces; catalysis; lanthanum hexaboride; solar cell

1. Introduction

Plasmonics (or nanoplasmonics) is a young topic of research, which is part of nanophotonics and nano-optics. Plasmonics concerns to the investigation of electron oscillations in metallic nanostructures and nanoparticles (NPs). Surface plasmons have optical properties, which are very interesting. For instance, surface plasmons have the unique capacity to confine light at the nanoscale [1–3]. Moreover, surface plasmons are very sensitive to the surrounding medium and the properties of the materials on which they propagate. In addition to the above, the surface plasmon resonances can be controlled by adjusting the size, shape, periodicity, and materials nature. Indeed, the technological progress allows researchers to produce new plasmonic systems by controlling all the parameters described previously [4–14]. Moreover, theoretical, computational, and numerical simulation tools have been developed in this last decade, allowing for a better understanding of the optical properties of plasmonic systems [1]. Thus, all these optical properties of plasmonic systems can enable a great number of applications, such as biosensors [15–20], optical devices [21–24], and photovoltaic devices [25–28]

2. Synopsis

This special issue is composed of five review articles, five research articles, and two communications. The first part of the latter is devoted to the applications of plasmonics to physics and engineering [29–33]. Concerning the applications to physics, such as non-linear optics, Mattox et al. demonstrated the control of plasmonic properties of LaB_6 via Eu-doping on a spectral range from near-infrared to infrared [29]. Then, Mattox et al. presented a review on the tuning of the plasmonic resonance of lanthanum hexaboride for a potential application to solar heat absorption [30]. Besides, Ogata et al. investigated the effect of the plasmonic resonance of metallic nanostructures on the optical third harmonic generation (THG) enhancement of nickel nanostructure-covered microcubes [31]. For the application to photovoltaics, Hajjiah et al. presented a simulation study of the efficiency enhancement of peroskite solar cells by using plasmonic nanoparticles [32]. To finish this first part dedicated to physics with the application to metasurfaces, Li et al. proposed a novel computational method in order to optimize the coupling of the electric fields of a metasurface consisting of nanorod plasmonic antennas. This novel computational method is based on the coupling of the decomposition into several orders [33].

In the second and last part, the discussed topics are devoted to chemistry and sensing, such as surface-enhanced fluorescence, surface-enhanced Raman scattering (SERS), sum-frequency generation (SFG) spectroscopy, and electrocatalysis by using plasmonics [34–40]. Concerning the surface-enhanced fluorescence, Lu et al. numerically demonstrated a high enhancement effect of the fluorescence signal obtained with a hybrid metal-dielectric nano-aperture antenna consisting of silicon and gold layers [34]. Besides, for the SERS topic, Magno et al. showed excellent analytical enhancement factors of the SERS signal obtained with hybrid Al/Si nanopillars for the detection of thiophenol molecules. These hybrid Al/Si nanopillars have been realized with a simple and quick fabrication technique [35]. Moreover, Sarychev et al. presented a review on the light concentration by metal-dielectric micro/nano-resonators for efficient SERS sensing. In this review, the recent advances in this topic of metal-dielectric micro/nano-resonators for SERS are exposed [36]. Furthermore, D'Orlando et al. showed the feasibility to carry out and control nanostructures of gold nanoparticles, which can be seen as plasmonic molecules whose optical resonances are tuned by modifying the shape, symmetry, and interparticle distances with an AFM (Atomic Force Microscope) device coupled with an optical spectrometer [37]. To complete the sensing part, Han et al. presented a short review on plasmonic biosensing based on the design of nanovoids in thin films by reviewing resonance modes, materials, and hybrid functions using simultaneously electrical conductivity [38]. In addition, Humbert et al. presented a review on the sum-frequency generation (SFG) spectroscopy of plasmonic nanomaterials. In this review, the authors introduced the fundamentals of SFG spectroscopy. Then, they presented an overview of studies of plasmonic nanomaterials by this SFG spectroscopy over the last five years [39]. To conclude this part, as well as this special issue dedicated to *"Plasmonics and its Applications"*, Subramanian et al. presented a review on the electrocatalysis induced by plasmon with multi-component nanostructures. Indeed, the authors highlight the recent progress obtained in the synthesis of these multi-component nanostructures, especially for the plasmonic electrocatalysis of major fuel-forming and fuel cell reactions [40].

3. Conclusions

In making this special issue on plasmonics and its applications, I had the pleasure of obtaining contributions from high-quality authors worldwide, and I thank them for that. To conclude, I hope that this special issue dedicated to plasmonics and its applications will be read with interest by the students or researchers who wish to be involved in this topic or to gain an advanced understanding of it.

Funding: This research received no external funding.

Conflicts of Interest: The authors declare no conflict of interest.

References

1. Shahbazyan, T.V.; Stockman, M.I. *Plasmonics: Theory and Applications*; Springer: Dordrecht, The Netherlands, 2013; pp. 1–577.
2. Maier, S.A. *Plasmonics: Fundamentals and Applications*; Springer: New York, NY, USA, 2007; pp. 3–220.
3. Barbillon, G. *Nanoplasmonics-Fundamentals and Applications*; InTech: Rijeka, Croatia, 2017; pp. 3–481.
4. Barbillon, G.; Faure, A.-C.; El Kork, N.; Moretti, P.; Roux, S.; Tillement, O.; Ou, M. G.; Descamps, A.; Perriat, P.; Vial, A.; et al. How nanoparticles encapsulating fluorophores allow a double detection of biomolecules by localized surface plasmon resonance and luminescence. *Nanotechnology* **2008**, *19*, 035705. [CrossRef] [PubMed]
5. Barbillon, G.; Bijeon, J.-L.; Lérondel, G.; Plain, J.; Royer, P. Detection of chemical molecules with integrated plasmonic glass nanotips. *Surf. Sci.* **2008**, *602*, L119–L122. [CrossRef]

6. Faure, A.-C.; Barbillon, G.; Ou, M.; Ledoux, G.; Tillement, O.; Roux, S.; Fabregue, D.; Descamps, A.; Bijeon, J.-L.; Marquette, C.A.; et al. Core/shell nanoparticles for multiple biological detection with enhanced sensitivity and kinetics. *Nanotechnology* **2008**, *19*, 485103. [CrossRef] [PubMed]

7. Bryche, J.-F.; Gillibert, R.; Barbillon, G.; Sarkar, M.; Coutrot, A.-L.; Hamouda, F.; Aassime, A.; Moreau, J.; Lamy de la Chapelle, M.; Bartenlian, B.; et al. Density effect of gold nanodisks on the SERS intensity for a highly sensitive detection of chemical molecules. *J. Mater. Sci.* **2015**, *50*, 6601–6607. [CrossRef]

8. Bryche, J.-F.; Gillibert, R.; Barbillon, G.; Gogol, P.; Moreau, J.; Lamy de la Chapelle, M.; Bartenlian, B.; Canva, M. Plasmonic enhancement by a continuous gold underlayer: application to SERS sensing. *Plasmonics* **2016**, *11*, 601–608. [CrossRef]

9. Bryche, J.-F.; Tsigara, A.; Bélier, B.; Lamy de la Chapelle, M.; Canva, M.; Bartenlian, B.; Barbillon, G. Surface enhanced Raman scattering improvement of gold triangular nanoprisms by a gold reflective underlayer for chemical sensing. *Sens. Actuator B* **2016**, *228*, 31–35. [CrossRef]

10. Masson, J.-F.; Gibson, K.F.; Provencher-Girard, A. Surface-enhanced Raman spectroscopy amplification with film over etched nanospheres. *J. Phys. Chem. C* **2010**, *114*, 22406–22412. [CrossRef]

11. Lee, Y.; Lee, J.; Lee, T.K.; Park, J.; Ha, M.; Kwak, S.K.; Ko, H. Particle-on-Film Gap Plasmons on Antireflective ZnO Nanocone Arrays for Molecular-Level Surface-Enhanced Raman Scattering Sensors. *ACS Appl. Mater. Interfaces* **2015**, *7*, 26421–26429. [CrossRef] [PubMed]

12. Bryche, J.-F.; Bélier, B.; Bartenlian, B.; Barbillon, G. Low-cost SERS substrates composed of hybrid nanoskittles for a highly sensitive sensing of chemical molecules. *Sens. Actuator B* **2017**, *239*, 795–799. [CrossRef]

13. Magno, G.; Bélier, B.; Barbillon, G. Gold thickness impact on the enhancement of SERS detection in low-cost Au/Si nanosensors. *J. Mater. Sci.* **2017**, *52*, 13650–13656. [CrossRef]

14. Bryche, J.-F.; Barbillon, G.; Bartenlian, B.; Dujardin, G.; Boer-Duchemin, E.; Le Moal, E. k-space optical microscopy of nanoparticle arrays: Opportunities and artifacts. *J. Appl. Phys.* **2018**, *124*, 043102. [CrossRef]

15. Barbillon, G.; Noblet, T.; Busson, B.; Tadjeddine, A.; Humbert, C. Localised detection of thiophenol with gold nanotriangles highly structured as honeycombs by nonlinear sum frequency generation spectroscopy. *J. Mater. Sci.* **2018**, *53*, 4554–4562. [CrossRef]

16. Dolci, M.; Bryche, J.-F.; Leuvrey, C.; Zafeiratos, S.; Gree, S.; Begin-Colin, S.; Barbillon, G.; Pichon, B.P. Robust clicked assembly based on iron oxide nanoparticles for a new type of SPR biosensor. *J. Mater. Chem. C* **2018**, *6*, 9102–9110. [CrossRef]

17. Pichon, B.P.; Barbillon, G.; Marie, P.; Pauly, M.; Begin-Colin, S. Iron oxide magnetic nanoparticles used as probing agents to study the nanostructure of mixed self-assembled monolayers. *Nanoscale* **2011**, *3*, 4696–4705. [CrossRef] [PubMed]

18. Barbillon, G. Fabrication and SERS Performances of Metal/Si and Metal/ZnO Nanosensors: A Review. *Coatings* **2019**, *9*, 86. [CrossRef]

19. He, Y.; Su, S.; Xu, T.T.; Zhong, Y.L.; Zapien, J.A.; Li, J.; Fan, C.H.; Lee, S.T. Silicon nanowires-based highly-efficient SERS-active platform for ultrasensitive DNA detection. *Nano Today* **2011**, *6*, 122–130. [CrossRef]

20. Huang, J.-A.; Zhao, Y.-Q; Zhang, X.-J.; He, L.-F.; Wong, T.-L.; Chui, Y.-S.; Zhang W.-J.; Lee, S.-T. Ordered Ag/Si Nanowires Array: Wide-Range Surface-Enhanced Raman Spectroscopy for Reproducible Biomolecule Detection. *Nano Lett.* **2013**, *13*, 5039–5045. [CrossRef]

21. Salamin, Y.; Ma, P.; Baeuerle, B.; Emboras, A.; Fedoryshyn, Y.; Heni, W.; Cheng, B.; Josten, A.; Leuthold, J. 100 GHz Plasmonic Photodetector. *ACS Photonics* **2018**, *5*, 3291–3297. [CrossRef]

22. Thomaschewski, M.; Yang, Y.Q.; Bozhevolnyi, S.I. Ultra-compact branchless plasmonic interferometers. *Nanoscale* **2018**, *10*, 16178–16183. [CrossRef] [PubMed]

23. Ayata, M.; Fedoryshyn, Y.; Heni, W.; Baeuerle, B.; Josten, A.; Zahner, M.; Koch, U.; Salamin, Y.; Hoessbacher, C.; Haffner, C.; et al. High-speed plasmonic modulator in a single metal layer. *Science* **2017**, *358*, 630–632. [CrossRef] [PubMed]

24. Haffner, C.; Heni, W.; Fedoryshyn, Y.; Niegemann, J.; Melikyan, A.; Elder, D.L.; Baeuerle, B.; Salamin, Y.; Josten, A.; Koch, U.; et al. All-plasmonic Mach-Zehnder modulator enabling optical high-speed communication at the microscale. *Nat. Photonics* **2015**, *9*, 525–528. [CrossRef]

25. Shao, W.J.; Liang, Z.Q.; Guan, T.F.; Chen, J.M.; Wang, Z.F.; Wu, H.H.; Zheng, J.Z.; Abdulhalim, I.; Jiang, L. One-step integration of a multiple-morphology gold nanoparticle array on a TiO$_2$ film via a facile sonochemical method for highly efficient organic photovoltaics. *J. Mater. Chem. A* **2018**, *6*, 8419–8429. [CrossRef]

26. Vangelidis, I.; Theodosi, A.; Beliatis, M.J.; Gandhi, K.K.; Laskarakis, A.; Patsalas, P.; Logothetidis, S.; Silva, S.R.P.; Lidorikis, E. Plasmonic Organic Photovoltaics: Unraveling Plasmonic Enhancement for Realistic Cell Geometries. *ACS Photonics* **2018**, *5*, 1440–1452. [CrossRef]

27. Li, M.Z.; Guler, U.; Li, Y.A.; Rea, A.; Tanyi, E.K.; Kim, Y.; Noginov, M.A.; Song, Y.L.; Boltasseva, A.; Shalaev, V.M.; et al. Plasmonic Biomimetic Nanocomposite with Spontaneous Subwavelength Structuring as Broadband Absorbers. *ACS Energy Lett.* **2018**, *3*, 1578–1583. [CrossRef]

28. Chen, X.; Fang, J.; Zhang, X.D.; Zhao, Y.; Gu, M. Optical/Electrical Integrated Design of Core-Shell Aluminum-Based Plasmonic Nanostructures for Record-Breaking Efficiency Enhancements in Photovoltaic Devices. *ACS Photonics* **2017**, *4*, 2102–2110. [CrossRef]

29. Mattox, T.M.; Coffman, D.K.; Roh, I.; Sims, C.; Urban, J.J. Moving the Plasmon of LaB$_6$ from IR to Near-IR via Eu-Doping. *Materials* **2018**, *11*, 226. [CrossRef] [PubMed]

30. Mattox, T.M.; Urban, J.J. Tuning the Surface Plasmon Resonance of Lanthanum Hexaboride to Absorb Solar Heat: A Review. *Materials* **2018**, *11*, 2473. [CrossRef]

31. Ogata, Y.; Vorobyev, A.; Guo, C. Optical Third Harmonic Generation Using Nickel Nanostructure-Covered Microcube Structures. *Materials* **2018**, *11*, 501. [CrossRef]

32. Hajjiah, A.; Kandas, I.; Shehata, N. Efficiency Enhancement of Perovskite Solar Cells with Plasmonic Nanoparticles: A Simulation Study. *Materials* **2018**, *11*, 1626. [CrossRef]

33. Li, Y.; Hong, M. Diffractive Efficiency Optimization in Metasurface Design via Electromagnetic Coupling Compensation. *Materials* **2019**, *12*, 1005. [CrossRef] [PubMed]

34. Lu, G.; Xu, J.; Wen, T.; Zhang, W.; Zhao, J.; Hu, A.; Barbillon, G.; Gong, Q. Hybrid Metal-Dielectric Nano-Aperture Antenna for Surface Enhanced Fluorescence. *Materials* **2018**, *11*, 1435. [CrossRef]

35. Magno, G.; Bélier, B.; Barbillon, G. Al/Si Nanopillars as Very Sensitive SERS Substrates. *Materials* **2018**, *11*, 1534. [CrossRef] [PubMed]

36. Sarychev, A.K.; Ivanov, A.; Lagarkov, A.; Barbillon, G. Light Concentration by Metal-Dielectric Micro-Resonators for SERS Sensing. *Materials* **2019**, *12*, 103. [CrossRef] [PubMed]

37. D'Orlando, A.; Bayle, M.; Louarn, G.; Humbert, B. AFM-Nano Manipulation of Plasmonic Molecules Used as "Nano-Lens" to Enhance Raman of Individual Nano-Objects. *Materials* **2019**, *12*, 1372. [CrossRef] [PubMed]

38. Han, X.; Liu, K.; Sun, C.-S. Plasmonics for Biosensing. *Materials* **2019**, *12*, 1411. [CrossRef] [PubMed]

39. Humbert, C.; Noblet, T.; Dalstein, L.; Busson, B.; Barbillon, G. Sum-Frequency Generation Spectroscopy of Plasmonic Nanomaterials: A Review. *Materials* **2019**, *12*, 836. [CrossRef]

40. Subramanian, P.; Meziane, D.; Wojcieszak, R.; Dumeignil, F.; Boukherroub, R.; Szunerits, S. Plasmon-Induced Electrocatalysis with Multi-Component Nanostructures. *Materials* **2019**, *12*, 43. [CrossRef] [PubMed]

materials

MDPI

Communication

Moving the Plasmon of LaB$_6$ from IR to Near-IR via Eu-Doping

Tracy M. Mattox *[ID], D. Keith Coffman, Inwhan Roh, Christopher Sims and Jeffrey J. Urban *

Molecular Foundry, Lawrence Berkeley National Laboratory, One Cyclotron Rd., Berkeley, CA 94720, USA; dcoffm5261@gatech.edu (D.K.C.); noinhwan@gmail.com (I.R.); christophersims2017@u.northwestern.edu (C.S.)
* Correspondence: tmmattox@lbl.gov (T.M.M.); jjurban@lbl.gov (J.J.U.)

Received: 9 January 2018; Accepted: 26 January 2018; Published: 1 February 2018

Abstract: Lanthanum hexaboride (LaB$_6$) has become a material of intense interest in recent years due to its low work function, thermal stability and intriguing optical properties. LaB$_6$ is also a semiconductor plasmonic material with the ability to support strong plasmon modes. Some of these modes uniquely stretch into the infrared, allowing the material to absorb around 1000 nm, which is of great interest to the window industry. It is well known that the plasmon of LaB$_6$ can be tuned by controlling particle size and shape. In this work, we explore the options available to further tune the optical properties by describing how metal vacancies and Eu doping concentrations are additional knobs for tuning the absorbance from the near-IR to far-IR in La$_{1-x}$Eu$_x$B$_6$ (x = 0, 0.2, 0.5, 0.8, and 1.0). We also report that there is a direct correlation between Eu concentration and metal vacancies within the Eu$_{1-x}$La$_x$B$_6$.

Keywords: plasmon; hexaboride; doping; lanthanum hexaboride; LaB$_6$

1. Introduction

Plasmonic nanoparticles are well known for their intriguing properties [1], and are being explored in a variety of fields such as photovoltaics [2], nanosensors [3], drug delivery devices [4], and quantum optics [5]. The physical properties of plasmonic materials are typically easy to tune because of their high carrier concentration and small size, where seemingly minor adjustments such as altering the particle shape or size have a substantial influence on the absorbance spectrum [1]. Vacancies also play a large role in tuning the optical properties of such materials, having a significant influence on free carrier density and doping constraints [6,7]. It's even possible to fully tune the plasmon independent of dopant concentration in core-shell indium-tin-oxide nanoparticles [8] and by reducing holes in the valence band in copper sulfide [9].

Plasmonic materials are highly sought after in the windows industry. The ability to design a material to selectively transmit in the visible region while absorbing the most intense radiative heat in the IR (about 750 nm–1250 nm) is important for smarter window design, especially in hot climates. [10–12] Metal hexaborides (MB$_6$) are being sought after for these applications, and with lanthanum hexaboride (LaB$_6$) absorbing in the middle of this range (1000 nm) [13,14] we focus our efforts here on the tuning of LaB$_6$. It has already been shown that changing the particle size of LaB$_6$ nanoparticles offers a means of controlling the plasmon [15,16] and that these particles may be incorporated into polymers to make films [17,18]. Though some work has been done on LaB$_6$ to study how La vacancies influence vibrational energies [19] and how doping impacts the thermionic power [20,21], there is a potential link between doping content and vacancies in LaB$_6$ that has gone unexplored. Given the ability of doping levels and metal vacancies to alter free electron concentrations and thus the optical properties in Eu$_{1-x}$La$_x$B$_6$, we wished to explore the possible connection between doping concentration and metal vacancies as an additional means of controlling the plasmon.

In this work we demonstrate the possibility of alloying LaB_6 nanoparticles with Eu using, for the first time, a low temperature solid state technique with varying ratios of Eu to La. Interestingly, we report there is a direct correlation between Eu concentration and metal (M) vacancies within the $Eu_{1-x}La_xB_6$ system. Furthermore, this method allows the plasmon to be tuned across an incredibly large absorbance range from 1100 nm to 2050 nm, which may open doors to new optoelectronic applications.

2. Experimental Procedures

Anhydrous lanthanum (III) chloride (99.9% pure, Strem Chemical), anhydrous europium (III) chloride (99.99% pure, Strem Chemical) and sodium borohydride (EMD) were used as received and stored in an argon atmosphere glove box until use. Reactant powders were a stoichiometric 6:1 ratio of $NaBH_4$ to metal chloride, where the metal chloride content was a mixture of $EuCl_3$ and $LaCl_3$ with varying ratios of (Eu:La). The mixtures were transferred to alumina boats approximately two inches long and 1 cm wide and the reactions run in a one-inch diameter quartz tube in a Lindberg tube furnace. The reaction was purged with argon at 200 cc/min for 20 min prior to heating. Gas flow was reduced to 100 cc/min and the furnace heated to 450 °C at a rate of 10 °C/min. The reaction was held at 450 °C for 60 min and then cooled to room temperature under argon. The black solid was cleaned in air using methanol to react excess $NaBH_4$, HCl to convert residual sodium into sodium chloride and, finally, water to remove the sodium chloride. With each washing step, the solution was centrifuged at 10,000 rpm for ten minutes and the solvent removed. Severe aggregation of these ligand-free particles rendered electron-microscopy imaging infeasible. However, diffraction data suggest that the particles were approximately 17 nm, with the Scherrer equation giving calculated sizes of 17.46, 16.84, 17.62 and 17.21 nm, respectively, for x = 0.2, 0.5, 0.8 and 1.0.

Samples were analyzed by powder X-ray diffraction on a D8 Discover diffractometer (Bruker AXS Inc., Madison, WI, USA) operated at 35 kV and 40 mA using CoKα radiation. Samples were prepared for optical measurements by drop casting onto quartz slides. Raman spectra were collected on a LabRAM ARAMIS (HORIBA Jobin Yvon, Edison, NJ, USA) automated scanning confocal Raman microscope using a 532-nm excitation laser. Elemental analysis was performed by EDX spectroscopy on a Gemini Ultra-55 scanning electron microscope (Zeiss, Thornwood, NY, USA), and FTIR spectroscopy was performed on a Spectrum One equipped with an HATR assembly (PerkinElmer, Santa Clara, CA, USA). The absorbance was collected on a Cary-5000 UV-Vis-NIR (Agilent Technologies, Santa Clara, CA, USA). Samples were prepared for optical measurements by drop casting from water onto quartz slides, and the films were allowed to dry naturally in air.

3. Results and Discussion

The success of the incorporation of a Eu into LaB_6 was evident in changes to the XRD pattern of $La_xEu_{1-x}B_6$ (Figure 1A). Note that the small peak at ~33° is from an unidentified impurity in the $EuCl_3$. Increasing the concentration of Eu in the $La_xEu_{1-x}B_6$ synthesis caused a shift of the diffraction pattern to higher 2-Theta (Figure 1B), which is indicative of increased compressive lattice strain. This seems counterintuitive since incorporating larger atoms typically expands a crystal lattice. For instance, in $Eu_{1-x}Ca_xB_6$ the larger Eu atom replaces Ca and the lattice expands [21]. There is a possibility that increasing the amount of Eu in $La_xEu_{1-x}B_6$ may produce two phases, as reported for the $(Ba_xCa_{1-x})B_6$ system which has a mixture of both Ba-rich and Ca-rich particles in the final product [22]. Though this could account for the unexpected change to the lattice strain in our system, the diffraction peaks of $La_xEu_{1-x}B_6$ are symmetric, which is indicative of a single phase (Figure 1C). In $La_xEu_{1-x}B_6$, there appears to be a decrease in lattice spacing with increasing Eu content (Figure 1D), even though Eu is larger than La. The B_6 network, like all boron lattices, is electron-deficient and is only stable because of electron transfer from the metals [23]. Though Eu^{2+} and Ca^{2+} in $Eu_{1-x}Ca_xB_6$ are different sizes they are also both divalent, so the free electron density does not change when increasing the Ca content. By contrast, in $La_xEu_{1-x}B_6$ there is a mix of trivalent La^{3+} and divalent Eu^{2+}. This and the metal (M) vacancies within the system are likely responsible for the increasing lattice strain with increasing Eu concentration.

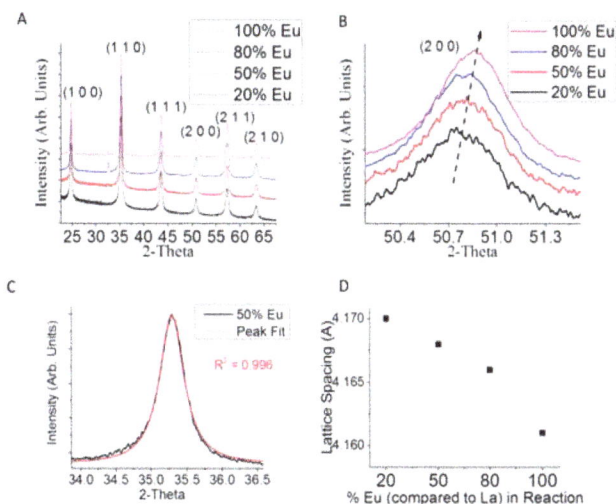

Figure 1. X-ray diffraction of (**A**) $La_xEu_{1-x}B_6$; (**B**) a magnified image of the (2 0 0) diffraction plane with $La_xEu_{1-x}B_6$ where x = 0.0, 0.2, 0.5 and 0.8; (**C**) Pearson VII peak fit of $La_xEu_{1-x}B_6$ with x = 0.5; and (**D**) lattice spacing versus atomic % Eu in the $La_xEu_{1-x}B_6$ reaction (calculated using Bragg's law).

EDS confirmed the presence of all three elements (La, B, and Eu) in $La_xEu_{1-x}B_6$ samples (Figure 2A; $La_xEu_{1-x}B_6$ with x = 0.2). Intriguingly, LaB_6 synthesized under this method contained about 97% B, which indicates a huge amount of M vacancies with x = 0.19 (equivalent to about 80% M vacancies). M vacancies are common in LaB_6, but it is understood that the lattice constant is unaffected by these voids [19,24,25]. he stability of the crystal structure is dictated by the bonds in the boron framework and not by the metal content so long as the electronic requirements of the structure are met [26]. However, if there is too much void space then MB_6 becomes unstable. Though there is a lot of disagreement surrounding La-B phase diagrams, a B content above 90% [24,27] is expected to contain both LaB_6 and an additional B phase [25,26,28,29], which suggests that any excess boron in our system may not lie within the MB_6 structure. However, we see no indication of a separate B phase beyond $La_xEu_{1-x}B_6$ by XRD. The phase diagrams of La-B were developed under the assumption that high temperatures ($\geq$1500 °C) are required to make LaB_6, which was disproved only recently [15,30]. With low temperature reactions we recently reported the existence of bridging halogens between La atoms which are involved in the lattice formation of LaB_6 [31–33], so even though a sample containing 97% B may potentially have a massive amount of vacancies, it's possible that the structure was stable during formation because these halogens fulfilled the electronic requirements necessary to stabilize the material without the need of an additional B phase. Unfortunately, the amount of Cl in the materials reported here were either too low in concentration to be detected by EDS or the 450 °C reaction temperature was high enough to remove the bridging-Cl atoms as the final product formed. Work is ongoing understand exactly how halogen atoms enter into the reaction mechanism.

As the concentration of Eu in the $La_xEu_{1-x}B_6$ reaction is increased there is a clear trend of increasing amounts of B relative to M until the system becomes stoichiometric with EuB_6 (86% B or x = 1; Figure 2B), with a slightly higher Eu content in $La_xEu_{1-x}B_6$ than was expected with x < 1 (Figure 2C). It's possible that EuB_6 is more energetically favored than LaB_6 or that there are so many vacancies that at low concentrations the divalent Eu^{2+} has an easier time filling holes in addition to replacing La atoms. Regardless, there is a clear trend of decreasing vacancies with increasing Eu in the reaction (Figure 2D). Unfortunately, the ligand-free nature of these particles results in an aggregated product, rendering single-particle analysis on individual LaB_6 particles infeasible.

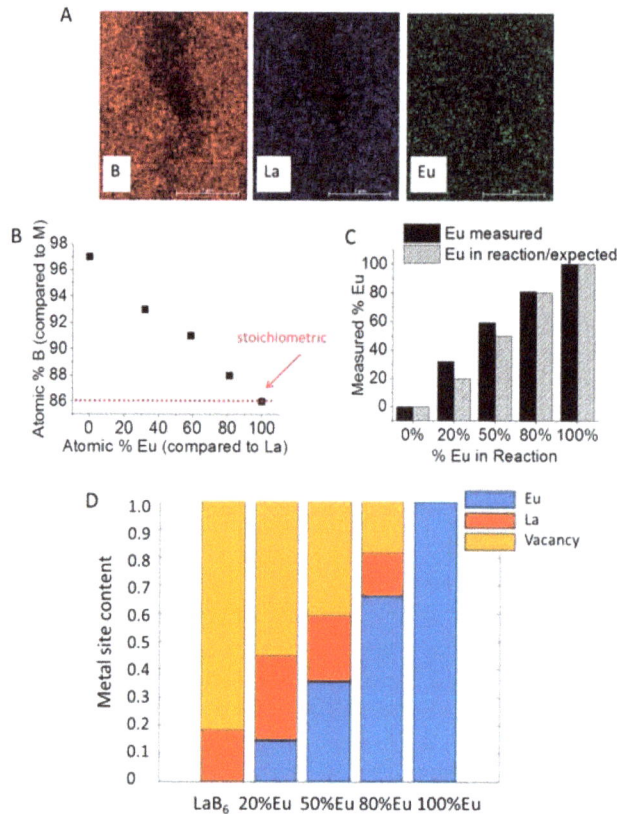

Figure 2. (**A**) View of the EDS map of $La_xEu_{1-x}B_6$ (x = 0.2) including B, La and Eu; (**B**) atomic % B versus atomic % Eu (the red dashed line is stoichiometric with 1M:6B); (**C**) measured versus expected % Eu (comparing Eu to La) in $La_xEu_{1-x}B_6$; and (**D**) metal content (Eu and La) and M void in $La_xEu_{1-x}B_6$.

There have been several publications studying the ability to tune the plasmon of LaB_6 to achieve desired optical properties [13–15], but little is yet known about how vacancies influence these properties. Research discussing vacancies related to optical and vibrational properties in LaB_6 are very recent [19,31], and though much has been done to study the magnetic and thermoelectric properties of $La_xEu_{1-x}B_6$ [34,35], no one until now has synthesized doped hexaborides at low temperatures. Furthermore, only in very recent years have the optical properties of doped MB_6 been explored [36–38]. In this work, we used absorbance spectroscopy to determine how the Eu concentration and M vacancies in $La_xEu_{1-x}B_6$ nanocrystals can be used to tune the plasmonic properties (Figure 3A). When increasing the concentration of Eu the small absorbance peak in the visible region that is indicative of metal hexaborides shifts from ~380 nm in pure LaB_6 to 730 nm in pure EuB_6, while the larger absorbance peak red shifts from 1100 nm in pure LaB_6 to 2050 nm in pure EuB_6 (Figure 3B). Introducing Eu as a dopant causes a constant red shift of the absorbance peak from 1100 nm in pure LaB_6 to 2050 nm in pure EuB_6 (Figure 3B). This shift is a result of changes to the number of electrons in the conduction band as divalent Eu^{2+} replaces trivalent La^{3+}. The sudden broadening of the absorbance at 80% Eu is most likely due to the changing carrier concentration which results from Eu incorporation as well as from changing metal vacancies within the lattice.

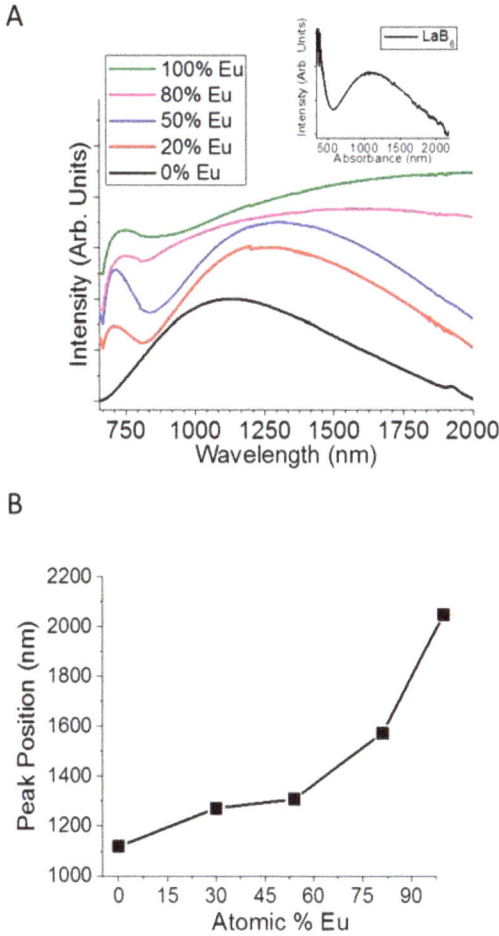

Figure 3. (**A**) Absorbance of $La_xEu_{1-x}B_6$ changing with Eu content (normalized) and (**B**) absorbance peak position versus atomic % Eu in $La_xEu_{1-x}B_6$.

The electron deficiency is calculated as vacancy content minus lanthanum content. Whatever the mechanism causing the change in lattice spacing (vacancies or changing Eu content), the shifting absorption peak is indicative of an increase in carrier density with lanthanum content, and is impacted by vacancies within the system. Equation 1 gives the most basic model for the wavelength of the plasmon resonance [39],

$$\lambda = 2\pi c \sqrt{\frac{\varepsilon_0 m * (\varepsilon\infty + k\varepsilon_m)}{Ne^2}}, \tag{1}$$

where N is the number of charge carriers per unit volume, e is the charge of each carrier, m^* is the effective mass of the charge carriers, ε_0 is the permittivity of free space, ε_m is the dielectric function of the surrounding medium, ε_∞ is the dielectric limit for the material at high frequencies (accounting for bound charge), and k is a geometrical factor. The absorbance spectroscopy was performed in air, so ε_m is nearly unity. We treat the particles as spherical [15,19], which is associated with a constant of $k = 2$ and an effective electron mass of 0.225 m_0 in EuB_6 as reported based on optical measurements [40]. Finally, taking ε_∞ as unity, our absorption peaks translate to the charge concentrations in Figure 4. In short, Figure 4 illustrates qualitative agreement between increasing carrier concentration as inferred from plasmonic resonance and increasing carrier concentration as inferred from composition measurements. As the Eu content is increased the samples lose free electrons and the absorbance peak expands and broadens.

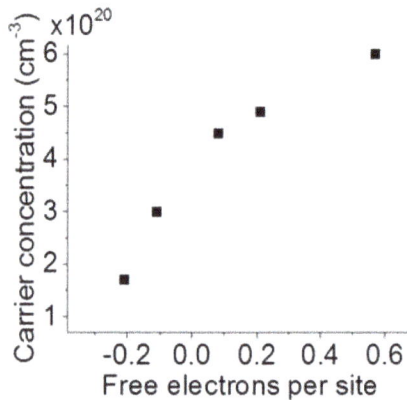

Figure 4. Localized surface plasmon resonance inferred carrier concentration versus number of free electrons per metal site in $La_xEu_{1-x}B_6$.

4. Conclusions

We have found that systematically increasing the amount of divalent Eu^{2+} compared to trivalent La^{3+} within $Eu_{1-x}La_xB_6$ not only decreases the lattice spacing but drastically changes the vacancies within the system. These vacancies have a large influence on the optical properties and allow the plasmon to be tuned across an incredibly large range from 1100 nm to 2050 nm. The true nature of these particles on the nanoscale is not fully understood (i.e., the influence of Cl bridging atoms), but we are making great strides to improve our knowledge of this system. It is our hope that this work will not only help to further our understanding of the MB_6 crystal structure, but may open new doors for developing new devices, optoelectronics, and more. Research is ongoing to study how this synthetic method may be used to alter the nanoparticle surface, bringing to light new properties which may become a vital aspect for biosensing applications.

Acknowledgments: This work was supported by the Molecular Foundry and the Advanced Light Source at Lawrence Berkeley National Laboratory, both user facilities supported by the Office of Science, Office of Basic Energy Sciences, of the U.S. Department of Energy (DOE) under Contract No. DE-AC02-05CH11231. Work was also supported in part by the DOE Office of Science, Office of Workforce Development for Teachers and Scientists (WDTS) under the Science Undergraduate Laboratory Internship (SULI) program.

Author Contributions: T.M.M. and J.J.U. conceived the idea; T.M.M. designed the experiments and wrote the paper; D.K.C., I.R., and C.S. performed experiments and analysis.

Conflicts of Interest: The authors declare no competing financial interests.

References

1. Mattox, T.M.; Ye, X.; Manthiram, K.; Schuck, P.J.; Alivisatos, A.P.; Urban, J.J. Chemical Control of Plasmons in Metal Chalcogenide and Metal Oxide Nanostructures. *Adv. Mater.* **2015**, *27*, 5830–5837. [CrossRef] [PubMed]
2. Stockman, M.I. Nanoplasmonics: The physics behind the applications. *Phys. Today* **2011**, *64*, 39–44. [CrossRef]

3.   Anker, J.N.; Hall, W.P.; Lyandres, O.; Shah, N.C.; Zhao, J.; Duyne, R.P.V. Biosensing with plasmonic nanosensors. *Nat. Mater.* **2008**, *7*, 442. [CrossRef] [PubMed]

4.   Lopatynskyi, A.M.; Lytvyn, V.K.; Mogylnyi, I.V.; Rachkov, O.E.; Soldatkin, O.P.; Chegel, V.I. Smart nanocarriers for drug delivery: Controllable LSPR tuning. *Semicond. Phys. Quantum Electron. Optoelectron.* **2016**, *19*, 358–365. [CrossRef]

5.   Tame, M.S.; McEnery, K.R.; Özdemir, S.K.; Lee, J.; Maier, S.A.; Kim, M.S. Quantum plasmonics. *Nat. Phys.* **2013**, *9*, 329–340. [CrossRef]

6.   Elimelech, O.; Liu, J.; Plonka, A.M.; Frenkel, A.I.; Banin, U. Size Dependence of Doping by a Vacancy Formation Reaction in Copper Sulfide Nanocrystals. *Angew. Chem.* **2017**, *129*, 10471–10476. [CrossRef]

7.   Luther, J.M.; Jain, P.K.; Ewers, T.; Alivisatos, A.P. Localized surface plasmon resonances arising from free carriers in doped quantum dots. *Nat. Mater.* **2011**, *10*, 361. [CrossRef] [PubMed]

8.   Crockett, B.M.; Jansons, A.W.; Koskela, K.M.; Johnson, D.W.; Hutchison, J.E. Radial Dopant Placement for Tuning Plasmonic Properties in Metal Oxide Nanocrystals. *ACS Nano* **2017**, *11*, 7719–7728. [CrossRef] [PubMed]

9.   Kalanur, S.S.; Seo, H. Tuning plasmonic properties of CuS thin films via valence band filling. *RSC Adv.* **2017**, *7*, 11118–11122. [CrossRef]

10. Granqvist, C.G. Electrochromics for smart windows: Oxide-based thin films and devices. *Thin Solid Films* **2014**, *564*, 1–38. [CrossRef]

11. Runnerstrom, E.L.; Llordés, A.; Lounis, S.D.; Milliron, D.J. Nanostructured electrochromic smart windows: Traditional materials and NIR-selective plasmonic nanocrystals. *Chem. Commun.* **2014**, *50*, 10555–10572. [CrossRef] [PubMed]

12. Zhou, Y.; Huang, A.; Li, Y.; Ji, S.; Gao, Y.; Jin, P. Surface plasmon resonance induced excellent solar control for VO$_2$ @SiO$_2$ nanorods -based thermochromic foils. *Nanoscale* **2013**, *5*, 9208–9213. [CrossRef] [PubMed]

13. Adachi, K.; Miratsu, M.; Asahi, T. Absorption and scattering of near-infrared light by dispersed lanthanum hexaboride nanoparticles for solar control filters. *J. Mater. Res.* **2010**, *25*, 510–521. [CrossRef]

14. Takeda, H.; Kuno, H.; Adachi, K. Solar Control Dispersions and Coatings with Rare-Earth Hexaboride Nanoparticles. *J. Am. Ceram. Soc.* **2008**, *91*, 2897–2902. [CrossRef]

15. Mattox, T.M.; Agrawal, A.; Milliron, D.J. Low Temperature Synthesis and Surface Plasmon Resonance of Colloidal Lanthanum Hexaboride (LaB6) Nanocrystals. *Chem. Mater.* **2015**, *27*, 6620–6624. [CrossRef]

16. Machida, K.; Adachi, K. Particle shape inhomogeneity and plasmon-band broadening of solar-control LaB6 nanoparticles. *J. Appl. Phys.* **2015**, *118*, 013013. [CrossRef]

17. Jiang, F.; Leong, Y.K.; Martyniuk, M.; Keating, A.; Dell, J.M. Dispersion of lanthanum hexaboride nanoparticles in water and in sol-gel silica arrays. In Proceedings of the 2010 Conference on Optoelectronic and Microelectronic Materials and Devices, Canberra, Australia, 12–15 December 2010; pp. 163–164.

18. Schelm, S.; Smith, G.B. Dilute LaB6 nanoparticles in polymer as optimized clear solar control glazing. *Appl. Phys. Lett.* **2003**, *82*, 4346–4348. [CrossRef]

19. Mattox, T.M.; Chockkalingam, S.; Roh, I.; Urban, J.J. Evolution of Vibrational Properties in Lanthanum Hexaboride Nanocrystals. *J. Phys. Chem. C* **2016**, *120*, 5188–5195. [CrossRef]

20. Zhou, S.; Zhang, J.; Liu, D.; Hu, Q.; Huang, Q. The effect of samarium doping on structure and enhanced thermionic emission properties of lanthanum hexaboride fabricated by spark plasma sintering. *Phys. Status Solid A* **2014**, *211*, 555–564. [CrossRef]

21. Jong-Soo Rhyee, B.K.C.; Kim, H.C. Possible adiabatic paloronic hopping in Ca1-xEuxB6 (x = 0.005, 0.01, and 0.05). *Phys. Rev. B* **2005**, *71*, 073104. [CrossRef]

22. Cahill, J.T.; Alberga, M.; Bahena, J.; Pisano, C.; Borja-Urby, R.; Vasquez, V.R.; Edwards, D.; Misture, S.T.; Graeve, O.A. Phase Stability of Mixed-Cation Alkaline-Earth Hexaborides. *Cryst. Growth Des.* **2017**, *17*, 3450–3461. [CrossRef]

23. Etourneau, J. Critical survey of rare-earth borides: Occurrence, crystal chemistry and physical properties. *J. Less-Common Met.* **1985**, *110*, 267–281. [CrossRef]

24. Otani, S.; Honma, S.; Ishizawa, Y. Preparation of LaB6 single crystals by the floating zone method. *J. Alloys Compd.* **1993**, *193*, 286–288. [CrossRef]

25. Korsukova, M.M.; Gurin, V.N. Physicochemical Problems in the Preparation of Defect-free Monocrystals of Lanthanum Hexaboride. *Russ. Chem. Rev.* **1987**, *56*, 1. [CrossRef]

26. Johnson, R.W.; Daane, A.H. The lanthanum-boron system. *J. Phys. Chem.* **1961**, *65*, 909–915. [CrossRef]

27. Otani, S.; Nakagawa, H.; Nishi, Y.; Kieda, N. Floating Zone Growth and High Temperature Hardness of Rare-Earth Hexaboride Crystals: LaB6, CeB6, PrB6, NdB6, and SmB6. *J. Solid State Chem.* **2000**, *154*, 238–241. [CrossRef]

28. Schlesinger, M.E.; Liao, P.K.; Spear, K.E. The B-La (Boron-Lanthanum) System. *J. Phase Equilib.* **1999**, *20*, 73–78. [CrossRef]

29. Lundstrom, T. The homogeneity range of LaB6—An instructive example of phase analytical techniques. *Z. Anorg. Allg. Chem.* **1986**, *540*, 163–168. [CrossRef]

30. Zhang, M.; Wang, X.; Zhang, X.; Wang, P.; Xiong, S.; Shi, L.; Qian, Y. Direct low-temperature synthesis of RB6 (R = Ce, Pr, Nd) nanocubes and nanoparticles. *J. Solid State Chem.* **2009**, *182*, 3098–3104. [CrossRef]

31. Groome, C.; Roh, I.; Mattox, T.M.; Urban, J.J. Effects of size and structural defects on the vibrational perperties of lanthanum hexaboride nanocrystals. *ACS Omega* **2017**, *2*, 2248–2254. [CrossRef]

32. Mattox, T.M.; Croome, G.; Doran, A.; Beavers, C.M.; Urban, J.J. Anion-mediated negative thermal expansion in lanthanum hexaboride. *Solid State Commun.* **2017**. accepted. [CrossRef]

33. Mattox, T.M.; Groome, C.; Doran, A.; Beavers, C.M.; Urban, J.J. Chloride Influence on the Formation of Lanthanum Hexaboride: An In-Situ Diffraction Study. *J. Cryst. Growth* **2018**, in press. [CrossRef]

34. Song, M.; Yang, I.-S.; Seo, C.W.; Cheong, H.; Kim, J.Y.; Cho, B.K. Local symmetry breaking in Eu1−xLaxB6. *J. Magn. Magn. Mater.* **2007**, *310*, 1019–1020. [CrossRef]

35. Zhitomirsky, M.E.; Rice, T.M.; Anisimov, V.I. Ferromagnetism in the hexaborides. *Nature* **1999**, *402*, 251–253. [CrossRef]

36. Chao, L.; Bao, L.; Shi, J.; Wei, W.; Tegus, O.; Zhang, Z. The effect of Sm-doping on optical properties of LaB6 nanoparticles. *J. Alloys Compd.* **2015**, *622*, 618–621. [CrossRef]

37. Chao, L.; Bao, L.; Wei, W.; Tegus, O. Optical properties of Yb-doped LaB6 from first-principles calculation. *Mod. Phys. Lett. B* **2016**, *30*, 1650091. [CrossRef]

38. Li, Q.; Zhao, Y.; Fan, Q.; Han, W. Synthesis of one-dimensional rare earth hexaborides nanostructures and their optical absorption properties. *Ceram. Int.* **2017**, *43*, 10715–10719. [CrossRef]

39. Willets, K.A.; Duyne, R.P.V. Localized Surface Plasmon Resonance Spectroscopy and Sensing. *Annu. Rev. Phys. Chem.* **2007**, *58*, 267–297. [CrossRef] [PubMed]

40. Gurin, V.N.; Korsukova, M.M.; Karin, M.G.; Sidorin, K.K.; Smirnov, I.A.; Shelikh, A.I. Optical Constants of EuB sub (6) and LaB sub (6). *Sov. Phys. Solid State Commun.* **1980**, *22*, 418–421.

materials

MDPI

Review

Tuning the Surface Plasmon Resonance of Lanthanum Hexaboride to Absorb Solar Heat: A Review

Tracy M. Mattox * and Jeffrey J. Urban *

Molecular Foundry, Lawrence Berkeley National Laboratory, Berkeley, CA 94720, USA
* Correspondence: tmmattox@lbl.gov (T.M.M.); jjurban@lbl.gov (J.J.U.); Tel.: +510-495-2649 (T.M.M.);
  +510-486-4526 (J.J.U.)

Received: 5 November 2018; Accepted: 30 November 2018; Published: 5 December 2018

Abstract: While traditional noble metal (Ag, Au, and Cu) nanoparticles are well known for their plasmonic properties, they typically only absorb in the ultraviolet and visible regions. The study of metal hexaborides, lanthanum hexaboride (LaB_6) in particular, expands the available absorbance range of these metals well into the near-infrared. As a result, LaB_6 has become a material of interest for its energy and heat absorption properties, most notably to those trying to absorb solar heat. Given the growing popularity of LaB_6, this review focuses on the advances made in the past decade with respect to controlling the plasmonic properties of LaB_6 nanoparticles. This review discusses the fundamental structure of LaB_6 and explains how decreasing the nanoparticle size changes the atomic vibrations on the surface and thus the plasmonic absorbance band. We explain how doping LaB_6 nanoparticles with lanthanide metals (Y, Sm, and Eu) red-shifts the absorbance band and describe research focusing on the correlation between size dependent and morphological effects on the surface plasmon resonance. This work also describes successes that have been made in dispersing LaB_6 nanoparticles for various optical applications, highlighting the most difficult challenges encountered in this field of study.

Keywords: lanthanum hexaboride; LaB_6; plasmon; nanoparticles; heat absorption

1. Introduction

Traditional plasmonic metals (Ag, Au, and Cu) possess enormous free carrier densities, and when confined on the nanoscale the quantized free electron oscillations result in sharp localized surface plasmon resonance (LSPR) modes. The LSPR properties enhance light-matter interactions, making these materials ideal for a wide variety of electronic and optical applications [1–4]. Furthermore, their sensitivity to small changes within their structure (e.g., size, morphology, atomic vacancies, etc.) makes the properties easy to tune by introducing defects or changing the surface by varying the size or shape.

The ability of plasmonic metals to convert solar light into electricity and chemical energy is well documented, but the absorbance is restricted to the ultraviolet and visible spectrums in traditional plasmonic metals (Figure 1). This leaves the near infrared (NIR) region mostly inaccessible to metal nanoparticles, with the exception of novel engineered geometries of some metals, such as the cases of Au nanowires and shells [5,6]. There is a growing need to find materials with a NIR absorbance for applications such as window coatings to absorb solar heat [7,8]. Researchers attempting to reduce heat entering buildings and automobiles through windows need a visibly transparent material that absorbs the most intense radiative heat from the sun, ideally in the range of 750–1200 nm. While some chalcogenides and metal oxides are potential candidates [9–14], the metal borides are often overlooked.

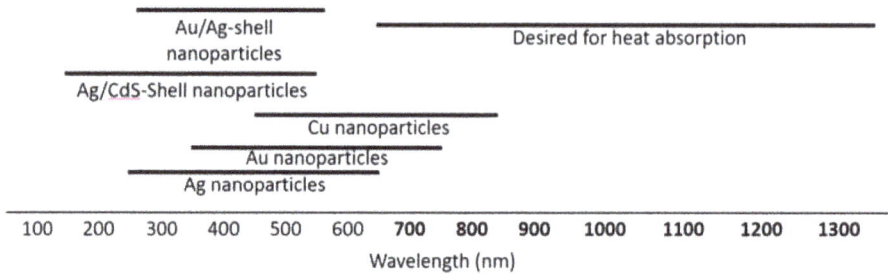

Figure 1. Typical absorbance range of plasmonic metal nanoparticles: Ag [15], Au [16], Cu [17], Ag/CdS-shell [18], Au/Ag-shell [19], and the desired range to absorb solar heat.

Lanthanum hexaboride (LaB_6) is a plasmonic metal with a large free carrier density [20–22] that is best known for its impressive thermionic emission properties and low work function [23–29]. However, LaB_6 also absorbs light very strongly at about 1000 nm, which falls well within the targeted range to absorb solar heat [30,31]. The optical properties of LaB_6 nanoparticles coupled with its incredible hardness [32–36] and high thermal stability [37–43] make it an excellent choice to include in alloys and composites for solar window applications.

The ability to directly synthesize LaB_6 on the nanoscale has only recently become a reality, so research focusing on plasmonic control in LaB_6 is relatively new. That said, LaB_6 does offer the same wide range of methods for optical tuning as other LSPR particles, including controlling the carrier density through vacancies and doping, particle size, morphology, and the surrounding media (i.e. ligands and polymer matrices) [10,44–47]. There is still much to be learned about the optical properties of LaB_6 nanoparticles, but what has been discovered thus far has been quite exciting and has made LaB_6 a very popular material of interest in recent years.

This review focuses on the advances that have been made in the last decade with respect to tuning the plasmon of LaB_6 nanoparticles. Though this field is still relatively new, we feel that the growing demand for optically tunable LaB_6 warrants a comprehensive review to explain the nuances of controlling the plasmon of LaB_6. This work describes how some of the fundamentals behind this intriguing material correspond with the optical properties, and how doping, size, and morphology are contributing factors when attempting to meet requirements of desired plasmonic applications.

2. Relating LaB_6 Fundamentals to Plasmonics

2.1. Crystal Structure of LaB_6

LaB_6 is composed of interconnecting hexaboride clusters with lanthanum (La) atoms residing in the interstitial spaces (Figure 2A) [48]. It is well known that the boron network is responsible for the rigidity of the structure, and consequently the thermal and mechanical stability. Though La is not bound within the system it does provide electrons to stabilize the structure. Interestingly, this rigid material is more relaxed on the particle surface (Figure 2B) [49–51]. On the outermost surface, lanthanum atoms relax closer to the B_6 framework while the B_6 octahedral clusters relax and expand slightly outwards. Though this relaxation is insignificant to the optical properties of bulk sized LaB_6, the surface vibrations of nanoparticles are incredibly important when attempting to control the LSPR.

Researchers are still exploring the surface chemistry of LaB_6, and remain uncertain of the mechanistic details of crystal formation and growth. Until recently, it was assumed that high temperatures (>1500 °C) were required to make LaB_6, most often by reacting lanthanum salt (e.g., $LaCl_3$) with sodium borohydride ($NaBH_4$). Unfortunately, the quick nucleation and growth at such high temperatures prevents the direct synthesis of nano-sized LaB_6. In typical nanoparticle syntheses, it is important to control the rate of nucleation and growth in order to make particles of a uniform size. This is often accomplished by varying temperature, concentration, or including additives (ligands) [52,53]. Unfortunately, the assumed high temperature requirements make small adjustments traditionally used in nanoparticle work impossible, especially the incorporation of organic ligands that cannot withstand the extreme heat. As a result, researchers have had to rely on ball milling techniques to reduce the size of LaB_6, which introduces contaminants by the very nature of the process [54–56]. In recent years, researchers have found that significantly lowering the heat can yield phase-pure LaB_6 nanoparticles using a variety of methods, including tube furnaces [57,58], autoclaves [59,60], and vapor deposition [61,62]. In reducing the reaction temperature, the nucleation and growth rate is significantly reduced, making it possible to observe crystal lattice formation through in-situ diffraction measurements [48]. Researchers have recently discovered that in low temperature reactions the halogen on the lanthanum salt acts as a bridging ligand between La atoms [48,58,63]. When these LaB_6 nanoparticles are heated, the halogen atoms force the lattice to contract until the halogen is removed (Figure 2C). This has important implications to LSPR studies, which are incredibly sensitive to the particle surface.

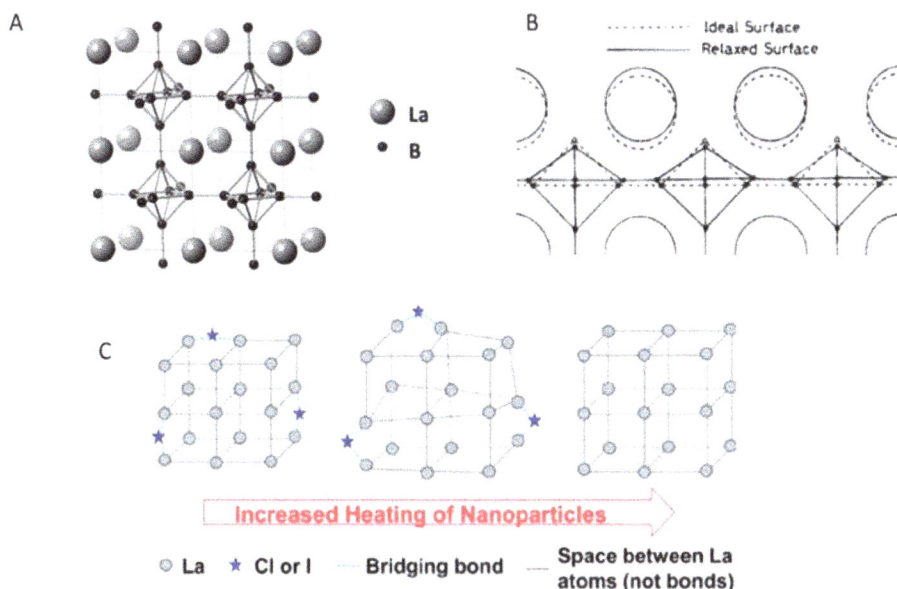

Figure 2. (**A**) Four unit cells of lanthanum hexaboride (LaB_6) with a lattice constant of approximately 4.15 Å (reprinted from [48]). (**B**) Position of La (circle) and B_6 (diamond) in relaxed and ideal states on LaB_6 surface (reprinted from [49]). (**C**) Removal of halogen bridge between La atoms with continued heating of LaB_6 nanoparticle (reprinted from [58]).

2.1.1. Vibrational Structure of LaB_6

Decreasing the particle size of LaB_6 increases the surface area, and in the case of extremely small sizes it becomes possible for the nanoparticles to be composed almost entirely of surface atoms. As a result, minor changes to the surface composition can have a significant impact on the lattice and consequently the LSPR. In order to use this concept to fine tune the position of the plasmon, it is essential to first describe the vibrational modes of LaB_6, which are displayed in Equation (1).

$$\Gamma = A_{1g} + E_g + T_{1g} + T_{2g} + 3T_{1u} + T_{2u} \tag{1}$$

As seen in Figure 3A, the vibrational modes of the B_6 cluster in LaB_6 include the bending (T_{2g}) and stretching (E_g and A_{1g}) modes, while La includes vibrations from moving within the boron cage ("rattling mode") and movement with respect to the boron cage (T_{1u}). When LaB_6 particles decrease in size from bulk to 2.5 nm, there is a shift to higher energy for all of the B_6 cluster vibrational modes due to the increased surface area which has a larger number of relaxed B_6 on the surface (Figure 3B). Furthermore, different sized halogens from the La salt precursor also impact the vibration modes, where larger atoms take up more space as bridging atoms that increase the vibrational energy. More details on the direct influence of particle size on the plasmon position will be discussed below.

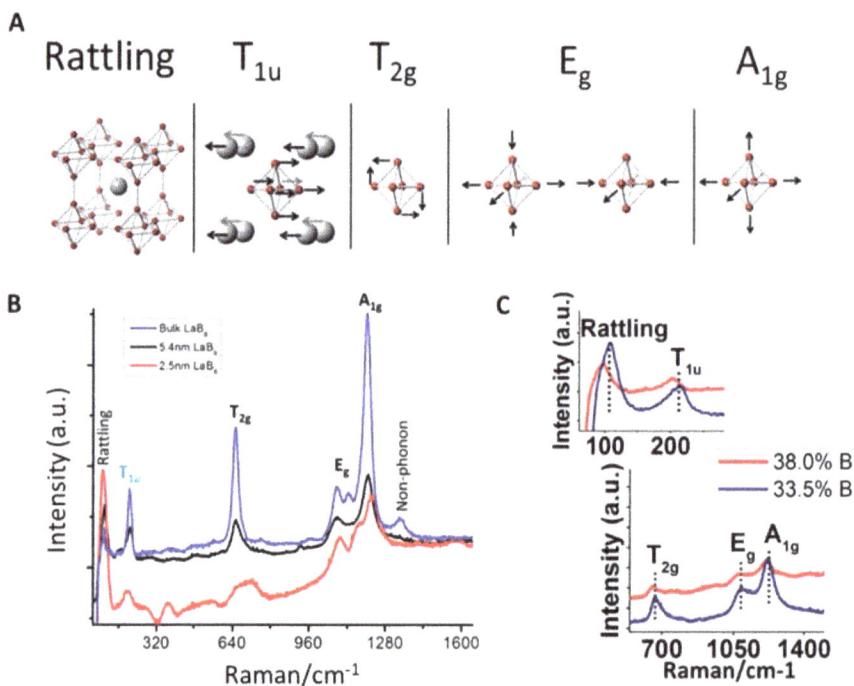

Figure 3. (**A**) Raman active vibrational modes of LaB_6; (**B**) vibrational mode changes with changing nanoparticle sizes; and (**C**) shifting vibrational modes of 6.2 nm particles with varying B content (reprinted from [63]).

2.1.2. Structural Defects

LaB_6 is well known to have defects within the structure, and is often non-stoichiometric because La atoms are missing or excess boron is residing in the interstitial spaces. As with other plasmonic systems, these defects can be advantageous to scientists when tuning the optical properties so long as the defects can be predicted and controlled. Only recently have researchers published on the ability to control La defects simply through reaction temperature and heating rate [64]. When the system is not stoichiometric (6B:1La) and the %B is increased (La atoms are missing) the Raman vibrations shift to lower energy. This is very clear in Figure 3C when comparing 6.2 nm particles containing 38.0% and 33.5% B (note that stoichiometric LaB_6 contains 31.8% B). The vacant lanthanum positions of the structure lower the overall energy of the system, and with less La available there are fewer electrons contributing to the structure, which causes the plasmon to shift.

3. Controlling the Plasmon of LaB$_6$

The structure of LaB$_6$ is exceptionally robust. Although vacancies and slight changes to the crystal structure clearly influence the position of the plasmon, it is possible to move beyond the fundamental structure to find other knobs to control the plasmonic properties. Similar to chalcogenides and metal oxides, LaB$_6$ is significantly influenced by the particle size and the shape of the surface. Unlike traditional noble-metal based materials, LaB$_6$ has the ability to incorporate dopants within its structure, which offers an additional means to increase the free carrier density and red-shift the frequency of the plasmon.

3.1. Doping

LaB$_6$ is well known for its stability, even when La vacancies leave unfilled holes within the crystal lattice. Like all boron frameworks, the B$_6$ network is electron-deficient and stable only as a result of electron transfer from the surrounding metals [65]. The natural defects or holes found in the LaB$_6$ framework make this material ideal for incorporating other metals via doping to tune the optical properties. However, few people have been successful in producing phase-pure hexaboride nanoparticles containing more than one type of metal, and fewer still have considered plasmonic changes in these doped systems. Fortunately, the introduction of dopants into LaB$_6$ offers a promising route to plasmon control.

Part of the difficulty arises from the nature of the synthesis. The mechanism of formation is not understood, and it is only in very recent years that researchers have found ways to control nucleation and growth in LaB$_6$ by significantly reducing the reaction temperature. The other difficulty in doping LaB$_6$ is that the metal does not form a direct bond with the rigid B$_6$ framework. As a result, LaB$_6$ is not influenced much by differing sizes of metal atoms that could alter the structure and shift the plasmon. The contribution of electrons from the metal to the structure also means that attempting to dope the trivalent La^{3+} containing LaB$_6$ with other trivalent metals will not shift the plasmon because the free electron density will not change [66]. Consequently, researchers must be very selective when designing doped LaB$_6$ nanoparticles when intending to tune the plasmon. Doping this system also changes the vacancies within the crystal lattice, so careful attention must be paid to determine whether the dopants or vacancies are responsible for changes to the optical properties.

Studies of the Fermi level may help provide insight into the movement of electrons within the system. The spatial distribution of electrons near the Fermi level was recently reported for trivalent LaB$_6$ and divalent BaB$_6$ [67], and weak electron lobes were found around the interior-B$_6$ octahedral bond (Figure 4A). Comparing theoretical and experimental work, it was determined that these electron lobes were responsible for the conductive π-electrons in LaB$_6$. Trivalent LaB$_6$ behaves as a metal while divalent hexaborides (i.e. BaB$_6$) are semiconductors [68], but how do these and the optical properties change when LaB$_6$ becomes a mixed valence system? Below are the few examples in the literature that tune the plasmon of LaB$_6$ nanoparticles by doping the material with divalent lanthanide metals, including some comments on how the changed Fermi surface influences the LSPR. In all cases, introducing a lanthanide metal as a dopant in the system causes a red-shift of the plasmon.

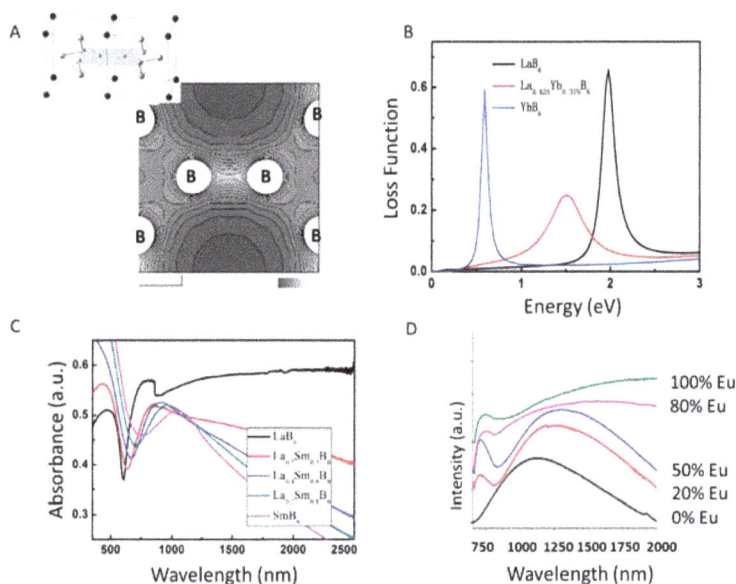

Figure 4. (**A**) Total charge density map of LaB$_6$ (grey scale shows increasing charge density) (reprinted from [67]). (**B**) Energy loss function of LaB$_6$, La$_{0.625}$Yb$_{0.375}$B$_6$ and YbB$_6$ in the low energy region (reprinted from [69]). (**C**) Absorbance spectra of SmB$_6$, La$_{0.2}$Sm$_{0.8}$B$_6$, La$_{0.4}$Sm$_{0.6}$B$_6$, La$_{0.7}$Sm$_{0.3}$B$_6$ and LaB$_6$ (reprinted from [70]). (**D**) Absorbance spectra of La$_x$Eu$_{1-x}$B$_6$ changing with Eu concentration (normalized) (reprinted from [71]).

3.1.1. Yb-Doped LaB$_6$ Nanoparticles: Theory

A recent report used density functional theory (DFT) first-principal calculations to study the potential optical effects of LaB$_6$ nanoparticles doped with ytterbium (Yb), and found that the 4f states of the Yb dopant at the Fermi surface have the potential to uniquely influence the optical properties, due to the participation of f-orbitals [69]. This is clear in Figure 4B, which compares phase pure YbB$_6$, LaB$_6$, and La$_{0.625}$Yb$_{0.375}$B$_6$ in the energy loss spectra (another method for measuring plasmon energies). Doping LaB$_6$ with Yb is expected to reduce the plasmon energy of the system, coinciding with a change of the LSPR. The concentration of charge carriers in the system decreases and reduces the plasma frequency when some of the La atoms are replace with Yb, which causing the minimum plasma absorption to shift to higher wavelengths. Though the calculations suggest that Yb-doping LaB$_6$ should indeed tune the plasmon, such an experiment has yet to be reported.

3.1.2. Sm-Doped LaB$_6$ Nanoparticles

The first synthetic example of LaB$_6$ nanoparticles doped with a trivalent metal was reported using samarium (Sm) [70]. With the aid of DFT calculations, it was determined that the shift of the optical properties was due to the changing Sm 4f states near the Fermi surface of LaB$_6$ after doping. The same as was predicted for Yb doping, with a reduced number of conduction electrons causing the shift of absorption spectra. When increasing the Sm content in SmB$_6$ the plasmon peak shifts to higher wavelengths, which is clear as the onset of the absorbance peak shifts from 603 nm for LaB$_6$ to 756 nm for SmB$_6$ (Figure 4C).

3.1.3. Eu-Doped LaB$_6$ Nanoparticles

A low temperature solid-state synthesis was employed to dope LaB$_6$ with europium (Eu), forming a single phase of La$_x$Eu$_{1-x}$B$_6$ nanoparticles [71]. Combining divalent Eu^{2+} and trivalent La^{3+} changed the free electron density of the system, resulting in a shift of the plasmon when changing the ratios of the two metals (Figure 4D). Increasing the concentration of Eu^{2+} significantly increased the number of metal vacancies in the structure, which allowed the plasmon to be tuned across a very wide range (1100 nm to 2050 nm).

3.2. Nanoparticle Size

In addition to doping, controlling the size of phase-pure LaB_6 also offers a means of control when tuning the position of the plasmon, where shifts of size and shape change the particle surface enough to influence the resonant frequency [72]. However, due to past difficulties in synthesizing LaB_6 nanoparticles directly, it has been difficult to directly synthesize a series of pure LaB_6 with varying nanoparticle sizes.

Theoretical models using DFT clearly show how particle sizes ranging from 10 nm to 100 nm can shift the LSPR, where the larger the particle the longer the wavelength absorbed so long as the particles are smaller than 80 nm (Figure 5A) [73]. In contrast, LaB_6 nanoparticles larger than 80 nm are too big to influence the plasmon and have a lower efficiency for NIR absorbance. Experiments of particles below 5nm have shown that even a small difference in size has a notable impact on the position of the plasmon. For example, increasing the particle size of phase pure LaB_6 from 2.1 nm to 4.7 nm red-shifts the absorbance from 1080 nm to 1250 nm, and the same red shift trend is observed when a diol-based ligand is included, where changing the size from 2.5 nm to 2.8 nm moves the plasmon from 1190 nm to 1220 nm (Figure 5B) [74]. Interestingly, with particles well below 10 nm in size the DFT calculations predicting the position of the plasmon are lower than the actual values, and further work is needed to understand what additional contributions from such small sized particles influence the LSPR theoretical calculations.

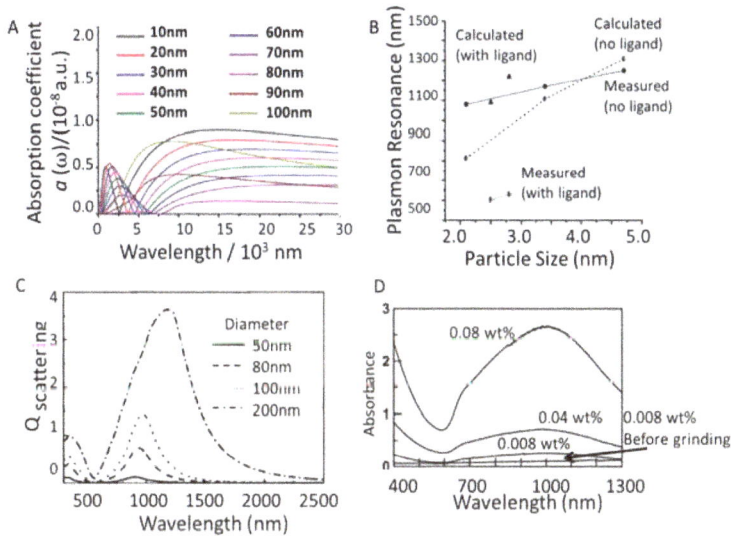

Figure 5. (**A**) Modeled absorption spectrum for LaB_6 nanoparticles ranging from 10 nm to 100 nm (reprinted from [73]). (**B**) Calculated and measured (via diffuse reflectance) LSPR peak positions of ligand-free and ligand-bound LaB_6 nanoparticles (reprinted from [74]). (**C**) Calculated scattering efficiency of single spherical LaB_6 nanoparticles embedded in a polymer for sizes ranging from 50 nm to 200 nm (inset shows fraction of scattering to total extinction efficiency at 500 nm) (reprinted from [30]). (**D**) Absorption spectra for ethylene glycol dispersion of LaB_6 powders before and after grinding at different concentrations (reprinted from [75]).

Calculations have also found that the scattering efficiency changes when adding different sizes of LaB$_6$ to a polymer matrix (Figure 5C), which ought to be considered when developing films for window coatings. Decreasing the nanoparticle size from 200 nm to 50 nm significantly red-shifts the plasmon, and not only increases the intensity of the scattering but broadens the scatter as well. It should also be noted that the shape of the scattering coefficient curve changes in particles larger than 80 nm, which may be indicative of size-dependent LSPR behavior. When attempting to design a material that will capture a wide wavelength range, like the solar heat from 750 nm to 1200 nm mentioned above, it is ideal to use nanomaterials with a broad and intense absorbance peak [30].

Bulk sized LaB$_6$ may also be reduced in size through grinding in the presence of a surfactant such as dodecylbenzene-sulfonic acid (DBS). The DBS makes it easier to disperse ~100 nm particles into ethylene glycol, with the absorbance intensity increasing in higher particle concentrations (Figure 5D) [75]. This makes LaB$_6$ a feasible NIR photothermal conversion material.

3.3. Morphology

In addition to size, the plasmonics of nanoparticles can also be tuned by altering the morphology of the system, where introducing different facets has a large impact on how free electrons behave on the surface [76]. Unfortunately, publications describing more than one nanoparticle shape of LaB$_6$ are uncommon due to synthetic limitations, and those connecting shape and LSPR are even more rare. Whether morphological limitations are due to the chemistry itself or if low temperature methods simply need additional time to study has yet to be determined. The shape of LaB$_6$ nanoparticles tends to be described as generic nanoparticles (assumed spheres) versus cubes [56,77] (Figure 6A) or as nanowires [78–80], but to our knowledge only two publications to date tie plasmonics to varying particle shape [81,82].

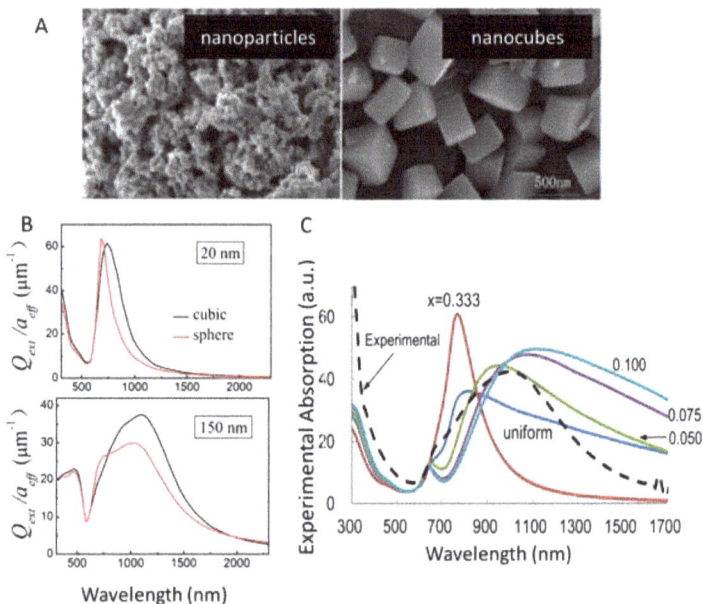

Figure 6. (**A**) SEM images of LaB$_6$ nanoparticles and nanocubes (reprinted from [57]). (**B**) Comparison of extinction efficiencies divided by effective radius between cubic and spherical LaB$_6$ particles (reprinted from [82]). (**C**) Absorption cross sections calculated for 100,000 LaB$_6$ oblate particles with different aspect ratios and a standard deviation of 0.1 (reprinted from [81]).

The optical response of LaB$_6$ nanoparticles of varying sizes and shapes was compared using the discrete dipole approximation and experimental results were in agreement with the findings [82]. Comparing the extinction efficiencies of cubic and spherical particles of the same diameter (Figure 6B), it is clear that the optical properties are significantly influenced by the shape of the particles and that the nanocubes exhibit stronger NIR extinction. The Mie integration method is also an effective means of estimating optical properties of nanoparticles of changing sizes and shapes [83]. For LaB$_6$, it was found that increasing the aspect ratio of spheroids can enhance the LSPR properties in the NIR [81]. In oblate spheroids, decreasing the aspect ratio has the effect of red-shifting the LSPR, and can significantly broaden the absorbance peak (Figure 6C).

4. Improving LaB$_6$ Plasmonic Applications

In order to make LaB$_6$ nanoparticles feasible as coatings for various solar energy and window applications, researchers must gain an understanding of plasmonic behaviors when incorporating the particles into films. Once the surrounding media is changed, the absorbance and intensity of the LSPR signal is also altered. To make cleaner films, LaB$_6$ nanoparticles can be put into solutions using various polymer matrixes [84,85]. As with the changing particle concentrations mentioned above, varying the size of LaB$_6$ nanoparticles (while maintaining the same concentration) in polymethyl methacrylate (PMMA) composites changes the absorbance intensity (Figure 7A), and the ideal particle size to achieve maximum intensity is just below 100 nm [86]. A similar effect has been observed in the transmittance and reflectance profiles of LaB$_6$ nanoparticles dispersed in acrylic coatings, where decreasing the LaB$_6$ concentration increases the transmittance intensity (Figure 7B) [87].

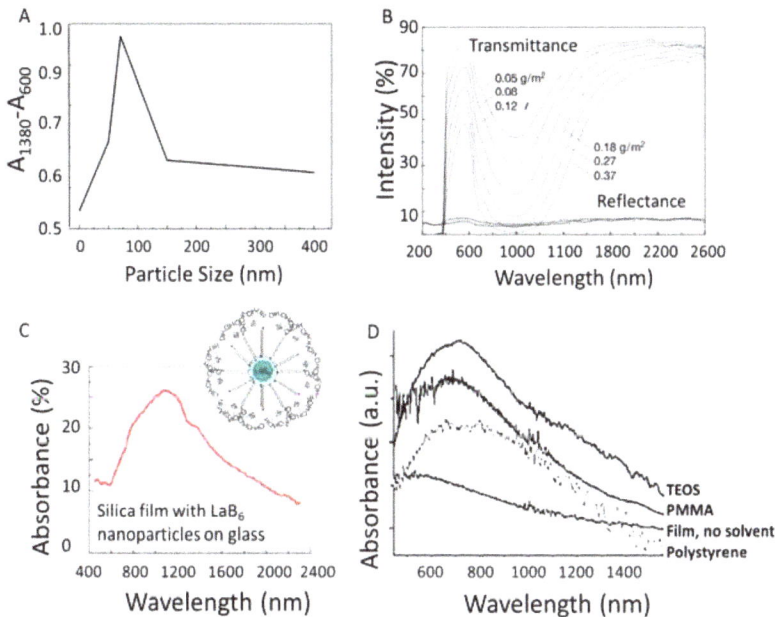

Figure 7. (**A**) Relative strength of absorption in composites containing LaB$_6$ particles of different sizes; 0 nm represents pure polymer without LaB$_6$. (reprinted from [86]). (**B**) Transmittance and reflectance profiles of LaB$_6$ nanoparticle-dispersed acrylic coatings with different thicknesses on PET films (reprinted from [87]). (**C**) Absorbance spectra of cetyltrimethyl ammonium bromide (CTAB)-stabilized LaB$_6$ nanoparticles (reprinted from [84]). (**D**) Absorbance spectra of 2.8nm ligand-bound LaB$_6$ nanoparticles embedded in polymethyl methacrylate (PMMA), polystyrene, and tetraethoxyorthosilane (TEOS)-derived glass (reprinted from [74]).

Creating visibly clear films with LaB$_6$ remains an ongoing challenge. Since LaB$_6$ nanoparticles are typically made with no ligands, they tend to aggregate with no physical barrier keeping them apart. To improve upon this, researchers are starting to develop methods to incorporate surfactants directly on the surface of LaB$_6$. For example, using cetyltrimethyl ammonium bromide (CTAB) as a surfactant coating makes it possible to stabilize and reduce the agglomeration of LaB$_6$ nanoparticles in water (Figure 7C). There has also been a report of incorporating a ligand during the LaB$_6$ reaction using a low temperature method to convert isophthalic acid to 1,3-phenylenedimethanol in-situ, with the diol binding to the LaB$_6$ surface as soon as it is formed in-situ [74]. As a result, the ligand-bound LaB$_6$ nanoparticles can be embedded directly into various polymer matrixes without the need for additional treatment, producing unaggregated and clear films composed of LaB$_6$ in PMMA, polystyrene, or tetraethoxyorthosilane (TEOS) glass (Figure 7D). While this review focuses on tuning the plasmon of LaB$_6$, it is also important to note that there are many more applications for this material beyond building specialized windows for heat control. Many new applications of LaB$_6$ nanoparticles have come to light in recent years. For example, the photothermal conversion properties of LaB$_6$ are being explored for biomedical applications like cancer [75,88–90]. LaB$_6$ nanoparticles also interact with bacteria, and studies focusing on ablation and hydrogen production are ongoing [91,92]. Looking beyond nanoparticles to nanowires, LaB$_6$ is now known to be an efficient field emitter [78–80]. There are many more potential applications, and this list offers you a small sampling of how this material is advancing multiple fields of research.

Future Challenges of Plasmonic LaB$_6$

Much has been learned about the optical behavior and tunability of LaB$_6$, and work is ongoing. Perhaps the unspoken secret of this research space is how challenging it is to develop the synthetic procedures to make small phase-pure LaB$_6$ nanoparticles. Ball milling bulk particles to reduce the particle size introduces contaminants and does not maintain uniform morphology. Cutting single crystals into nano-sized particles is difficult and does not translate easily to larger scale [78]; plasma methods do not offer a means of easy control for tuning [93], etc.

To further complicate matters, the mechanism behind the formation of the B$_6$ cluster is still unknown. For decades it was assumed that high temperatures (>1200 °C) and pressures were energetically required to make LaB$_6$, but that is not the case [94,95]. It has also been assumed that a reducing agent such as magnesium was required to drive the reaction [59,96], but new reports find this to be untrue [74,95]. Only recently was it found that the halogen of the lanthanum salt influences the reaction to expand or contract the crystal lattice, and can act as a bridging ligand between La atoms in low temperature reactions [48,58].

In order advance this field it is becoming increasingly important to build a better understanding of the chemistry involved in the formation of LaB$_6$. In learning more about the mechanism, it will be much easier to tune these particles for optical applications as well as electronics, thermionics, and solar energy. With a better understanding of the formation of the robust boron clusters in LaB$_6$ and the potential interplay of boron and halogens, we will likely discover new applications previously unexplored.

5. Conclusions

LaB$_6$ is an intriguing plasmonic material that has gained much attention in the last decade for its ability to intensely absorb in the NIR region of the electromagnetic spectrum. Large advances have been made in developing methods to finely tune the position of the plasmon, making the shift from the visible region to the NIR possible. This review explains how some of the fundamentals of the LaB$_6$ framework influence the optical properties, and how the plasmon can be tuned by controlling lattice defects, particle size and shape, and through doping. As advances continue to be made in this field, it is hoped that a there will be a better understanding of how the structure forms, enabling better control over this system and potentially opening the door to new applications in need of stable plasmonic nanoparticles.

Funding: This research received no external funding.

Acknowledgments: This work was supported by the Molecular Foundry at Lawrence Berkeley National Laboratory, a user facility supported by the Office of Science, Office of Basic Energy Sciences, of the U.S. Department of Energy (DOE) under Contract No. DE-AC02-05CH11231.

Conflicts of Interest: The authors declare no conflict of interest.

References

1. Aslam, U.; Rao, V.G.; Chavez, S.; Linic, S. Catalytic conversion of solar to chemical energy on plasmonic metal nanostructures. *Nat. Catal.* **2018**, *1*, 656–665. [CrossRef]
2. Byers, C.P.; Zhang, H.; Swearer, D.F.; Yorulmaz, M.; Hoener, B.S.; Huang, D.; Hoggard, A.; Chang, W.-S.; Mulvaney, P.; Ringe, E.; et al. From tunable core-shell nanoparticles to plasmonic drawbridges: Active control of nanoparticle optical properties. *Sci. Adv.* **2015**, *1*, e1500988. [CrossRef] [PubMed]
3. Wen, J.; Wang, W.J.; Li, N.; Li, Z.F.; Lu, W. Plasmonic optical convergence microcavity based on the metal-insulator-metal microstructure. *Appl. Phys. Lett.* **2017**, *110*, 231105. [CrossRef]
4. Zhang, J.Z.; Noguez, C. Plasmonic Optical Properties and Applications of Metal Nanostructures. *Plasmonics* **2008**, *3*, 127–150. [CrossRef]
5. Takahata, R.; Yamazoe, S.; Koyasu, K.; Tsukuda, T. Surface Plasmon Resonance in Gold Ultrathin Nanorods and Nanowires. *J. Am. Chem. Soc.* **2014**, *136*, 8489–8491. [CrossRef] [PubMed]
6. Lee, K.-S.; El-Sayed, M.A. Gold and Silver Nanoparticles in Sensing and Imaging: Sensitivity of Plasmon Response to Size, Shape, and Metal Composition. *J. Phys. Chem. B* **2006**, *110*, 19220–19225. [CrossRef] [PubMed]
7. Xiaoa, L.; Sub, Y.; Qiud, W.; Liua, Y.; Rand, J.; Wua, J.; Lua, F.; Shaod, F.; Tangc, D.; Peng, P. Solar radiation shielding properties of transparent LaB6 filters through experimental and first-principles calculation methods. *Ceram. Int.* **2016**, *42*, 14278–14281. [CrossRef]
8. Bao, L.; Chao, L.; Wei, W.; Tegus, O. Tunable transmission light in nanocrystalline La$_{1-x}$Eu$_x$B$_6$. *Mater. Lett.* **2015**, *139*, 187–190. [CrossRef]
9. Xiao, L.; Su, Y.; Zhou, X.; Chen, H.; Tan, J.; Hu, T.; Yan, J.; Peng, P. Origins of high visible light transparency and solar heat-shielding performance in LaB6. *Appl. Phys. Lett.* **2012**, *101*, 041913. [CrossRef]
10. Mattox, T.M.; Ye, X.; Manthiram, K.; Schuck, P.J.; Alivisatos, A.P.; Urban, J.J. Chemical Control of Plasmons in Metal Chalcogenide and Metal Oxide Nanostructures. *Adv. Mater.* **2015**, *27*, 5830–5837. [CrossRef]
11. Lounis, S.D.; Runnerstrom, E.L.; Llordés, A.; Milliron, D.J. Defect Chemistry and Plasmon Physics of Colloidal Metal Oxide Nanocrystals. *J. Phys. Chem. Lett.* **2014**, *5*, 1564–1574. [CrossRef] [PubMed]
12. Wolf, A.; Kodanek, T.; Dorfs, D. Tuning the LSPR in copper chalcogenide nanoparticles by cation intercalation, cation exchange and metal growth. *Nanoscale* **2015**, *7*, 19519–19527. [CrossRef] [PubMed]
13. Mattox, T.M.; Bergerud, A.; Agrawal, A.; Milliron, D.J. Influence of Shape on the Surface Plasmon Resonance of Tungsten Bronze Nanocrystals. *Chem. Mater.* **2014**, *26*, 1779–1784. [CrossRef]
14. Zhao, Y.; Pan, H.; Lou, Y.; Qiu, X.; Zhu, J.; Burda, C. Plasmonic Cu$_{2-x}$S Nanocrystals: Optical and Structural Properties of Copper-Deficient Copper(I) Sulfides. *J. Am. Chem. Soc.* **2009**, *131*, 4253–4261. [CrossRef] [PubMed]
15. González, A.L.; Noguez, C.; Beránek, J.; Barnard, A.S. Size, Shape, Stability, and Color of Plasmonic Silver Nanoparticles. *J. Phys. Chem. C* **2014**, *118*, 9128–9136. [CrossRef]
16. Amendola, V.; Pilot, R.; Frasconi, M.; Maragò, O.M.; Iati, M.A. Surface plasmon resonance in gold nanoparticles: A review. *J. Phys. Condens. Matter* **2017**, *29*, 203002. [CrossRef]
17. Liu, P.; Wang, H.; Li, X.; Rui, M.; Zeng, H. Localized surface plasmon resonance of Cu nanoparticles by laser ablation in liquid media. *RSC Adv.* **2015**, *5*, 79738–79745. [CrossRef]
18. Duan, H.; Xuan, Y. Synthesis and optical absorption of Ag/CdS core/shell plasmonic nanostructure. *Sol. Energy Mater. Sol. Cells* **2014**, *121*, 8–13. [CrossRef]
19. Chen, Y.; Wu, H.; Li, Z.; Wang, P.; Yang, L.; Fang, Y. The Study of Surface Plasmon in Au/Ag Core/Shell Compound Nanoparticles. *Plasmonics* **2012**, *7*, 509–513. [CrossRef]
20. Schmidt, P.H.; Joy, D.C.; Longinotti, L.D.; Leamy, H.J.; Ferris, S.D.; Fisk, Z. Anisotropy of thermionic electron emission values for LaB6 single-crystal emitter cathodes. *Appl. Phys. Lett.* **1976**, *29*, 400–401. [CrossRef]

21. Ahmed, H.; Broers, A.N. Lanthanum Hexaboride Electron Emitter. *J. Appl. Phys.* **1972**, *43*, 2185–2192. [CrossRef]

22. Goebel, D.M.; Hirooka, Y.; Sketchley, T.A. Large-area lanthanum hexaboride electron emitter. *Rev. Sci. Instrum.* **1985**, *56*, 1717–1722. [CrossRef]

23. Zhang, H.; Tang, J.; Zhang, Q.; Zhao, G.; Yang, G.; Zhang, J.; Zhou, O.; Qin, L.-C. Field emission of electrons from single LaB₆ nanowires. *Adv. Mater.* **2006**, *18*, 87–91. [CrossRef]

24. Zhou, S.; Zhang, J.; Liu, D.; Lin, Z.; Huang, Q.; Bao, L.; Ma, R.; Wei, Y. Synthesis and properties of nanostructured dense LaB6 cathodes by arc plasma and reactive spark plasma sintering. *Acta Mater.* **2010**, *58*, 4978–4985. [CrossRef]

25. Back, T.C.; Schmid, A.K.; Fairchild, S.B.; Boeckl, J.J.; Cahay, M.; Derkink, F.; Chen, G.; Sayir, A. Work function characterization of directionally solidified LaB₆–VB₂ eutectic. *Ultramicroscopy* **2017**, *183*, 67–71. [CrossRef] [PubMed]

26. Liu, H.; Zhang, X.; Ning, S.; Xiao, Y.; Zhang, J. The electronic structure and work functions of single crystal LaB6 typical crystal surfaces. *Vacuum* **2017**, *143*, 245–250. [CrossRef]

27. Ning, S.-Y.; Iitaka, T.; Yang, X.-Y.; Wang, Y.; Zhao, J.-J.; Li, Z.; Zhang, J.-X. Enhanced thermionic emission performance of LaB6 by Ce doping. *J. Alloys Compd.* **2018**, *760*, 1–5. [CrossRef]

28. Torgasin, K.; Morita, K.; Zen, H.; Masuda, K.; Katsurayama, T.; Murata, T.; Suphakul, S.; Yamashita, H.; Nogi, T.; Kii, T.; et al. Thermally assisted photoemission effect on CeB₆ and LaB₆ for application as photocathodes. *Phys. Rev. Accel. Beams* **2017**, *20*, 073401. [CrossRef]

29. Voss, J.; Vojvodic, A.; Chou, S.H.; Howe, R.T.; Abild-Pedersen, F. Inherent Enhancement of Electronic Emission from Hexaboride Heterostructure. *Phys. Rev. Appl.* **2014**, *2*, 024004. [CrossRef]

30. Schelm, S.; Smith, G.B.; Garrett, P.D.; Fisher, W.K. Tuning the surface-plasmon resonance in nanoparticles for glazing applications. *J. Appl. Phys.* **2005**, *97*, 124314. [CrossRef]

31. Schelm, S.; Smith, G.B. Dilute LaB6 nanoparticles in polymer as optimized clear solar control glazing. *Appl. Phys. Lett.* **2003**, *82*, 4346–4348. [CrossRef]

32. Bogomol, I.; Nishimura, T.; Nesterenko, Y.; Vasylkiv, O.; Sakka, Y.; Loboda, P. The bending strength temperature dependence of the directionally solidified eutectic LaB₆–ZrB₂ composite. *J. Alloys Compd.* **2011**, *509*, 6123–6129. [CrossRef]

33. Chen, D.; Min, G.; Wu, Y.; Yu, H.; Zhang, L. The preparation and composition design of boron-rich lanthanum hexaboride target for sputtering. *J. Alloys Compd.* **2015**, *638*, 380–386. [CrossRef]

34. Gridneva, I.V.; Lazorenko, V.I.; Lotsko, D.V.; Mil'man, Y.V.; Paderno, Y.B.; Chugunova, S.I. Mechanical properties of single-crystal lanthanum hexaboride with local loading. *Sov. Powder Metall. Met. Ceram.* **1990**, *29*, 967–972. [CrossRef]

35. Loboda, P.I.; Kysla, H.P.; Dub, S.M.; Karasevs'ka, O.P. Mechanical properties of the monocrystals of lanthanum hexaboride. *Mater. Sci* **2009**, *45*, 108–113. [CrossRef]

36. Mitterer, C.; Komenda-Stallmaier, J.; Losbichler, P.; Schmölz, P.; Störi, H. Decorative boride coatings based on LaB₆. *Surf. Coat. Technol.* **1995**, *74–75*, 1020–1027. [CrossRef]

37. Bogomol, I.; Nishimura, T.; Vasylkiv, O.; Sakka, Y.; Loboda, P. High-temperature strength of directionally reinforced LaB₆–TiB₂ composite. *J. Alloys Compd.* **2010**, *505*, 130–134. [CrossRef]

38. Buckingham, J.D. Thermionic emission properties of a lanthanum hexaboride/rhenium cathode. *Br. J. Appl. Phys.* **1965**, *16*, 1821. [CrossRef]

39. Xu, G.-L.; Chen, J.-D.; Xia, Y.-Z.; Liu, X.-F.; Liu, Y.-F.; Zhou, X.-Z. First-Principles Calculations of Elastic and Thermal Properties of Lanthanum Hexaboride. *Chin. Phys. Lett.* **2009**, *26*, 056201.

40. Hu, P.; Zhang, X.-H.; Han, J.-C.; Luo, X.-G.; Du, S.-Y. Effect of Various Additives on the Oxidation Behavior of ZrB₂-Based Ultra-High-Temperature Ceramics at 1800 °C. *J. Am. Ceram. Soc.* **2010**, *93*, 345–349. [CrossRef]

41. Kelly, J.P.; Kanakala, R.; Graeve, O.A. A Solvothermal Approach for the Preparation of Nanostructured Carbide and Boride Ultra-High-Temperature Ceramics. *J. Am. Ceram. Soc.* **2010**, *93*, 3035–3038. [CrossRef]

42. Marchenko, A.A.; Cherepanov, V.V.; Tarashchenko, D.T.; Kazantseva, Z.I.; Naumovets, A.G. A low work function substrate for STM studies of objects with poor tunneling transparency: Lanthanum hexaboride (100). *Surf. Sci.* **1998**, *416*, 460–465. [CrossRef]

43. Monteverde, F.; Alfano, D.; Savino, R. Effects of LaB6 addition on arc-jet convectively heated SiC-containing ZrB₂-based ultra-high temperature ceramics in high enthalpy supersonic airflows. *Corros. Sci.* **2013**, *75*, 443–453. [CrossRef]

44. Cortie, M.B.; McDonagh, A.M. Synthesis and Optical Properties of Hybrid and Alloy Plasmonic Nanoparticles. *Chem. Rev.* **2011**, *111*, 3713–3735. [CrossRef] [PubMed]

45. Hou, W.; Cronin, S.B. A Review of Surface Plasmon Resonance-Enhanced Photocatalysis. *Adv. Funct. Mater.* **2013**, *23*, 1612–1619. [CrossRef]

46. Linic, S.; Christopher, P.; Ingram, D.B. Plasmonic-metal nanostructures for efficient conversion of solar to chemical energy. *Nat. Mater.* **2011**, *10*, 911–921. [CrossRef] [PubMed]

47. Lu, X.; Rycenga, M.; Skrabalak, S.E.; Wiley, B.; Xia, Y. Chemical Synthesis of Novel Plasmonic Nanoparticles. *Annu. Rev. Phys. Chem.* **2009**, *60*, 167–192. [CrossRef] [PubMed]

48. Mattox, T.M.; Groome, C.; Doran, A.; Beavers, C.M.; Urban, J.J. Chloride influence on the formation of lanthanum hexaboride: An in-situ diffraction study. *J. Cryst. Growth* **2018**, *486*, 60–65. [CrossRef]

49. Nagao, T.; Kitamura, K.; Iizuka, Y.; Oshima, C. Surface phonons of LaB$_6$(100): Deformation of boron octahedra at the surface. *Surf. Sci.* **1993**, *290*, 436–444. [CrossRef]

50. Nagao, T.; Kitamura, T.; Iizuka, T.; Umeuchi, M.; Oshima, C. Deformation of octahedra at LaB$_6$(100) surface studied by HREELS. *Surf. Sci.* **1993**, *287–288*, 391–395. [CrossRef]

51. Rokuta, E.; Yamamoto, N.; Hasegawa, Y.; Trenary, M.; Nagao, T.; Oshima, C.; Otani, S. Deformation of boron networks at the LaB$_6$(111) surface. *Surf. Sci.* **1998**, *416*, 363–370. [CrossRef]

52. Li, Y.; Cao, Y.; Jia, D. A general strategy for synthesis of metal nanoparticles by a solid-state redox route under ambient conditions. *J. Mater. Chem. A* **2014**, *2*, 3761–3765. [CrossRef]

53. Polte, J. Fundamental growth principles of colloidal metal nanoparticles—A new perspective. *CrystEngComm* **2015**, *17*, 6809–6830. [CrossRef]

54. Adachi, K.; Miratsu, M.; Asahi, T. Absorption and scattering of near-infrared light by dispersed lanthanum hexaboride nanoparticles for solar control filters. *J. Mater. Res.* **2010**, *25*, 510–521. [CrossRef]

55. Sato, Y.; Terauchi, M.; Mukai, M.; Kaneyama, T.; Adachi, K. High energy-resolution electron energy-loss spectroscopy study of the dielectric properties of bulk and nanoparticle LaB$_6$ in the near-infrared region. *Ultramicroscopy* **2011**, *111*, 1381–1387. [CrossRef] [PubMed]

56. Zhang, L.; He, W.J.; Tolochko, O.V.; Polzik, L.; Min, G.H. Morphology Characterization and Optical Properties Analysis for Nanostructured Lanthanum Hexaboride Powders. *Adv. Mater. Res.* **2009**, *79–82*, 107–110. [CrossRef]

57. Hang, C.-L.; Yang, L.-X.; Wang, F.; Xu, Y.-B.; Yi, C.-Y. Melt-assisted synthesis to lanthanum hexaboride nanoparticles and cubes. *Bull. Mater. Sci.* **2017**, *40*, 1241–1245. [CrossRef]

58. Mattox, T.M.; Groome, C.; Doran, A.; Beavers, C.M.; Urban, J.J. Anion-mediated negative thermal expansion in lanthanum hexaboride. *Solid State Commun.* **2017**, *265*, 47–51. [CrossRef]

59. Zhang, M.; Yuan, L.; Wang, X.; Fan, H.; Wang, X.; Wu, X.; Wang, H.; Qian, Y. A low-temperature route for the synthesis of nanocrystalline LaB6. *J. Solid State Chem.* **2008**, *181*, 294–297. [CrossRef]

60. Yuan, Y.; Zhang, L.; Liang, L.; He, K.; Liu, R.; Min, G. A solid-state reaction route to prepare LaB$_6$ nanocrystals in vacuum. *Ceram. Int.* **2011**, *37*, 2891–2896. [CrossRef]

61. Brewer, J.R.; Deo, N.; Wang, Y.M.; Cheung, C.L. Lanthanum Hexaboride Nanoobelisks. *Chem. Mater.* **2007**, *19*, 6379–6381. [CrossRef]

62. Zhang, H.; Shang, Q.; Tang, J.; Qin, L.-C. Single-crystalline LaB$_6$ nanowires. *J. Am. Chem. Soc.* **2005**, *127*, 2862–2863. [CrossRef] [PubMed]

63. Mattox, T.M.; Chockkalingam, S.; Roh, I.; Urban, J.J. Evolution of Vibrational Properties in Lanthanum Hexaboride Nanocrystals. *J. Phys. Chem. C* **2016**, *120*, 5188–5195. [CrossRef]

64. Groome, C.; Roh, I.; Mattox, T.M.; Urban, J.J. Effects of size and structural defects on the vibrational perperties of lanthanum hexaboride nanocrystals. *ACS Omega* **2017**, *2*, 2248–2254. [CrossRef]

65. Etourneau, J. Critical survey of rare-earth borides: Occurrence, crystal chemistry and physical properties. *J. Less-Common Met.* **1985**, *110*, 267–281. [CrossRef]

66. Cahill, J.T.; Alberga, M.; Bahena, J.; Pisano, C.; Borja-Urby, R.; Vasquez, V.R.; Edwards, D.; Misture, S.T.; Graeve, O.A. Phase Stability of Mixed-Cation Alkaline-Earth Hexaborides. *Cryst. Growth Des.* **2017**, *17*, 3450–3461. [CrossRef]

67. Kasai, H.; Nishibori, E. Spatial distribution of electrons near the Fermi level in the metallic LaB$_6$ through accurate X-ray charge density study. *Sci. Rep.* **2017**, *7*, 41375. [CrossRef]

68. Etourneau, J.; Hagenmuller, P. Structure and physical features of the rare-earth borides. *Philos. Mag. B* **1985**, *52*, 589–610. [CrossRef]

69. Chao, L.; Bao, L.; Wei, W.; Tegus, O. Optical properties of Yb-doped LaB$_6$ from first-principles calculation. *Mod. Phys. Lett. B* **2016**, *30*, 1650091. [CrossRef]
70. Chao, L.; Bao, L.; Shi, J.; Wei, W.; Tegus, O.; Zhang, Z. The effect of Sm-doping on optical properties of LaB6 nanoparticles. *J. Alloys Compd.* **2015**, *622* (Suppl. C), 618–621. [CrossRef]
71. Mattox, T.M.; Coffman, D.K.; Roh, I.; Sims, C.; Urban, J.J. Moving the Plasmon of LaB$_6$ from IR to Near-IR via Eu-Doping. *Materials* **2018**, *11*, 226. [CrossRef] [PubMed]
72. Noguez, C. Surface Plasmons on Metal Nanoparticles: The Influence of Shape and Physical Environment. *J. Phys. Chem. C* **2007**, *111*, 3806–3819. [CrossRef]
73. Hong, Y.; Zhang, X.; Li, B.; Li, M.; Shi, Q.; Wang, Y.; Li, L. Size dependent optical properties of LaB$_6$ nanoparticles enhanced by localized surface plasmon resonance. *J. Rare Earth* **2013**, *31*, 1096–1101. [CrossRef]
74. Mattox, T.M.; Agrawal, A.; Milliron, D.J. Low Temperature Synthesis and Surface Plasmon Resonance of Colloidal Lanthanum Hexaboride (LaB$_6$) Nanocrystals. *Chem. Mater.* **2015**, *27*, 6620–6624. [CrossRef]
75. Chen, C.-J.; Chen, D.-H. Preparation of LaB$_6$ nanoparticles as a novel and effective near-infrared photothermal conversion material. *Chem. Eng. J.* **2012**, *180*, 337–342. [CrossRef]
76. Rivera, V.A.G.; Ferri, A.F.; Marega, E., Jr. Localized Surface Plasmon Resonances: Noble Metal Nanoparticle Interaction with Rare-Earth Ions. In *Plasmonics: Principles and Applications*; IntechOpen: Rijeka, Croatia, 2012.
77. Kanakala, R.; Rojas-George, G.; Graeve, O.A. Unique Preparation of Hexaboride Nanocubes: A First Example of Boride Formation by Combustion Synthesis. *J. Am. Ceram. Soc.* **2010**, *93*, 3136–3141. [CrossRef]
78. Zhang, H.; Tang, J.; Yuan, J.; Ma, J.; Shinya, N.; Nakajima, K.; Murakami, H.; Ohkubo, T.; Qin, L.-C. Nanostructured LaB6 Field Emitter with Lowest Apical Work Function. *Nano Lett.* **2010**, *10*, 3539–3544. [CrossRef] [PubMed]
79. Xu, J.; Hou, G.; Li, H.; Zhai, T.; Dong, B.; Yan, H.; Wang, Y.; Yu, B.; Bando, Y.; Golberg, D. Fabrication of vertically aligned single-crystalline lanthanum hexaboride nanowire arrays and investigation of their field emission. *NPG Asia Mater.* **2013**, *5*, e53. [CrossRef]
80. Xu, J.Q.; Zhao, Y.M.; Zhang, Q.Y. Enhanced electron field emission from single-crystalline LaB$_6$ nanowires with ambient temperature. *J. Appl. Phys.* **2008**, *104*, 124306. [CrossRef]
81. Machida, K.; Adachi, K. Particle shape inhomogeneity and plasmon-band broadening of solar-control LaB$_6$ nanoparticles. *J. Appl. Phys.* **2015**, *118*, 013103. [CrossRef]
82. Chao, L.; Bao, L.; Wei, W.; Tegus, O.; Zhang, Z. Effects of nanoparticle shape and size on optical properties of LaB$_6$. *Plasmonics* **2016**, *11*, 697–701. [CrossRef]
83. Kelly, K.L.; Coronado, E.; Zhao, L.L.; Schatz, G.C. The Optical Properties of Metal Nanoparticles: The Influence of Size, Shape, and Dielectric Environment. *J. Phys. Chem. B* **2003**, *107*, 668–677. [CrossRef]
84. Jiang, F.; Leong, Y.; Saunders, M.; Martyniuk, M.; Faraone, L.; Keating, A.; Dell, J. Uniform Dispersion of Lanthanum Hexaboride Nanoparticles in a Silica Thin Film: Synthesis and Optical Properties. *ACS Appl. Mater. Interfaces* **2012**, *4*, 5833–5838. [CrossRef]
85. Seeboth, A.; Ruhmann, R.; Mühling, O. Thermotropic and Thermochromic Polymer Based Materials for Adaptive Solar Control. *Materials* **2010**, *3*, 5143–5168. [CrossRef] [PubMed]
86. Yuan, Y.; Zhang, L.; Hu, L.; Wang, W.; Min, G. Size effect of added LaB6 particles on optical properties of LaB$_6$/Polymer composites. *J. Solid State Chem.* **2011**, *184*, 3364–3367. [CrossRef]
87. Takeda, H.; Kuno, H.; Adachi, K. Solar Control Dispersions and Coatings with Rare-Earth Hexaboride Nanoparticles. *J. Am. Ceram. Soc.* **2008**, *91*, 2897–2902. [CrossRef]
88. Lai, B.H.; Chen, D.H. LaB6 nanoparticles with carbon-doped silica coating for fluorescence imaging and near-IR photothermal therapy of cancer cells. *Acta Biomater.* **2013**, *9*, 7556–7563. [CrossRef] [PubMed]
89. Chen, M.-C.; Lin, Z.-W.; Ling, M.-H. Near-Infrared Light-Activatable Microneedle System for Treating Superficial Tumors by Combination of Chemotherapy and Photothermal Therapy. *ACS Nano* **2016**, *10*, 93–101. [CrossRef] [PubMed]
90. Cheng, F.-Y.; Chen, C.-T.; Yeh, C.-S. Comparative efficiencies of photothermal destruction of malignant cells using antibody-coated silica@Au nanoshells, hollow Au/Ag nanospheres and Au nanorods. *Nanotechnology* **2009**, *20*, 425104. [CrossRef]
91. Lai, B.-H.; Chen, D.-H. Vancomycin-modified LaB$_6$@SiO$_2$/Fe$_3$O$_4$ composite nanoparticles for near-infrared photothermal ablation of bacteria. *Acta Biomater.* **2013**, *9*, 7573–7579. [CrossRef]

92. Li, Y.; Zhong, N.; Liao, Q.; Fu, Q.; Huang, Y.; Zhu, X.; Li, Q. A biomaterial doped with LaB$_6$ nanoparticles as photothermal media for enhancing biofilm growth and hydrogen production in photosynthetic bacteria. *Int. J. Hydrogen Energy* **2017**, *42*, 5794–5803.

93. Szépvölgyi, J.; Mohai, I.; Károly, Z.; Gál, L. Synthesis of nanosized ceramic powders in a radiofrequency thermal plasma reactor. *J. Eur. Ceram. Soc.* **2008**, *28*, 895–899. [CrossRef]

94. Selvan, R.K.; Genish, I.; Perelshtein, I.; Calderon Moreno, J.M.; Gedanken, A. Single Step, Low-Temperature Synthesis of Submicron-Sized Rare Earth Hexaborides. *J. Phys. Chem. C* **2008**, *112*, 1795–1802. [CrossRef]

95. Yu, Y.; Wang, S.; Li, W.; Chen, Z. Low temperature synthesis of LaB$_6$ nanoparticles by a molten salt route. *Powder Technol.* **2018**, *323*, 203–207. [CrossRef]

96. Wang, L.; Xu, L.; Ju, Z.; Qian, Y. A versatile route for the convenient synthesis of rare-earth and alkaline-earth hexaborides at mild temperatures. *CrystEngComm* **2010**, *12*, 3923–3928. [CrossRef]

materials

MDPI

Communication

Optical Third Harmonic Generation Using Nickel Nanostructure-Covered Microcube Structures

Yoichi Ogata *, Anatoliy Vorobyev and Chunlei Guo *

The Institute of Optics, University of Rochester, Hutchison Road 275, Rochester, NY 14627, USA;
avorobye@ur.rochester.edu
* Correspondences: ogachi.yo@gmail.com or ogata@ipr-ctr.t.u-tokyo.ac.jp (Y.O.);
chunlei.guo@rochester.edu (C.G.); Tel.: +1-585-275-2134 (C.G.)

Received: 10 February 2018; Accepted: 23 March 2018; Published: 27 March 2018

Abstract: We investigated the optical third harmonic generation (THG) signal from nanostructure-covered microcubes on Ni. We found that the hierarchical structures greatly change the third-order optical nonlinearity of the metallic surface. While the symmetry and lightning rod (LR) effects on microstructures did not significantly influence the THG, the localized surface plasmon (LSP) effect on the nanostructures enhanced it. By removing the nanostructures on the hierarchical structures, THG intensity could be strongly suppressed. In the present paper, we also discuss the mechanism that enhances THG in nano/micro structures.

Keywords: Ni; THG; nonlinearity; plasmon; lightning rod

1. Introduction

Optical third harmonic generation (THG) is a complicated coherent nonlinear optical process [1]. In general, nonlinear waves are forbidden for centrosymmetric systems [1–4]. However, it is clear that the magnitude of the THG signal is dependent on the specific structure of the sample. THG processes are useful for various applications, such as autocorrelators [5] and nonlinear imaging [6]. Since nonlinear processes through photon–photon interactions are intrinsically weak, studies on the possibility of enhancing the nonlinear efficiency are essential [1,3,4].

The enhancement of THG responses by surface plasmons (SPs) has been reported among many researchers [7–9]. SPs are coherent electron oscillations that exist at the metal surface [10]. Since THG intensity is proportional to the third power of the incident light intensity [1], the SP-assisted electric field enhancement at the surface areas greatly enhances THG light strength. The magnitude of this enhancement is different from that of optical second harmonic generation (SHG).

Recently, we produced nanostructure-covered laser-induced periodic surface structures (NC-LIPSSs) [11,12] and investigated their SHG [13–15]. We found that NC-LIPSSs could significantly modify the second-order nonlinearity of the metallic surfaces. The micro-scale grooves and the nanostructures on the NC-LIPSSs could influence not only symmetry but also SPs [13–15].

In spite of the systematic study on the SP-assisted SHG processes in nano/micro hierarchical structures [13–15], a systematic study on the SP-assisted THG processes of nano/micro hierarchical structures is currently lacking. Nano- and micro structures both play important roles for generating SPs. Therefore, a sufficient investigation of the dual enhancement effect of nano/micro hierarchical structures is needed to fully understand the potential applications these structures might have in nonlinear plasmon devices. One interesting component of the recently obtained structure, the nanostructure-covered microcube (NC-MC) [14], is their dominant feature. The different sizes of the surface structure excite different SPs. The micro-cubes (MCs) on the NC-MC surface lead to lightning rod (LR) effects. Meanwhile, the nanostructures developed on surface layer of the hierarchical structure can also excite localized SPs (LSPs). Here, LR effects induced at the corners of cubes lead to large local field enhancements, evidenced by much electromagnetic (EM) wave simulation [16–18]. Strong local fields at the tips (corners) enhance the SHG by over two orders of magnitude compared with a nonsharp reference [19]. By sufficiently studying the SPs excited by structures of different sizes and the SPs combined with hierarchical structures, we can control the THG efficiency in nano/micro structures. In this study, we observe THG signals from Ni NC-MCs created via the laser ablation method. We will discuss how their THG properties relate to their nano/micro structure.

2. Materials and Methods

THG measurements on the NC-MCs on Ni, as seen in Figure 1a, were performed using an amplified femtosecond (fs) laser. Using an amplified fs laser system, two laser beams with different polarization directions (i.e., x and y directions) were focused on the Ni surface. Then, the two beams were controlled by a mechanical chopper and the number of pulses per pulse burst was adjusted to be 1. Namely, it can be said that the two beams alternately passed through the chopper pulse by pulse. Here, the laser fluence and incidence angle were set at 0.137 J/cm^2 and 1–1.5° ($\fallingdotseq$ normal to the substrate), respectively. By doing so, we succeeded in producing an NC-MCs sample on Ni. After fabrication, the surface morphologies of NC-MCs were monitored by a scanning electron microscope (SEM). The depth of the microstructures was measured by a ultra-violet (UV) laser microscope that has a lateral resolution of 200 nm and transverse resolution as high as 1 nm. Figure 1a shows the SEM image of the Ni NC-MCs. According to Figure 1a, the average periodicity Λ of the micro-cubes on Ni was 600 nm, and their depth d was ~100 nm. Moreover, the micro-cubes were covered with nanostructures. The fabrication method and other details of the sample have been reported elsewhere [14,15]. Figure 1b shows the arrangement of the THG intensity measurements performed on the NC-MC sample. An amplified fs laser (60 fs, 1.2 W, 1 kHz repetition rate) emitting at ~800 nm with the s-polarization was focused onto the NC-MCs sample with a 45° incidence. The s-polarization is parallel to the y-direction. We observed the dependence of the THG intensity on laser polarization. THG with 266 nm was directed through a color filter and a band-pass filter (FGUV5) to cut the surplus fundamental light. An output light was set to the s-polarization. By acknowledging the influence of hyper-Rayleigh scattering (HRS) [20], the generated THG photons were collected through a large N.A. lens. Then, the THG signal was detected using a photomultiplier tube (PMT).

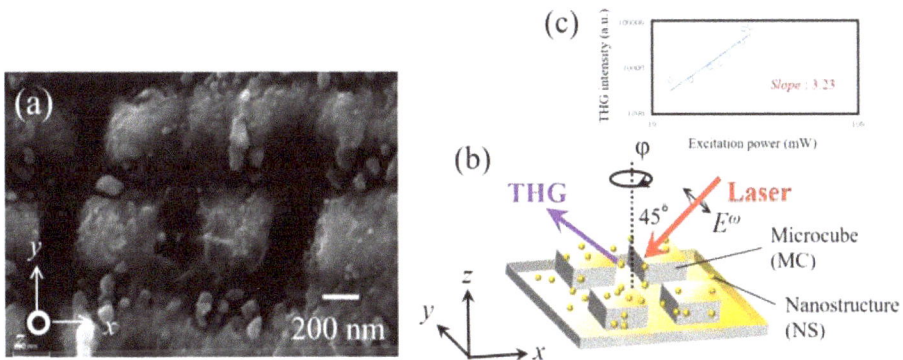

Figure 1. (a) A scanning electron microscope (SEM) image of the nanostructure-covered microcube (NC-MC) structures on Ni from a planar view normal to the surface; (b) Schematic detailing of measurement conditions. The rotation angle φ is defined as the angle between the incident plane and the travel direction of beam. When $\varphi = 0°$, the incident light propagates along the x-axis direction; (c) Excitation power dependence of the third harmonic generation (THG) intensity from the NC-MC sample.

3. Results

3.1. Definition of THG Intensity for Ni NC-MCs

The s-polarized THG intensity $|E_s(3\omega)|^2 \{= I_s(3\omega)\}$ with the $\chi^{(3)}_{YYYY}$ element is given as:

$$\left|E_s^R(3\omega)\right|_2 \propto \left|F_Y(3\omega)\chi^{(3)}_{YYYY}F_Y(\omega)^3 E_Y(\omega)^3\right|^2 \tag{1}$$

where third-order nonlinear susceptibility $\chi^{(3)}{}_{ijkl}$ is a fourth rank tensor. The indices *ijkl* are summed over the three directions of laser polarization *x*, *y*, and *z*. The coordinate system is oriented so that the *x* and *y* coordinates are in the plane and the *z* coordinate is in the direction normal to the substrate surface [14]. $F_Y(\omega)$ is the Fresnel factor at the fundamental frequency for the *y* direction, and $F_Y(3\omega)$ is the Fresnel factor at the THG frequency along the *y* direction. The relation as $E_{Y,loc} = F_Y(\omega)E_Y(\omega)$ holds for the local electric field E_{loc}. In the case of nanostructures, $F_Y(\omega)$ can depend on both LSP and LR effects [21]. On the other hand, $F_Y(3\omega)$ is expected to originate only from the LR effect because of no resonance. Either way, Equation (1) expresses that the s-polarized THG intensity $|E_s(3\omega)|^2$ depends on the nonlinear susceptibility χ and the *y*-directed local field $E_{Y,loc}$ of the electric field component.

3.2. THG Signal Intensity for Ni NC-MCs

We first investigated the output THG power dependence on fundamental frequency in the NC-MC sample. The power dependence of the THG intensity fits a slope of 3.23 when plotted on a log-log scale as shown in Figure 1c, which copes well with the third-order nature of the light emission.

In order to investigate the origin of the THG, the $\chi^{(3)}$ of the microstructure surface was analyzed. The $\chi^{(3)}$ is sensitive to the symmetry of the structure. Since the structure of MCs in a tetragonal system is characterized by C_{4v} symmetry with four mirror planes, 21 or 11 nonlinear susceptibility elements are permitted [1]. Under s-polarization configuration, the $\chi^{(3)}{}_{YYYY}$ should be accepted. If we assume that the $\chi^{(3)}{}_{YYYY}$ is dominated to the s-in/s-out polarization combination, the THG should be forbidden for $\varphi = 0°$, $45°$, and $90°$ when φ ranges from $0°$ to $90°$. In conclusion, we found that the $\chi^{(3)}{}_{YYYY}$ dependence on a $\sin(4\varphi)$ function is the main factor that modulates the THG emission from the MCs. The THG related to the symmetry for the MCs was then fitted by a $\sin^2(4\varphi)$ function, and this trend is the same as the one found for the SHG case [14,15]. On the other hand, the nanostructures contain nanoparticles with strong quadrupolar resonances, and they should provide the fixed nonlinear intensities independent of the φ. By combining these symmetry elements χ and the local field E_{loc}, the simulated curves shown in Figure 2 can be obtained.

We have also observed the sample rotation angle dependence of the THG from the NC-MC sample at a 45° incidence, as shown in Figure 1b. Figure 2a illustrates the THG intensity from the Ni NC-MCs for the s-in/s-out combination, as a function of the φ. The curve simulated in the previous section is patterned in Figure 2. In fact, the THG in the s-in/s-out combination showed an isotropic pattern. According to the literature [9], this enhancement is mainly caused by the LSPs from nanostructures on the microcubes. As a consequence, the THG intensity pattern for the s-in/s-out combination does not depend on the φ, and the THG intensity is sensitive to LSP effect on the nanostructures rather than depending on the symmetry and LR effects on microstructures. Figure 2b illustrates the SHG emission intensity in the s-in/s-out combination, as a function of the φ from the Ni NC-MCs [14]. The SHG pattern in the s-in/s-out polarization configuration exhibits eight dim minima at $\varphi = 0°$, $45°$, $90°$, $135°$, $180°$, $225°$, $270°$, and $315°$ [14]. Thus, the SHG intensity pattern for the s-in/s-out polarization combination strongly depends on the rotation angle φ, and the SHG intensity is sensitive to symmetry and LR effects on the microstructures rather than the LSPs' effect on the nanostructures [14].

Figure 2 shows that there is a large difference between second- and third- harmonic patterns. We hypothesize that the main reason is the magnitude of the contribution of the nanostructures to the LSPs. As Equation (1) points out, only the local field $E_{Y,loc}$ (including Fresnel factor $F_Y(\omega)$)

relates to the plasmon excitation if we exclude Fresnel factors $F_Y(3\omega)$ and $F_Y(2\omega)$. Due to their weak off-resonance coupling to the LR effect with nanostructures, the field relation would be $|E^{LSP}{}_{Y,loc}| > |E^{LR}{}_{Y,loc}|$. Since the THG signal intensity $I_s(3\omega)$ depends on the sixth power of the local field $|E_{Y,loc}|^6$, it will receive a contribution from the nanostructures that is more significant than in the case of SHG, depending on the fourth power of the local field $|E_{Y,loc}|^4$.

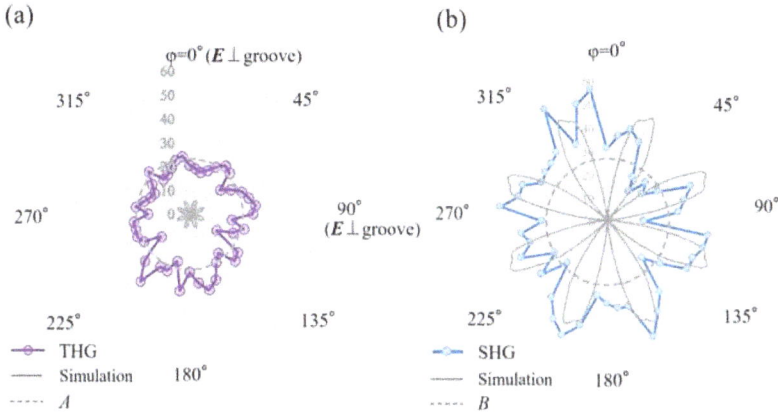

Figure 2. Angular (**a**) THG and (**b**) second harmonic generation (SHG) signal intensity. The angular nonlinear intensity of the NC-MCs as a function of the φ. The solid curve results from a simulation curve implanting a calculation by symmetrical parameters and local electric fields.

In order to control THG enhancement, we measured the THG signal using the MC sample. In Figure 1a, many nanodots were ablated from the MCs, leading to the isotropic THG as seen in Figure 2a. The SEM image of the surface of the MC sample is shown in Figure 3a. Indeed, several nanodots on microcubes were removed. Figure 3b shows the relative THG and SHG intensities in the s-in/s-out combination for NC-MCs and MCs on the Ni at $\varphi = 22.5°$. According to the data, the THG intensity decreases to a greater degree than the SHG intensity. The reason for this is related to the different magnitudes of contribution of THG and SHG to LSPs, as mentioned above.

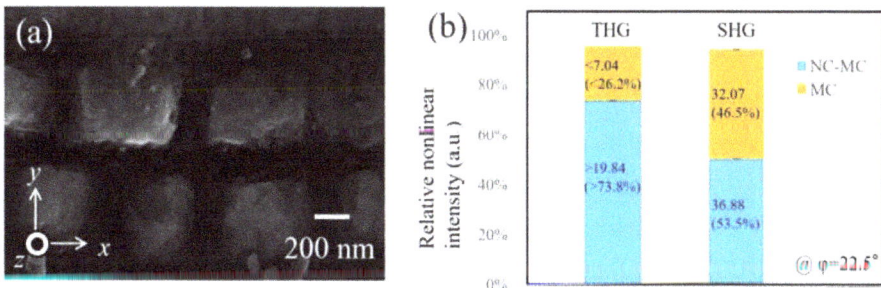

Figure 3. (**a**) SEM image of the microcube (MC) surface; (**b**) relative nonlinear optical intensities for s-in/s-out at $\varphi = 22.5°$ for NC-MCs and MCs, respectively.

We describe here the experimental studies for the different magnitudes of the contribution of the THG and SHG to LSPs. Microcube structures created via the laser-ablation method do not have sharper structure at their corners (as seen in Figure 1a), and they are like "dorm-style tents". Due to this, the magnitude of the LR effect on the cubes should be 10 times weaker than for the perfect cubes [19]. On the other hand, a lot of nanostructures generated via ablation lead to a large LSP effect, as expected [16]. Namely, the magnitude of the electric field enhanced by the LR on the microstructure might be weaker than that enhanced by LSPs on nanostructures. If so, LSP enhancement should result in greater SHG than LR enhancement. However, the magnitude of enhancements cannot be distinguished in Figure 2b. LSP enhancement resulted in greater THG than LR enhancement, as seen in Figure 2a. This is because THG intensity is proportional to the sixth power of electric fields, and thus the difference between their two effects appeared significantly. These facts suggest that LSP-enhanced THG screened LR-enhanced THG. In short, the gap may widen even more if higher-order nonlinear processes are taken into account when the module of the local field on the nanostructures ($E^{LSP}_{Y,loc}$) is greater than module of the local field on the microcubes ($E^{LR}_{Y,loc}$) (i.e., $|E^{LSP}_{Y,loc}| > |E^{LR}_{Y,loc}|$). THG intensity is resonantly enhanced by the LSPs from the nanostructures, and the enhancement due to the LR effect from the microstructures is somehow hidden. These facts show a clear relation between LSP and LR enhancements on the nano/micro structure.

The nano/micro hierarchical structure consisting of differently-sized structures can induce cascaded plasmon field enhancement [14,22]. If the nanostructures are in contact with other nanostructures (as seen in Figure 1a), a large enhancement of the local field $E_{Y,loc}(\omega)$ will be obtained thanks to the cascading *E*-field enhancement effect [22] due to the contentious LSP effect. Then, the *E*-field enhancement occurring due to the cascading effect may boost the THG intensity isotropically at the φ, as seen in Figure 2a.

4. Conclusions

We investigated the THG signal intensity from NC-MC nano/micro structures on Ni. The influence of LSPs on the nanostructures clearly enhanced THG. By removing the nanostructures on the hierarchical structures, we could extract the THG supported by the symmetry and LR effects on microstructures. We found that the enhancement due to the LR effect from the microstructures was hidden by resonant enhancement due to the LSPs from the nanostructures. These interesting physical phenomena could have applications in nonlinear plasmonic devices.

Acknowledgments: Support from the URnano is gratefully acknowledged.

Author Contributions: Yoichi Ogata designed the experiments and analyzed data. Anatoliy Vorobyev and Chunlei Guo joined the discussions and supported my work.

Conflicts of Interest: The authors declare no conflict of interest.

References

1. Boyd, R.W. *Nonlinear Optics*, 3rd ed.; Academic Elsevier: Amsterdam, The Netherlands, 2008.
2. Shen, Y.R. *The Principles of Nonlinear Optics*; John Wiley & Sons: New York, NY, USA, 1984.
3. Kauranen, M.; Zayats, A.V. Nonlinear plasmonics. *Nat. Photonics* **2012**, *6*, 737–748. [CrossRef]
4. Ogata, Y.; Mizutani, G. Control of Cross-Sections and Optical Nonlinearity of Pt Nanowires and the Roughness Effect. *Phys. Res. Int.* **2012**, *2012*, 969835. [CrossRef]
5. Monat, C.; Grillet, C.; Collins, M.; Clark, A.; Schroeder, J.; Xiong, C.; Li, J.; O'faolain, L.; Krauss, T.F.; Eggleton, B.J.; et al. Integrated optical auto-correlator based on third-harmonic generation in a silicon photonic crystal waveguide. *Nat. Commun.* **2014**, *5*, 3246. [CrossRef] [PubMed]
6. Olivier, N.; Aptel, F.; Plamann, K.; Schanne-Klein, M.C.; Beaurepaire, E. Harmonic microscopy of isotropic and anisotropic microstructure of the human cornea. *Opt. Express* **2010**, *18*, 5028–5040. [CrossRef] [PubMed]
7. Tsang, T.Y. Surface-plasmon-enhanced third-harmonic generation in thin silver films. *Opt. Lett.* **1996**, *21*, 245–247. [CrossRef] [PubMed]
8. Niu, J.; Luo, M.; Liu, Q.H. Enhancement of graphene's third-harmonic generation with localized surface plasmon resonance under optical/electro-optic Kerr effects. *JOSA B* **2016**, *33*, 615–621. [CrossRef]
9. Nezami, M.S.; Gordon, R. Localized and propagating surface plasmon resonances in aperture-based third harmonic generation. *Opt. Express* **2015**, *23*, 32006–32014. [CrossRef] [PubMed]

10. Piccione, B.; Aspetti, C.O.; Cho, C.H.; Agarwal, R. Tailoring light–matter coupling in semiconductor and hybrid-plasmonic nanowires. *Rep. Prog. Phys.* **2014**, *77*, 086401. [CrossRef] [PubMed]

11. Vorobyev, A.Y.; Guo, C. Effects of nanostructure-covered femtosecond laser-induced periodic surface structures on optical absorptance of metals. *Appl. Phys. A* **2007**, *86*, 321–324. [CrossRef]

12. Hwang, T.Y.; Vorobyev, A.Y.; Guo, C. Surface-plasmon-enhanced photoelectron emission from nanostructure-covered periodic grooves on metals. *Phys. Rev. B* **2009**, *79*, 085425. [CrossRef]

13. Ogata, Y.; Vorobyev, A.; Guo, C. Symmetry-sensitive plasmonic enhancement of nonlinear optical intensity in nano-micro hierarchical structures on silver. *Surf. Interface Anal.* **2016**, *48*, 1108–1113. [CrossRef]

14. Ogata, Y. Optical second harmonic generation from nanostructure-covered micro-cubes on nickel. *Opt. Mater. Express* **2016**, *6*, 1520–1529. [CrossRef]

15. Ogata, Y.; Guo, C. Nonlinear optics on nano/microhierarchical structures on metals: Focus on symmetric and plasmonic effects. *Nano Rev. Exp.* **2017**, *8*, 1339545. [CrossRef]

16. Gangadharan, D.T.; Xu, Z.; Liu, Y.; Izquierdo, R.; Ma, D. Recent advancements in plasmon-enhanced promising third-generation solar cells. *Nanophotonics* **2016**, *6*, 153–175.

17. Chen, S.; Li, G.; Zeuner, F.; Wong, W.H.; Pun, E.Y.B.; Zentgraf, T.; Cheah, K.W.; Zhang, S. Symmetry-selective third-harmonic generation from plasmonic metacrystals. *Phys. Rev. Lett.* **2014**, *113*, 033901. [CrossRef] [PubMed]

18. Ashok, A.; Arackal, A.; Jacob, G. Study of Surface Plasmon Excitation on Different Structures of Gold and Silver. *Nanosci. Nanotechnol.* **2015**, *5*, 71–81.

19. Kontio, J.M.; Husu, H.; Simonen, J.; Huttunen, M.J.; Tommila, J.; Pessa, M.; Kauranen, M. Nanoimprint fabrication of gold nanocones with ~10 nm tips for enhanced optical interactions. *Opt. Lett.* **2009**, *34*, 1979–1981. [CrossRef] [PubMed]

20. Gomopoulos, N.; Lütgebaucks, C.; Sun, Q.; Macias-Romero, C.; Roke, S. Label-free second harmonic and hyper Rayleigh scattering with high efficiency. *Opt. Express* **2013**, *21*, 815–821. [CrossRef] [PubMed]

21. Hubert, C.; Billot, L.; Adam, P.M.; Bachelot, R.; Royer, P.; Grand, J.; Gindre, D.; Dorkenoo, K.D.; Fort, A. Role of surface plasmon in second harmonic generation from gold nanorods. *Appl. Phys. Lett.* **2007**, *90*, 181105. [CrossRef]

22. Toroghi, S.; Kik, P.G. Cascaded plasmon resonant field enhancement in nanoparticle dimers in the point dipole limit. *Appl. Phys. Lett.* **2012**, *100*, 183105. [CrossRef]

materials   MDPI

Article

Efficiency Enhancement of Perovskite Solar Cells with Plasmonic Nanoparticles: A Simulation Study

Ali Hajjiah [1,*], Ishac Kandas [2,3,4] and Nader Shehata [2,3,4,5] ![ORCID]

[1]   Electrical Engineering Department, College of Engineering and Petroleum, Kuwait University, Safat 13113, Kuwait
[2]   Department of Engineering Mathematics and Physics, Faculty of Engineering, Alexandria University, Alexandria 21544, Egypt; ishac@vt.edu (I.K.); n.shehata@kcst.edu.kw (N.S.)
[3]   Center of Smart Nanotechnology and Photonics (CSNP), SmartCI Research Center, Alexandria University, Alexandria 21544, Egypt
[4]   Kuwait College of Science and Technology (KCST), Doha Spur Rd., Safat 13113, Kuwait
[5]   USTAR Bioinnovation Center, Faculty of Science, Utah State University, Logan, UT 84341, USA
*   Correspondence: ali.hajjiah@ku.edu.kw; Tel.: +965-9966-2243

Received: 4 August 2018; Accepted: 3 September 2018; Published: 5 September 2018

Abstract: Recently, hybrid organic-inorganic perovskites have been extensively studied due to their promising optical properties with relatively low-cost and simple processing. However, the perovskite solar cells have some low optical absorption in the visible spectrum, especially around the red region. In this paper, an improvement of perovskite solar cell efficiency is studied via simulations through adding plasmonic nanoparticles (NPs) at the rear side of the solar cell. The plasmonic resonance wavelength is selected to be very close to the spectrum range of lower absorption of the perovskite: around 600 nm. Both gold and silver nanoparticles (Au and Ag NPs) are selected to introduce the plasmonic effect with diameters above 40 nm, to get an overlap between the plasmonic resonance spectrum and the requested lower absorption spectrum of the perovskite layer. Simulations show the increase in the short circuit current density (J_{sc}) as a result of adding Au and Ag NPs, respectively. Enhancement in J_{sc} is observed as the diameter of both Au and Ag NPs is increased beyond 40 nm. Furthermore, there is a slight increase in the reflection loss as the thickness of the plasmonic nanoparticles at the rear side of the solar cell is increased. A significant decrease in the current loss due to transmission is achieved as the size of the nanoparticles increases. As a comparison, slightly higher enhancement in external quantum efficiency (EQE) can be achieved in case of adding Ag NPs rather than Au NPs.

Keywords: Perovskites; solar cell; plasmonic nanoparticles; short circuit current; quantum efficiency

---

1. Introduction

One of the hottest topics in materials science in the past few years has been hybrid organic-inorganic perovskites due to their superb properties in optoelectronic applications. These organo-metal halide materials have emerged as an excellent absorber material for thin-film photovoltaics with spectacular achievements in power conversion efficiencies that compete with silicon and other established thin-film technologies (i.e., CdTe and CIGS). The power conversion efficiency (PCE) of perovskite based solar cells has increased from 3.8% upon its inception in 2009 to a certified 22.1% in early 2016 [1,2]. The material possesses the ABX3 crystal structure, where A is a small organic cation, B is a cationic group 14 metal, and X is a halide anion. The most commonly used perovskite semiconductor material in solar cells is methylammonium-lead (II)-iodide with the chemical formula $CH_3NH_3PbI_3$ (MAPbI3), owing to its excellent material properties for photovoltaic applications. MAPbI3 is an inorganic-organic hybrid perovskite that forms a tetragonal crystal structure and is compatible with both solution processing [3] and evaporation techniques [4,5]. This material is a direct bandgap semiconductor [6] with a bandgap around 1.6 eV and a large open circuit voltage (Voc) of 1.07 V [4], only 0.53 V less than the perovskite bandgap potential (Eg/q) [7]. The bandgap of the MAPbI$_3$ perovskite (1.6 eV) can be continuously tuned up to 2.25 eV by substituting Br for I to make MAPb(I$_{1-x}$Br$_x$)$_3$ [8], which makes perovskite solar cells especially attractive for tandem applications. Furthermore, it is also an intrinsic material with high carrier mobilities [9], high absorption coefficient [10,11], shallow defect levels [12], and a long charge-carrier diffusion length [13–15], which are important metrics for highly performing solar cells.

One method for achieving light trapping in thin film solar cells is the use of metallic NPs [16–18]. Metallic NPs exhibits the phenomenon of surface plasmon resonance when illuminated with light of suitable frequency [19]. Metallic NPs show potential for enhancing light absorption and photocurrent, therefore, plasmonic resonances in metal NPs have attracted the attention in sensors and other applications such as solar cells [20–22]. Plasmonic structure can be integrated with solar cell in many ways [23]. Metal NPs can be deposited on the front surface of the solar cell. Also, they can be embedded inside the cell [24]. However, it was found that locating the particles on the rear side of the absorber layer is more effective in enhancing photocurrent [25,26]. From literature, metal NPs of different size, shape, and composition were used as absorption enhancers in methylammonium lead iodide perovskite solar cells. The absorption enhancement is the key point to reduce the thickness of the perovskite solar cell. Integration plasmonic gold nanostars (Au NSs) into mesoporous TiO$_2$ photoelectrodes for perovskite solar cells (PSCs) increased the efficiency from 15.19 up to 17.72% [27,28]. Also, size has been shown to play a pivotal role in performance enhancement. Previous work has systemically screened different AuNPs sizes in photoelectrodes to find the champion devices contained 8 nm plasmonic Au NPs [29]. Incorporation of Au NPs into titanium dioxide (TiO$_2$) photoelectrodes showed 20% improvement in average. The refractive index of metal NPs is complex. The permittivity is the square of the refractive index and consequently it is a complex quantity. In optics, the permittivity depends strongly on the frequency. Optical properties of NPs are different from the bulk specimen [30]. Noble NPs can resonate with light, which gives it a great importance. Localized surface plasmons (LSP) can be excited and cause resonance with the incident frequency under certain conditions. The resonant frequency is strongly affected by NPs size, nanoparticle shape, and surrounding medium. Near field enhancement can be exploited in many applications such as solar cell applications [31–34].

In this paper, we focus on the improvement of perovskite solar cell through the addition of plasmonic NPs using a simulation study. The perovskite solar cells may have some lower optical absorption in the visible spectrum around the red region. Therefore, our contribution is to prove the concept that the overall efficiency can be enhanced through adding metallic NPs whose plasmonic resonance wavelength is close to the spectrum range of lower absorption of the perovskite. In more details, Au and Ag NPs are selected to introduce the plasmonic effect with diameters above 40 nm, to get an overlap between the plasmonic resonance spectrum and the lower absorption spectrum of the perovskite layer around 600 nm wavelength. Therefore, this coupling can enhance the quantum efficiency in this spectrum region. In this work, both Au and Ag NPs are selected to be added at the rear-side as an additional layer of perovskite solar cell. Simulated optical properties and quantum efficiency calculations are presented along with a comparison of the impact of both Au and Ag additives.

2. Literature Background

Our targeted device in this work is the regular structure of n-i-p semi-transparent CH$_3$NH$_3$PbI$_3$ perovskite solar cell (area = 0.1–1 cm^2) with an architecture of glass/ITO-front/SnO$_2$/PCBM/CH$_3$NH$_3$PbI$_3$/spiro-OMeTAD/ITO-rear. A schematic model of the semi-transparent perovskite solar cell used in our transfer-matrix-based optical simulations can be seen in Figure 1.

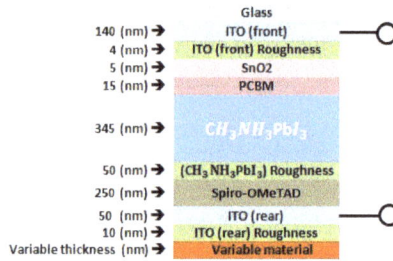

Figure 1. Schematic model of the semi-transparent n-i-p CH₃NH₃PbI₃ perovskite solar cells applied in the transfer-matrix-based optical simulations. The variable material layer is where gold and silver nanoparticles (Au and Ag NPs) are included in our simulations. Roughness is simulated as an effective medium of the adjacent media using the Bruggeman effective medium approximation.

To investigate the effect of both Au and Ag NPs on the performance of the perovskite solar cell, the Au and Ag NPs were applied at the rear side of the perovskite solar cell with diameter above 40 nm. Then, we can obtain plasmonic resonance wavelength above 550 nm, which can compensate the external quantum efficiency losses in perovskite in this range of spectrum. Then, the Transfer-Matrix-Based Optical Simulation Method (TMM) was used in our investigation. This method allows modeling of the optical properties of thin-film layer stacks by solving Maxwell's equations at each interface through using the complex refractive index and layer thicknesses of all relevant materials as input [35–38]. Reflectance (R), Transmittance (T), Absorbance (A), and EQE spectra of the fabricated perovskite solar cell were measured in order to calibrate and underline the accuracy of our optical simulations. More information on the calibration of the TMM optical simulator and the details of the perovskite solar cell fabrication process can be reviewed from our references [39,40]. Surface roughness is considered as an effective medium according to the Bruggeman effective medium approximation (BEMA) [41]. Therefore, in our simulations, interface roughness is simulated using a BEMA layer consisting of a mixture of the optical constants for the adjacent media. The accuracy of our optical simulation is confirmed by showing excellent agreement with experimental data, as shown in Figure 2a,b. However, small offsets between experimental and simulation data for long wavelengths in our transmission and reflection plots can be explained by absorption and/or scattering in the substrate.

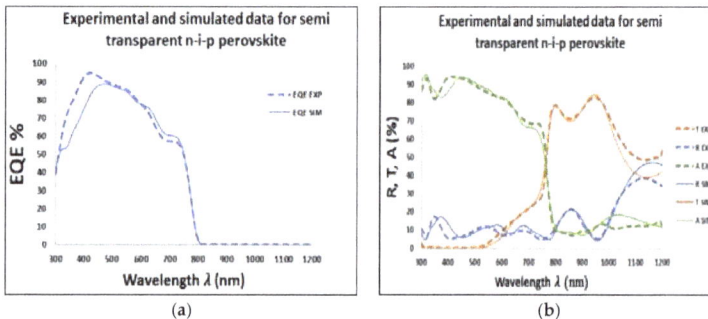

Figure 2. (a) Shows measured and simulated external quantum efficiency (EQE)% and (b) shows measured and simulated percentages of reflectance, transmittance, and absorbance spectra of semitransparent CH₃NH₃PbI₃ solar cell with the architecture glass/ITO-front/ITO-front-roughness/ SnO₂/PCBM/CH₃NH₃PbI₃/CH₃NH₃PbI₃-roughness/spiro-OMeTAD/ITO-rear/ITO-rear-roughness. The dotted line represents measurements on the actual device and solid line represents transfer-matrix-based simulations.

The model used to describe the permittivity of gold and silver is Drude-critical points model. The model can be given in Equation (1) as follows [42]

$$\varepsilon(\omega) = \varepsilon_\infty - \frac{\omega_D^2}{\omega^2 + i\gamma_{NP}\omega} + \sum_{p=1}^{2} A_p \Omega_p \left(\frac{e^{i\phi_p}}{\Omega_p - \omega - i\Gamma_p} + \frac{e^{-i\phi_p}}{\Omega_p + \omega + i\Gamma_p} \right) \tag{1}$$

where ε_∞ is the permittivity due to interband transitions, ω_D is the plasma frequency, γ_{NP} is the damping constant. A_p, Ω_p, ϕ_p, and Γ_p are constants and are summarized in Table 1 for gold and silver. The experimental data was taken from reference [43]. The model is valid for wavelengths in the range between 200 nm and 1000 nm [42]. The damping constant, γ_{NP} counts for absorption loss is size dependent and given by Equation (2) [44–46]

$$\gamma_{NP} = \gamma + \frac{Cv_F}{R} \tag{2}$$

where γ equals 1.0805×10^{14} rad/s and 4.5841×10^{13} rad/s for gold and silver respectively. The constant C is considered 1 for both of them [46], v_F is the Fermi velocity and equals 1.39×10^6 m/s and 1.38×10^6 m/s for gold and silver, respectively, and R is the radius of the spherical NPs [45,46]. The refractive index is complex and simply expressed as follows in Equation (3)

$$n = \sqrt{\varepsilon(\omega)} = n_r + ik \tag{3}$$

The real and imaginary parts were calculated for both Au and Ag to be used in the simulation part. To calculate the absorption, scattering, and extinction cross sections of noble NPs, Mie theory can be used. Mie theory is using Maxwell equations to calculate the fields in the vicinity of the nanoparticle. Equation (4) shows the cross sections which can be can be given as [47]

$$\sigma_{ext} = \frac{2\pi}{|k|^2} \sum_{n=1}^{\infty} (2n+1) Re[a_n + b_n] \tag{4a}$$

$$\sigma_{sca} = \frac{2\pi}{|k|^2} \sum_{n=1}^{\infty} (2n+1) [|a_n|^2 + |b_n|^2] \tag{4b}$$

$$\sigma_{abs} = \sigma_{ext} - \sigma_{sca} \tag{4c}$$

where σ_{ext}, σ_{sca}, and σ_{abs} are the extinction, scattering, and absorption cross sections, respectively, n is the multipole order, k is the wave number. Equation (5) shows the Mie coefficients a_n and b_n, which are given by [47]

$$a_n = \frac{m\psi_n(mx)\psi_n'(x) - \psi_n(x)\psi_n'(mx)}{m\psi_n(mx)\eta_n'(x) - \eta_n(x)\psi_n'(mx)} \tag{5a}$$

$$b_n = \frac{\psi_n(mx)\psi_n'(x) - m\psi_n(x)\psi_n'(mx)}{\psi_n(mx)\eta_n'(x) - m\eta_n(x)\psi_n'(mx)} \tag{5b}$$

where ψ_n, and η_n are Riccarti –Bessel functions, while ψ_n', and η_n' are their derivatives. m is the ration of the complex refractive indices of the nanoparticle and the surrounding medium. x is given by $2\pi R/\lambda$ where λ is the wavelength of the incident light.

Table 1. Parameters used in Drude model.

	ε_∞	ω_D (rad/s)	A1	A2	ϕ_1	ϕ_2	Ω_1 (rad/s)	Ω_2 (rad/s)	Γ_1 (rad/s)	Γ_2 (rad/s)
Au	1.1431	1.3202×10^{16}	0.2669	3.0834	-1.2371	-1.0968	3.8711×10^{15}	4.1684×10^{15}	4.4642×10^{14}	2.3555×10^{15}
Ag	15.833	1.3861×10^{16}	1.0171	15.797	-0.9394	1.8087	6.6327×10^{15}	9.2726×10^{17}	1.6666×10^{15}	2.3716×10^{17}

3. Simulation Procedure

In order to investigate the effect of Au and Ag NPs on the performance of the perovskite solar cell, optical loss analysis was considered by varying the thickness of both Au and Ag films at the rear side of the perovskite solar cell in our TMM simulations and calculating both internal current densities losses (J_{short}, J_{medium}, J_{long}), the external losses ($J_{escape-back}$, $J_{reflection}$), the average transmittance (800–1200 nm), and the J_{sc} in our perovskite solar cell. Here, we assume one homogeneous layer of NPs on the perovskite cell. Therefore, the thickness of the plasmonic layer is the same as the diameter of NPs. According to the limitations of the used coding we have assumed that Au and Ag NPs or even though the Au and Ag planar film is deposited as one layer over the perovskite with negligible grain boundaries problems that can be found in the real design.

The internal current losses are calculated using Equation (6) by integrating the area between the EQE and the absorbance curves over the AM1.5G solar spectrum for different wavelength regions corresponding to short (λ = 300–450 nm), medium (λ = 450–700 nm), and long (λ = 700–1000 nm) internal current losses. For the external current losses, however, we focused on both losses of $J_{escape,back}$ and $J_{reflection}$. Regarding the $J_{escape-back}$ external loss, the back side of the semi-transparent perovskite cell was considered where part of the long wavelength light is lost by transmission through the semi-transparent ITO-rear through the nanoparticle material. This external loss is calculated by integrating the transmission curve of the cell according to Equation (7). Moreover, at the device front, part of the light is lost due to external reflection. These reflection losses ($J_{reflection}$) are calculated by integrating the reflection spectrum of the front side of the cell according to Equation (8). Then, by using TMM method and Equation (9), J_{sc} for each specific nanoparticle thickness can be expressed as shown [48].

$$J_{internal-loss} = \frac{q}{hc} \int_{\lambda_1}^{\lambda_2} \lambda \cdot \Phi(\lambda) \cdot [A_{cell}(\lambda) - EQE(\lambda)] \cdot d\lambda \tag{6}$$

$$J_{escape-back\ external} = \frac{q}{hc} \int_{500}^{1000} \lambda \cdot \Phi(\lambda) \cdot T_{cell}(\lambda) \cdot d\lambda \tag{7}$$

$$J_{reflection} = \frac{q}{hc} \int_{\lambda_1}^{\lambda_2} \lambda \cdot \Phi(\lambda) \cdot R(\lambda) d\lambda \tag{8}$$

$$J_{sc} = \frac{q}{hc} \int_{300nm}^{1000nm} \lambda \cdot \Phi(\lambda) \cdot EQE(\lambda) d\lambda \tag{9}$$

where q is the elementary charge, h is Planck's constant, c is the speed of the light, λ is the wavelength, $\Phi(\lambda)$ is the AM1.5G solar spectrum, T_{cell} is the transmission of the cell, R is the reflection of the cell.

4. Results & Discussions

4.1. Plasmonic Resonances of Au and Ag NPs

The real and imaginary parts of Au and Ag NPs are shown in Figure 3. The radius of Ag or Au NPs is 50 nm. All the used parameters are summarized in Table below [36]. As noticed, the real part of Au and Ag permittivity is very close and negative for a wide range of wavelengths. The negative real refractive index is responsible for the appearance of resonance wavelength. The imaginary part is responsible for the absorption loss. As seen, the silver loss is less than the gold loss.

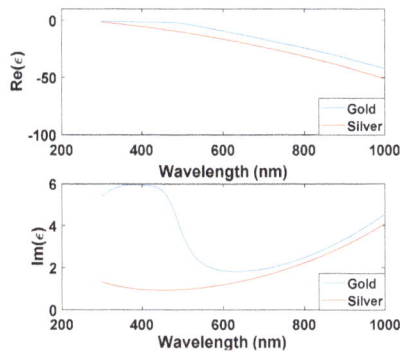

Figure 3. Real and imaginary parts of Au and Ag permittivity.

In Figure 4, the cross sections were simulated for particles of radius 50 nm. The surrounding medium has refractive index of value 1.5. For silver case, the cross section spectrum is much wide and may contain more than one peak. It is worthy to mention that, the resonance wavelength is size dependent. As the nanoparticle size increases, red shift occurs for Au and Ag cross sections.

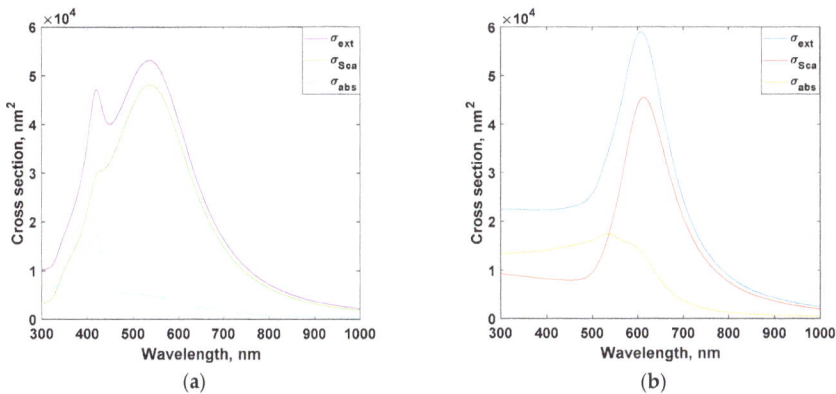

Figure 4. Cross sections of (**a**) Au, and (**b**) Ag plasmonic nanoparticles (NPs).

4.2. Impact of Ag NPs on Perovskite Cells

Figure 5 shows the optical characteristics of perovskite cell with added layer of Au NPs at different diameters. It can be noticed that the transmission is tremendously decreased with increasing the size of added Au NPs. However, the corresponding reflection and absorption have higher values with added Au NPs especially at wavelength around the plasmonic resonance wavelength. The resonance wavelength is varying from approximately ~550 to ~600 nm when the particle diameter is changing from 40 to 300 nm, respectively, for surrounding medium of refractive index of 1.5. Then, Figure 6 shows the improvement of EQE at wavelength range close to the plasmonic resonance frequency. The EQE enhancement can be explained due to the impact of reflected photoelectrons, which are trapped according to the added layer of plasmonic NPs, beside the optical enhancement of absorption due to plasmonic resonance of the added layer. There is a clear enhancement of EQE with lowest select size of Au NPs. By increasing Au NPs size, the EQE shows slight improvement with a saturation behavior at relatively higher NPs size up to 300 nm in diameter.

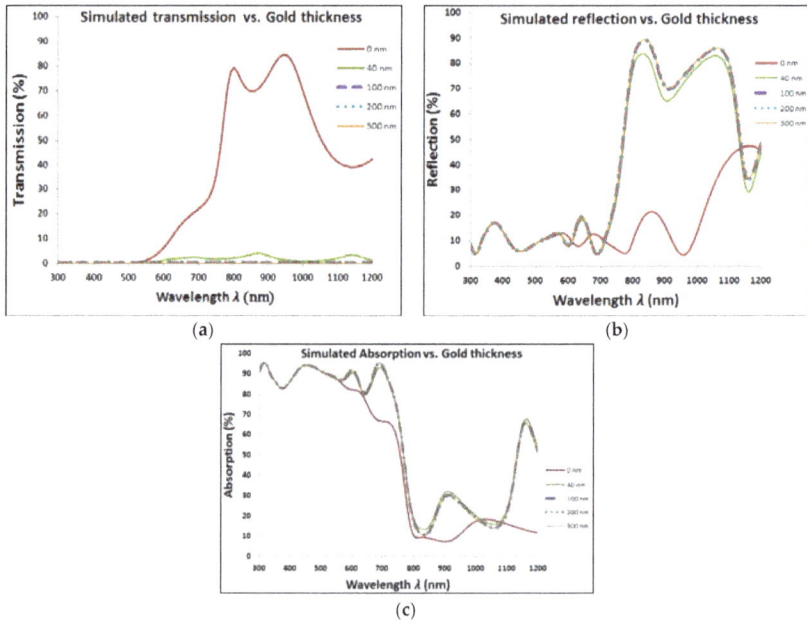

Figure 5. Optical characteristics for perovskite cell with different thickness of gold layer. These characteristics are (**a**) Transmission, (**b**) Reflection, and (**c**) Absorption.

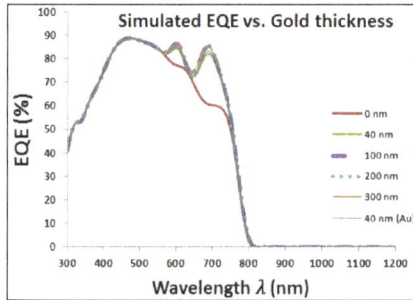

Figure 6. Enhancement of external quantum efficiency (EQE) due to added gold nanoparticles (Au NPs) in the wavelength range close to plasmonic resonance of gold.

In Table 2, we compare the internal losses, external losses, and the corresponding J_{sc} of our solar cell with uniform NPs radii at different sizes to show its effect on our simulated perovskite solar cell. Simulations show a 1.24 mA/cm^2 increase in the J_{sc} as a result of adding 40 nm Au NPs at the rear side of our perovskite solar cell. Further enhancement in J_{sc} is observed as the diameter of the Au NPs is increased beyond 40 nm. However, an insignificant difference in J_{sc} ($\approx$0.1 mA/cm^2) is observed when comparing the 40 nm Au NPs with a 40 nm Au planar thin film. The Au planar perovskite solar cell is displayed in order to better assess the performance of our Au NPs proposed solar cell structure. Furthermore, it can also be seen from the table a slight increase in the reflection loss as the thickness of the Au NPs at the rear side of the solar cell is increased. However, with increasing Au NPs thickness at the rear side of the perovskite solar cell, a noticeable decrease in the current loss due to transmission ($J_{escape,back}$) can be seen with nearly neglected value at 200 nm size or higher which shows a clear advantage of the added NPs.

Table 2. The effect of added gold nanoparticles (Au NPs) of different nanosize on the internal losses, external losses, and the short circuit current for our simulated Perovskite solar cell.

Perovskite Cell Losses (mA/cm^2)				J_{sc} (mA/cm^2)
Gold (nm)	$J_{Internal}$	$J_{escape,back}$	$J_{reflection}$	
0	4.15	16.49	6.55	18.69
40 (Au) planar	6.939	0.784	18.7	20.03
40 (Au NPs)	7.45	0.764	18.31	19.93
100 (Au NPs)	7.013	0.009	19.25	20.18
200 (Au NPs)	6.926	0.000	19.33	20.20
300 (Au NPs)	6.899	0.000	19.35	20.21

4.3. Impact of Ag NPs on Perovskite Cells

The same behavior of added Au NPs is repeated with another type of plasmonic nanostructure, which is silver. Figure 7 shows the optical characteristics of perovskite cell with added layer of Ag NPs at different diameters. As the case of gold, adding Ag NPs makes the transmission worse, but with improved reflection and absorption, especially at close wavelength-range to the plasmonic resonance of silver. The overall impact is the enhancement of EQE with added Ag NPs as shown in Figure 8. Table 3 shows the results of different types of current losses with different thickness of silver layer. Same behavior is existed as found in Table 1 in the case of Au NPs via enhancement of both reflection and short circuit currents. However, silver is showing higher enhancement in short circuit currents and faster decay of backside escape current with increasing the diameter of Ag NPs compared to gold.

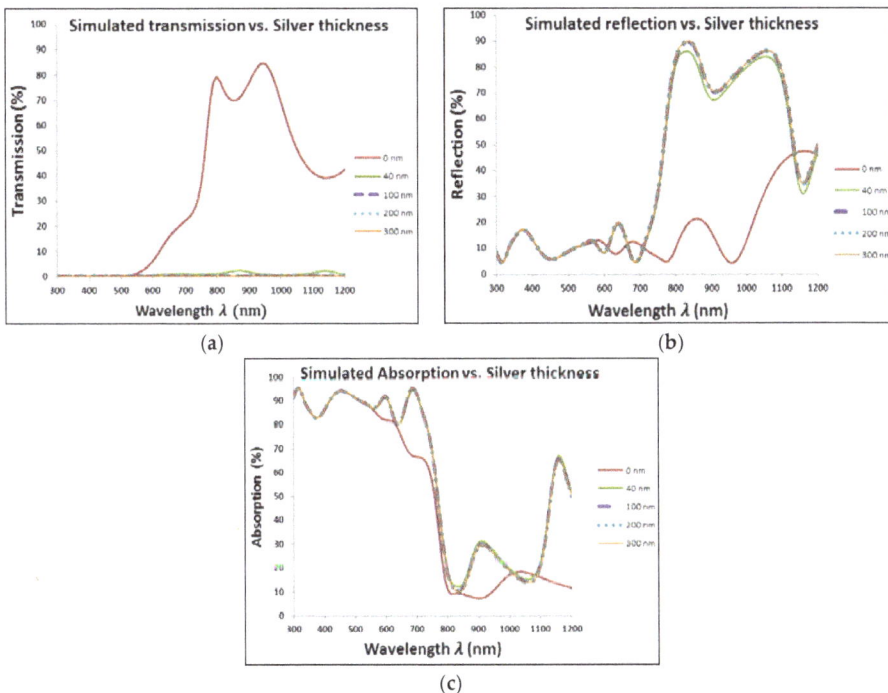

(a)

(b)

(c)

Figure 7. Optical characteristics for perovskite cell with different thicknesses of silver layer.

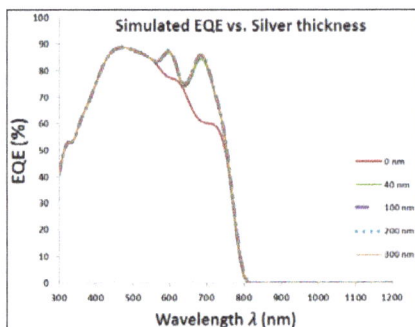

Figure 8. Enhancement of external quantum efficiency (EQE) due to added silver nanoparticles (Ag NPs) in the wavelength range close to plasmonic resonance of silver.

Table 3. The effect of added silver nanoparticles (Ag NPs) of different nanosize on the internal losses, external losses, and the short circuit current for our simulated Perovskite solar cell.

	Perovskite Cell Losses (mA/cm^2)			J_{sc} (mA/cm^2)
Silver (nm)	$J_{Internal}$	$J_{escape,back}$	$J_{reflection}$	
0	4.15	16.49	6.55	18.69
40	7.182	0.41	18.75	20.11
100	6.846	0.002	19.37	20.24
200	6.778	0.000	19.43	20.25
300	6.756	0.000	19.45	20.25

4.4. Comparison Between Gold and Silver Effects

In last two sections, it is proved the enhancement of EQE of perovskite solar cells with added plasmonic NPs. Here, we are going to compare between both added plasmonic Au and Ag NPs. It can be shown from Figure 9 that Ag NPs enhance the EQE of perovskite cell slightly more than the added gold for different thicknesses of plasmonic layer. It can be explained through the better reflection capability that silver can offer compared to gold, as shown in Figure 10. This consequently leads to higher enhancement of short-circuit current in the case of silver compared to gold, as shown in the comparison figure at different layer thicknesses in Figure 11.

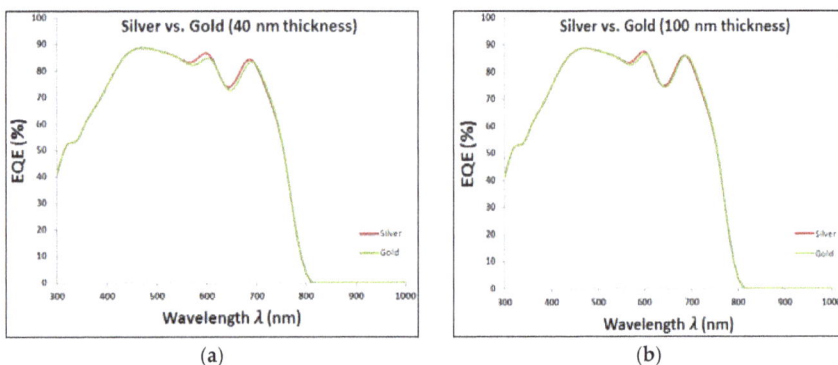

(a)

(b)

Figure 9. *Cont.*

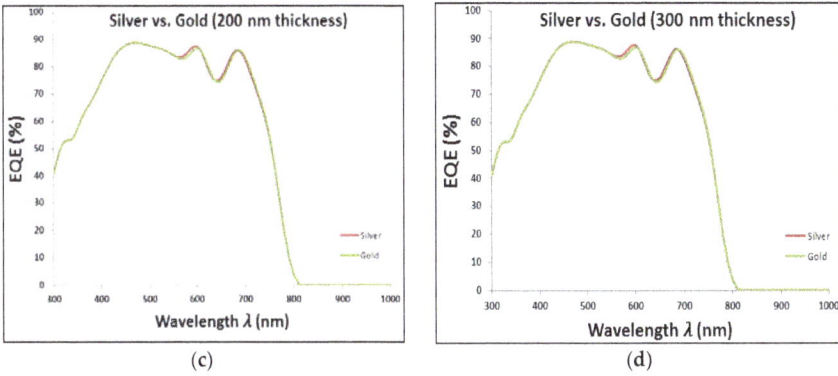

Figure 9. Comparison of external quantum efficiency (EQE) curves at added Au and Ag at different size of nanoparticles (NPs).

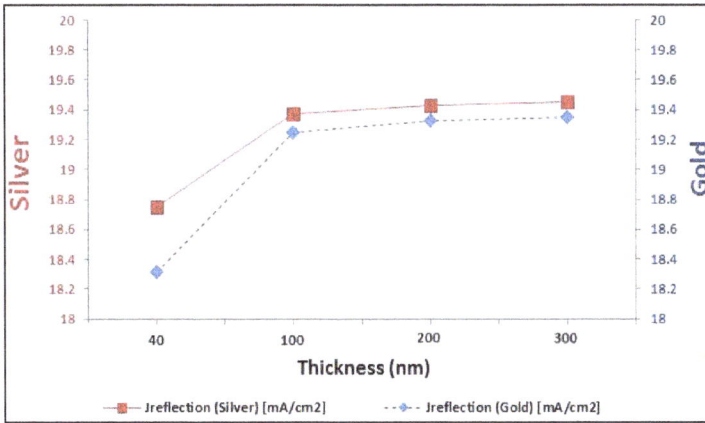

Figure 10. Optical reflection due to Au and Ag at different size of nanoparticles (NPs).

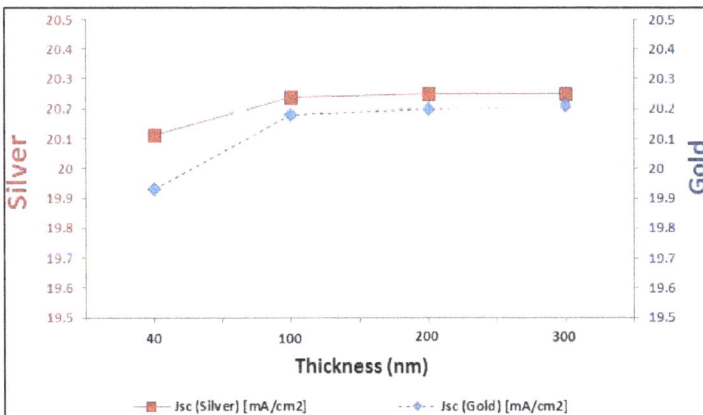

Figure 11. Comparison of J_{sc} values in the cell according to added Au and Ag at different size of nanoparticles (NPs).

5. Outlook

To fully benefit from the plasmonic effects of adding Au or Ag NPs at the rear-side of perovskite solar cells, the existing key optical losses need to be identified and addressed. These optical losses can be attributed to: a) overall reflection b) free carrier absorption in ITO electrodes, c) parasitic losses due to absorption in the substrate and diffuse scattering. To minimize the loss due to reflection, anti-reflective coating (ARC) or nanophotonic transparent electrodes can be used to improve the overall Power Conversion Efficiency (PCE) of the solar cell [49]. Furthermore, novel high-mobility Transparent Conducting Oxides (TCOs) such as hydrogenated indium oxide exhibit lower free carrier densities than the commonly used Indium Tin Oxide (ITO), hence offering the chance to minimize the parasitic absorption in ITO electrode at the front side [50,51]. Moreover, to completely mitigate the parasitic losses due to absorption in the substrate and diffuse scattering, ultra smooth $CH_3NH_3PbI_3$ films and non-absorbing substrates need to be used. More advanced light management concepts, trapping textures, and the optimization of the bandgap of the perovskite bear the potential to lead to very significant further improvements in light harvesting and current generation [52].

6. Conclusions

In this paper, the effect of adding plasmonic layer of Au and Ag NPs to the rear side of the perovskites solar cell is analytically studied. The resonance wavelength of the plasmonic NPs is adjusted to enhance the optical absorption of the solar cell in the visible range especially around the red wavelength. Both gold and silver lead to a promising enhancement in J_{sc} when the nanoparticle size becomes larger than 40 nm. Overall EQE enhancement is achieved. EQE improvement is slightly higher in case of adding silver compared to added gold.

Author Contributions: A.H. was responsible for simulations and analysis of perovskite solar cells. I.K. was responsible for simulation of the optical characteristics of gold and silver nanoparticles. N.S. had the main idea of the paper and helped in the analysis and explanation of the results of solar cells simulation.

Funding: The project was funded "partially" by Kuwait Foundation for the Advancement of Sciences under project code: P115-15EE-01.

Acknowledgments: The authors gratefully acknowledge the technical support of IMEC's Industrial Affiliation Program for Photovoltaics (IIAP-PV) for providing the measurements on the fabricated perovskite solar cell.

Conflicts of Interest: The authors declare no conflict of interest.

References

1. Kojima, A.; Teshima, K.; Shirai, Y.; Miyasaka, T. Organometal Halide Perovskites as Visible-Light Sensitizers for Photovoltaic Cells. *J. Am. Chem. Soc.* **2009**, *131*, 6050–6051. [CrossRef] [PubMed]
2. Contreras, M.A.; Mansfield, L.M.; Egaas, B.; Li, J.; Romero, M.; Noufi, R.; Rudiger-voigt, E.; Mannstadt, W. Wide bandgap Cu(In,Ga)Se2 solar cells with improved energy conversion efficiency. *Prog. Photovolt. Res. Appl.* **2007**, *15*, 659.
3. Burschka, J.; Pellet, N.; Moon, S.; Humphry-Baker, R.; Gao, P.; Nazeeruddin, M.; Grätzel, M. Sequential deposition as a route to high-performance perovskite-sensitized solar cells. *Nature* **2013**, *499*, 316–319. [CrossRef] [PubMed]
4. Liu, M.; Johnston, M.B.; Snaith, H. Efficient planar heterojunction perovskite solar cells by vapour deposition. *Nature* **2013**, *501*, 395–398. [CrossRef] [PubMed]
5. Chen, Q.; Zhou, H.; Hong, Z.; Luo, S.; Duan, H.; Wang, H.; Liu, Y.; Li, G.; Yang, Y. Planar heterojunction perovskite solar cells via vapor-assisted solution process. *J. Am. Chem. Soc.* **2013**, *136*, 622–625. [CrossRef] [PubMed]
6. Umebayashi, T.; Asai, K.; Kondo, T.; Nakao, A. Electronic structures of lead iodide based low-dimensional crystals. *Phys. Rev. B* **2003**, *67*, 155405. [CrossRef]
7. Snaith, H. Perovskites: The Emergence of a New Era for Low-Cost, High-Efficiency Solar Cells. *J. Phys. Chem. Lett.* **2013**, *4*, 3623. [CrossRef]

8. Noh, J.H.; Im, S.H.; Heo, J.H.; Mandal, T.N.; Seok, S.I. Chemical management for colorful, efficient, and stable inorganic-organic hybrid nanostructured solar cells. *Nano Lett.* **2013**, *13*, 1764–1769. [CrossRef] [PubMed]
9. Stoumpos, C.C.; Malliakas, C.D.; Kanatzidis, M.G. Semiconducting Tin and Lead Iodide Perovskites with Organic Cations: Phase Transitions, High Mobilities, and Near-Infrared Photoluminescent Properties. *Inorg. Chem.* **2013**, *52*, 9019–9038. [CrossRef] [PubMed]
10. De Wolf, S.; Holovsky, J.; Moon, S.J.; Löper, P.; Niesen, B.; Ledinsky, M.; Haug, F.; Yum, J.H.; Ballif, C. Organometallic Halide Perovskites: Sharp Optical Absorption Edge and Its Relation to Photovoltaic Performance. *J. Phys. Chem. Lett.* **2014**, *5*, 1035. [CrossRef] [PubMed]
11. Löper, P.; Stuckelberger, M.; Niesen, B.; Werner, J.; Filipic, M.; Moon, S.J.; Yum, J.H.; Topic, M.; De Wolf, S.; Ballif, C. Complex Refractive Index Spectra of CH3NH3PbI3 Perovskite Thin Films Determined by Spectroscopic Ellipsometry and Spectrophotometry. *J. Phys. Chem. Lett.* **2015**, *6*, 66. [CrossRef] [PubMed]
12. Yin, W.J.; Shi, T.; Yan, Y. Unusual defect physics in CH3NH3PbI3 perovskite solar cell absorber. *Appl. Phys. Lett.* **2014**, *104*, 063903. [CrossRef]
13. Xing, G.; Mathews, N.; Lim, S.; Lam, Y.; Mhaisalkar, S.; Sum, T.C. Long-range balanced electron- and hole-transport lengths in organic-inorganic CH$_3$NH$_3$PbI$_3$. *Science* **2013**, *6960*, 498. [CrossRef] [PubMed]
14. Stranks, S.; Eperon, G.; Grancini, G.; Menelaou, C.; Alcocer, M.; Leijtens, T.; Herz, L.; Petrozza, A.; Snaith, H. Electron-hole diffusion lengths exceeding 1 micrometer in an organometal trihalide perovskite absorber. *Science* **2014**, *342*, 341. [CrossRef] [PubMed]
15. Gonzalez-Pedro, V.; Juarez-Perez, E.; Arsyad, W.; Barea, E.; Fabregat-Santiago, F.; Mora-Sero, I.; Bisquert, J. General working principles of CH$_3$NH$_3$PbX$_3$ perovskite solar cells. *Nano Lett.* **2014**, *14*, 888–893. [CrossRef] [PubMed]
16. Krishnan, A.; Das, S.; Krishna, S.; Khan, M. Multilayer nanoparticle arrays for broad spectrum absorption enhancement in thin film solar cells. *Opt. Express* **2014**, *22*, A800–A811. [CrossRef] [PubMed]
17. Mendes, M.; Morawiec, S.; Simone, F.; Priolo, F.; Crupi, I. Colloidal plasmonic back reflectors for light trapping in solar cells. *Nanoscale* **2014**, *6*, 4796–4805. [CrossRef] [PubMed]
18. Dabirian, A.; Byranv, M.; Naqavi, A.; Kharat, A.; Taghavinia, N. Theoretical Study of Light Trapping in Nanostructured Thin Film Solar Cells Using Wavelength-Scale Silver Particles. *ACS Appl. Mater. Interfaces* **2016**, *8*, 247–255. [CrossRef] [PubMed]
19. Pillai, S.; Catchpole, K.; Trupke, T.; Green, M. Surface plasmon enhanced silicon solar cells. *J. Appl. Phys.* **2007**, *101*, 093105. [CrossRef]
20. Sannomiya, T.; Voros, J. Single plasmonic nanoparticles for biosensing. *Trends Biotechnol.* **2011**, *29*, 343–351. [CrossRef] [PubMed]
21. Zeng, S.; Yong, K.; Roy, I.; Dinh, X.; Yu, X.; Luan, F. A Review on Functionalized Gold Nanoparticles for Biosensing Applications. *Plasmonics* **2011**, *6*, 491–506. [CrossRef]
22. Sepulveda, B.; Angelome, P.; Lechuga, L.; Liz-Marzan, L. LSPR-based nanobiosensors. *Nano Today* **2009**, *4*, 244–251. [CrossRef]
23. Proise, F.; Joudrier, A.; Pardo, F.; Pelouard, J.; Guillemoles, J. Ultrathin mono-resonant nano photovoltaic device for broadband solar conversion. *Opt. Express* **2018**, *26*, A806. [CrossRef]
24. Atwater, H.; Polman, A. Plasmonics for improved photovoltaic devices. *Nat. Mater.* **2010**, *9*, 865. [CrossRef]
25. Beck, F.; Mokkapati, S.; Catchpole, K. Light trapping with plasmonic particles: beyond the dipole model. *Opt. Express* **2011**, *19*, 25230–25241. [CrossRef] [PubMed]
26. Saleh, Z.; Nasser, H.; Özkol, E.; Günöven, M.; Altuntas, B.; Bek, A.; Turan, R. Enhanced Optical Absorption and Spectral Photocurrent in a-Si: H by Single- and Double-Layer Silver Plasmonic Interfaces. *Plasmonics* **2014**, *9*, 357–365. [CrossRef]
27. Carretero-Palacios, S.; Jiménez-Solano, A.; Miguez, H. Plasmonic Nanoparticles as Light-Harvesting Enhancers in Perovskite Solar Cells: A User's Guide. *ACS Energy Lett.* **2016**, *1*, 323–331. [CrossRef] [PubMed]
28. Batmunkh, M.; Macdonald, T.; Peveler, W.; Bati, A.; Carmalt, C.; Parkin, I.; Shapter, J. Plasmonic Gold Nanostars Incorporated into High- Efficiency Perovskite Solar Cells. *ChemSusChem* **2017**, *10*, 3750–3753. [CrossRef] [PubMed]

29. Macdonald, T.; Ambroz, F.; Batmunkh, M.; Li, Y.; Kim, D.; Contini, C.; Poduval, R.; Liu, H.; Shapter, J.; Papakonstantinou, I.; et al. TiO$_2$ nanofiber photoelectrochemical cells loaded with sub-12 nm AuNPs: Size dependent performance evaluation. *Mater. Today Energy* **2018**, *9*, 254–263. [CrossRef]

30. Derkachova, A.; Kolwas, K.; Demchenko, I. Dielectric Function for Gold in Plasmonics Applications: Size Dependence of Plasmon Resonance Frequencies and Damping Rates for Nanospheres. *Plasmonics* **2016**, *11*, 941–951. [CrossRef] [PubMed]

31. Catchpole, K.; Polman, A. Design principles for particle plasmon enhanced solar cells. *Appl. Phys. Lett.* **2008**, *93*, 191113. [CrossRef]

32. Catchpole, K.; Polman, A. Plasmonic solar cells. *Opt. Express* **2008**, *16*, 21793–21800. [CrossRef] [PubMed]

33. Pillai, S.; Green, M. Plasmonics For Photovoltaic Applications. *Sol. Energy Mater. Sol. Cells* **2010**, *94*, 1481–1486. [CrossRef]

34. Zhu, J.; Xue, M.; Hoekstra, R.; Xiu, F.; Zeng, B.; Wang, K.L. Light concentration and redistribution in polymer solar cells by plasmonic nanoparticles. *Nanoscale* **2012**, *4*, 1978–1981. [CrossRef] [PubMed]

35. Anaya, M.; Lozano, G.; Calvo, M.; Zhang, W.; Johnston, M.; Snaith, H. Optical Description of Mesostructured Organic-Inorganic Halide Perovskite Solar Cells. *J. Phys. Chem. Lett.* **2015**, *6*, 48. [CrossRef] [PubMed]

36. Yang, D.; Yang, R.; Wang, K.; Wu, C.; Zhu, X.; Feng, J.; Ren, X.; Fang, G.; Priya, S.; Liu, S.F. High efficiency planar-type perovskite solar cells with negligible hysteresis using EDTA-complexed SnO$_2$. *Nat. Commun.* **2018**, *9*, 3239. [CrossRef] [PubMed]

37. Correa-Baena, J.; Anaya, M.; Lozano, G.; Tress, W.; Domanski, K.; Saliba, M.; Matsui, T.; Jacobsson, T.; Calvo, M.; Abate, A.; et al. Unbroken Perovskite: Interplay of Morphology, Electro-optical Properties, and Ionic Movement. *Adv. Mater.* **2016**, *1*, 5031. [CrossRef] [PubMed]

38. Paetzold, U.; Qiu, W.; Finger, F.; Poortmans, J.; Cheyns, D. Nanophotonic front electrodes for perovskite solar cells. *Appl. Phys. Lett.* **2015**, *106*, 173101. [CrossRef]

39. Eerden, M.; Jaysankar, M.; Hadipour, A.; Merckx, T.; Schermer, J.; Aernouts, T.; Poortmans, J.; Paetzold, U. Optical Analysis of Planar Multicrystalline Perovskite Solar Cells. *Adv. Opt. Mater.* **2017**, *5*, 1700151. [CrossRef]

40. Paetzold, U.W.; Gehlhaar, R.; Tait, J.; Qiu, W.; Bastos, J.; Debucquoy, M.; Poortmans, J. Optical loss analyses and energy yield modelling of perovskite/silicon multijunction solar cells. *Opt. Soc. Am.* **2016**, SoW2C-4. [CrossRef]

41. Bruggeman, D. Calculation of various physical constants of heterogeneous substances. I. dielectric constant and conductivity of the mixing body made of isotropic substances. *Ann. Phys.* **1935**, *416*, 636. [CrossRef]

42. Vial, A.; ·Laroche, T. Comparison of gold and silver dispersion laws suitable for FDTD simulations. *Appl. Phys. B* **2008**, *93*, 139–143. [CrossRef]

43. Johnson, P.B.; Christy, R.W. Optical Constants of the Noble Metals. *Phys. Rev. B* **1972**, *6*, 4370. [CrossRef]

44. Carminati, R.; Greffet, J.; Henkel, C.; Vigoureux, J. Radiative and non-radiative decay of a single molecule close to a metallic nanoparticles. *Opt. Commun.* **2006**, *261*, 368–375. [CrossRef]

45. Myroshnychenko, V.; Rodriıiguez-Fernandez, J.; Pastoriza-Santos, I.; Funston, A.; Novo, C.; Mulvaney, P.; Liz-Marzan, L.; Javier Garcıa de Abajo, F. Modelling the optical response of gold nanoparticles. *Chem. Soc. Rev.* **2008**, *37*, 1792–1805. [CrossRef] [PubMed]

46. Chung, H.; Leung, P.; Tsai1, D. Molecular fluorescence in the vicinity of a charged metallic nanoparticles. *Opt. Express* **2013**, *22*, 26483–26492. [CrossRef] [PubMed]

47. Craig, F.; Bohren, F.; Huffman, D. *Absorption and Scattering of Light by Small Particles*; John Wiley & Sons: New York, NY, USA, 1983.

48. Paviet-Salomon, B.; Tomasi, A.; Descoeudres, A.; Barraud, L.; Nicolay, S.; Despeisse, A.; De Wolf, S.; Ballif, C. Back-Contacted Silicon Heterojunction Solar Cells: Optical-Loss Analysis and Mitigation. *IEEE J. Photovolt.* **2015**, *5*, 1293–1303. [CrossRef]

49. Bailie, C.; Christoforo, M.; Mailoa, J.; Bowring, A.; Unger, E.; Nguyen, W.; Burschka, J.; Pellet, N.; Lee, J.; Grätzel, M.; et al. Semi-transparent perovskite solar cells for tandems with silicon and CIGS. *Energy Environ. Sci.* **2015**, *8*, 956. [CrossRef]

50. Fu, F.; Feurer, T.; Jäger, T.; Avancini, E.; Bissig, B.; Yoon, S.; Buecheler, S.; Tiwari, A. Low-temperature-processed efficient semi-transparent planar perovskite solar cells for bifacial and tandem applications. *Nat. Commun.* **2015**, *6*, 8932. [CrossRef] [PubMed]

51. Yin, G.; Steigert, A.; Manley, P.; Klenk, R.; Schmid, M. Enhanced absorption in tandem solar cells by applying hydrogenated In$_2$O$_3$ as electrode. *Appl. Phys. Lett.* **2015**, *107*, 211901. [CrossRef]
52. La, N.; White, T.; Catchpole, K. Optics and Light Trapping for Tandem Solar Cells on Silicon. *IEEE J. Photovolt.* **2014**, *4*, 1380.

materials

MDPI

Article

Diffractive Efficiency Optimization in Metasurface Design via Electromagnetic Coupling Compensation

Yang Li and Minghui Hong *

Department of Electrical and Computer Engineering, National University of Singapore, 4 Engineering Drive 3, Singapore 117576, Singapore; eleliy@nus.edu.sg
* Correspondence: elehmh@nus.edu.sg

Received: 21 February 2019; Accepted: 22 March 2019; Published: 27 March 2019

Abstract: Metasurface is an advanced flat optical component that can flexibly manipulate the electromagnetic wave in an ultrathin dimension. However, electromagnetic coupling among neighbored optical elements decreases the diffractive efficiency and increases the noise. In this paper, a novel computational method is proposed to optimize the coupling of the metasurface. The coupled electric fields in metasurface design are decomposed into various coupling orders and then restructured to replace the whole metasurface simulation. This method is applied to optimize a metasurface that consisted of conventional nanorod plasmonic antennas as a case study. The convergence of this method in calculation is demonstrated. The electric field intensity deviation of a nanoantenna array can be reduced from 112.2% to 0.5% by the second-order coupling correction. The diffractive efficiency of a three-level phase meta-deflector is optimized from 73% to 86% by optimized coupling compensation via particle swarm optimization (PSO). This process opens a new area of metasurface design by the detailed field distribution of optical elements.

Keywords: metasurface; coupling compensation; diffractive efficiency

1. Introduction

The electromagnetic couplings are important phenomena. They exist in many fields, including mutual inductance [1], plasmonics [2], and metamaterials [3]. In modern electromagnetic research, the coupling can lead to many unique effects, such as extraordinary optical transmission (EOT), negative refractive index, and high chirality. For the periodic design, the electromagnetic wave can transmit through subwavelength hole arrays on an opaque film. This EOT effect is due to the coupling of surface plasmons. However, for the non-periodic designs, too many parameters are involved in the design, which takes a long time in calculations for a complete scan of parameter optimization.

Metasurface is one of the most important optical designs involving non-periodic structures. It is an advanced flat optical device that consists of a large number of subwavelength optical antennas. In the development of a metasurface, the dimensions of the optical antennas become smaller, while more dielectric metasurfaces are designed, instead of metal metasurfaces [4]. These designs are applied for high efficiency and resolution light manipulation. On the other hand, it leads to weaker field localization. Thus, the performance of metasurface is obviously affected by electromagnetic coupling. To avoid this negative effect, a high refractive index of loss-free materials is desired for better field localization. In the infrared region, silicon is an ideal material with a high refractive index and low loss. However, when the metasurface is used in the visible spectrum, only 75% diffraction efficiency can be achieved due to the energy loss of the material [5]. Most optical materials with a high refractive index are not transparent. In this light spectrum, most materials with a high refractive index have high loss [6]. Silicon nitride (n ≈ 2), which has 90% transmission in the visible region, was used to design a metasurface. However, its diffraction efficiency is only about 40% [7]. Lower refractive index materials, such as glass, are barely selected in metasurface design due to lower field localization capability.

Due to the low diffraction efficiency caused by electromagnetic coupling, a series of negative effects appear. For instance, in a meta-hologram based on geometric phase, the noise of the unmodulated light (the zero-order noise) can be designed to be completely filtered out by the wave plate and polarizer. However, in practical holographic applications, the zero-order noise along the transmission direction cannot be completely removed in most cases due to electromagnetic coupling [6,8]. The designed phase shift is deviated randomly by the coupled field from the neighbor optical antennas, which leads to noise along the beam without modulation. The off-axis illumination is applied to improve the signal-to-noise ratio (SNR) in meta-holography [9–11]. However, this is only a compromise. The coupling also affects the symmetry of optical systems [12], which limits metasurface applications in optical integration, optical communication, and spatial multiplexing meta-holograms. Hence, it faces great challenges in the coupling compensation.

To overcome this issue, a novel design method of the metasurface is proposed in this paper by considering electromagnetic coupling. The element of the metasurface is simulated separately through coupling a group of optical antennas, instead of the traditional periodic design. The multiple orders of the coupling are modeled and simulated by field decomposition. This method is rigorous in theory, and its convergence is demonstrated. It gives an opportunity to consider the coupling in metasurface design. Combining particle swarm optimization (PSO) and our coupling compensation method, the diffractive efficiency of a meta-deflector is increased by 13%. It is found that the detailed light field through a metasurface can also be designed by this optimization method to achieve extraordinary improvement.

2. Principles

In traditional optics, there are some fundamental physics problems, such as optical aberration and multiple reflections [13,14]. The optical aberration can be reduced by a combination of multiple lenses with different radii and concave/convex designs in geometric shapes. In the simple cases, the chromatic aberration is modeled by the Abbe number of the lens materials. However, in a more complex system with multiple optical aberrations, the commercial software, such as Zemax (Zemax LLC, Bellevue, WA, USA), must be used to optimize the design. Similarly, for subwavelength optical design, such as metamaterial and photonic crystal [15–19], the parameters of periodic structures can be globally optimized via numerical simulation. However, for the non-periodic design, such as that of a metasurface, there are too many parameters. The elements of metasurface, which are called optical antennas, can modify the phase, amplitude, and polarization of light on a subwavelength scale [9,20]. For a complex optical function, there are thousands of optical antennas, but a lack of an effective method.

Conventionally, the optical antennas are optimized by a periodic design. Then, these antennas are arranged for light manipulation based on the optical property in periodic condition. However, the electromagnetic coupling among optical antennas is ignored. The complexity of the electromagnetic coupling makes the designed light manipulation difficult to be accurately realized. It needs time-consuming computations for optimization. Without proper design optimization, most functions of metasurface cannot be realized with high diffractive efficiencies (DEs).

To overcome this issue, the fundamentals are the basic electric field distribution and the contribution of electromagnetic coupling. The electric field distribution of a metasurface at a plane parallel to the surface can be expressed as $E_{meta}(x,y)$ illuminated by a plane wave $E_{in}(x,y)$. The plane wave normally illuminates the metasurface. The E_{in} can be considered a constant. The $E_{meta}(x,y)$ can be decomposed into the electric field contributions of each optical antenna and the coupling among the antennas:

$$
\begin{aligned}
E_{meta} &= E_{antenna} + E_{coupling} \\
&= \sum_i E_{antenna,i} + \sum_i E_{coupling,i} = \sum_i \left(E_{antenna,i} + E_{coupling,i} \right)
\end{aligned}
\tag{1}
$$

where $E_{antenna, i}$ and $E_{coupling, i}$ are the radiated and coupled electric fields of the ith optical antenna. The electric field distribution of the whole metasurface can be restructured by summing the radiated and coupled electric fields of all the antennas together. The radiated electric field distribution of the ith antenna can be simulated by local illumination on a single antenna. It is difficult to calculate the coupling for each antenna, as all the optical antennas are different. Hence, the electric field induced by the coupling is further decomposed by the number of the coupled optical antennas.

$$E_{0,i_1,i_2,\ldots,i_n} = E_0 + \sum_{m_1=i_1,i_2,\ldots,i_n} \Delta E_{0,m_1} + \sum_{m_1,m_2=i_1,i_2,\ldots,i_n} \Delta E_{0,m_1,m_2}$$
$$+ \ldots + \sum_{m_1,m_2,\ldots,m_n=i_1,i_2,\ldots,i_n} \Delta E_{0,m_1,m_2,\ldots,m_n} + \ldots \tag{2}$$

where $\sum_{m_1,m_2,\ldots,m_k=i_1,i_2,\ldots,i_n} \Delta E_{0,m_1,m_2,\ldots,m_k}$ is the kth order coupling. Figure 1 shows the schematic of the decomposition of the metasurface coupling. The radiated field of a single antenna is defined as E_0, which is generated by the resonance of the optical antenna itself. The coupling between the illuminated antenna and one coupled antenna is defined as the first-order coupling correction; the coupling among the illuminated antenna and two coupled antennas is defined as the second-order coupling. The higher order coupling can be defined in the similar way. Conventionally, the higher order coupling is weaker due to the increasing spatial separation and limited radiation efficiency. The coupling can be corrected by the finite order of calculation. In this way, the coupling effects for all the optical antennas are similar. A database of all the coupled electric field distributions can be built for the compensation. The irradiated and coupled electric fields of each antenna can be directly calculated from the database. Hence, the light distribution can be calculated accurately without whole model simulation. The accuracy of this coupling compensation method will be discussed, and then it will be applied for DE optimization.

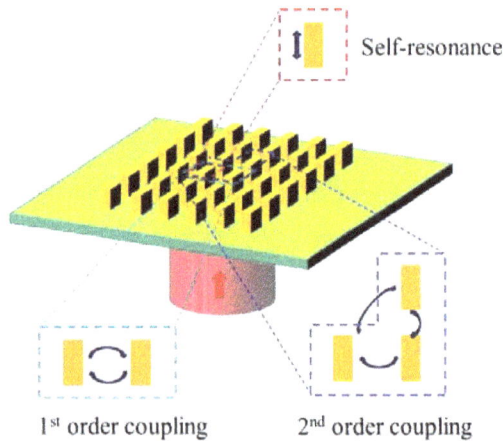

Figure 1. Schematic of the decomposition of electromagnetic coupling in metasurface design.

3. Results and Discussion

3.1. Coupling Compensation

To prove that the above-proposed method can calculate the field distribution efficiently for practical applications, it is important to investigate the calculation convergence. A traditional gold nanorod antenna is chosen as a case study. The lattice, length, and width of optical antennas are 200 nm, 140 nm, and 60 nm, respectively. The polarization of the incident light is parallel to the nanorod. The illumination area is 200 nm × 200 nm at 100 nm below the metasurface at an incident light wavelength of 800 nm. The grid size is set as 10 nm. By numerical simulation with a finite-difference time-domain (FDTD) software (Lumerical Inc, Canada), the electric field distributions of single and multiple antennas are shown in Figure 2a–d. Although the incident light only illuminates the optical antenna located at (0, 0), the field distributions of the multiple antennas are different from the field distribution of a single antenna E_0 (Figure 2a). The relative electric field deviation $\Delta E / E_0$ is as high as 30% to 40% without considering the coupling. Figure 2d,e show the restructured field distributions with the first-order and second-order coupling corrections. For a three-antenna system, the deviation can be decreased from 34.0% to 6.8% by the first-order coupling correction. For a four-antenna system, the deviation can be reduced to 10.9% and 2.25% by the first-order and second-order coupling corrections, respectively.

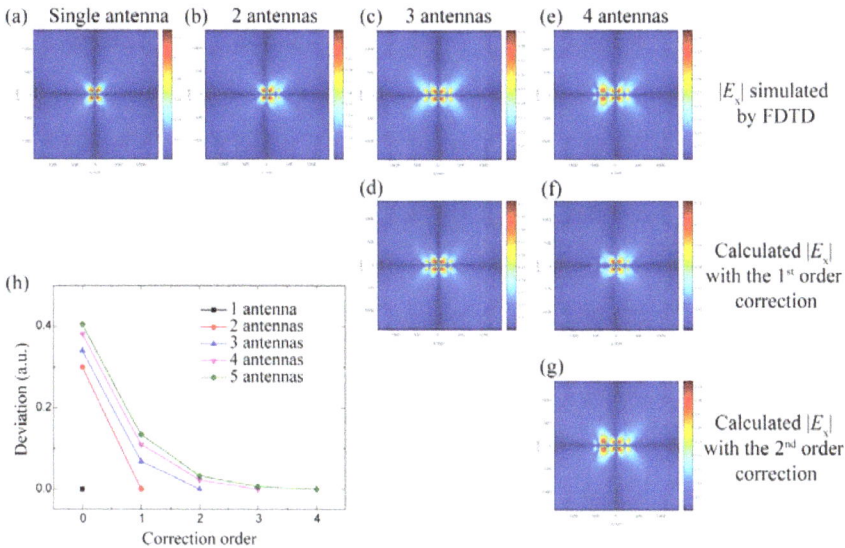

Figure 2. Simulated electrical filed distribution along x-axis (*Ex*) distributions simulated by FDTD (**a–d**) and calculated by the field of the elements with the first-order (**e,f**) and second-order (**g**) corrections (**h**) Field deviation for various correction orders.

These results indicate that the coupling correction is efficient, rigorous, and convergent in a calculable system. Similar results can be observed for a system with more optical antennas, as shown in Figure 2h. The deviation decreases inversely proportional to the correction order. It can be explained by the multiple scattering of the coupled optical antenna. The electric field excited by the scattering field from a neighbored antenna can be assumed as αE_0, where α is a coupling factor between zero and one. That is the first-order coupling. Considering the second-order coupling, the antenna with αE_0 can be treated as a subsource to excite another neighbored antenna that the intensity of the second-order coupling can be estimated as $\alpha^2 E_0$. In the calculation, it is difficult to get the accurate value of α due to the complexity of the coupling. It also depends on the orientations of the optical antennas; the value is different for different order corrections. In this case, α is in the range of 0.2 to 0.4, which is smaller than one. It is convergent in the calculation of the coupling corrections.

The coupling factor is one important parameter to the convergence of the coupling correction. The calculation time and the accuracy need to be properly balanced. The coupling intensity depends on the geometric parameters of the structure design, wavelength, and polarization. In the above cases, the geometric parameters include the length, width, height, and period of optical antennas. The performance of such a metallic rod design is not sensitive to the width and height [21]. The period of the optical antenna directly affects the distance between two antennas. The coupling is strong at a short distance and weak as the distance increases. It is due to the electric field radiated by subsources, which can be estimated as inversely proportional to the square of the distance. The antenna is considered as a point source. Figure 3a demonstrates that the coupling intensity reduces when the period increases. At the periods of 200 nm and 300 nm, the coupling intensities are almost the same, which is attributed to the near-field coupling at a deep subwavelength distance. Theoretically, when the period of the antenna is large enough, the coupling does not need to be considered. However, the efficiency and the maximum controllable special frequency would be affected at a large period. To fully control the wave factors in free space, the period of the optical antenna cannot be designed to be larger than the wavelength. Therefore, the coupling in the metasurface design cannot be avoided. The length of the optical antenna is another factor affecting the coupling, which affects the antenna resonance. The length of the optical antenna is usually at deep subwavelength scale for the metasurface design, which is shorter than the resonant length in the dipole model. Therefore, the increasing of the length of the antenna leads to stronger resonance and also affects its coupling. As shown in Figure 3b, the coupling of longer nanorods is stronger so that a higher order of correction is required. Figure 3c,d show the deviation caused by the coupling at different wavelengths and polarizations for comparison. The highest deviation always happens at the wavelength around 600 nm, which is at the plasmonic frequency of gold. Since the deviation in both polarization illuminations can be corrected with good convergence, the light field of common optical antennas can be restructured by this new coupling correction method.

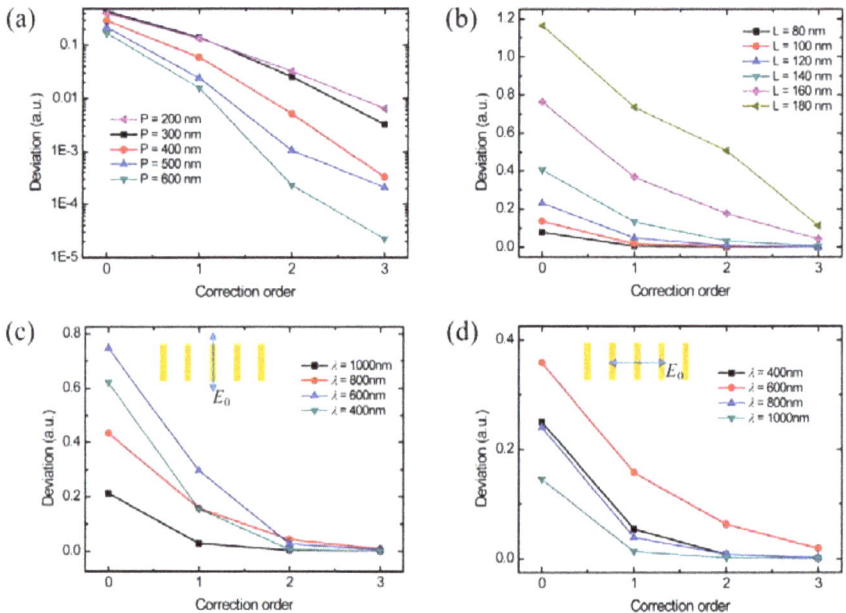

Figure 3. Calculated $|Ex|$ deviations compared with the simulated results at different (**a**) periods, (**b**) lengths of optical antenna, (**c**) wavelengths, and (**d**) polarizations.

In a more complex system with a large number of antennas, it is impossible to calculate all the coupled electric fields for all the antennas. However, the deviation of the electric field can be greatly reduced by the coupling correction via our proposed method to calculate only a limited area. For instance, a periodic optical antenna array is simulated. As shown in the insets of Figure 4, the electrical fields indicate the distribution through numerical simulation and the restructured field distribution via the field superposition. The results indicate that the field distributions calculated by different methods are similar, but the intensity is greatly changed. Without the correction, the radiated field is very weak. Hence, the deviation is as high as 112.2%. With the first-order correction, the deviation is slightly reduced to 98.7%. After the second-order correction, the deviation of the calculated electric field intensity is rapidly reduced to 0.5%. The restructured distribution and intensity of light match the simulated results very well. It is obvious that the coupling correction with the limited orders is efficient for metasurface design with a large number of elements.

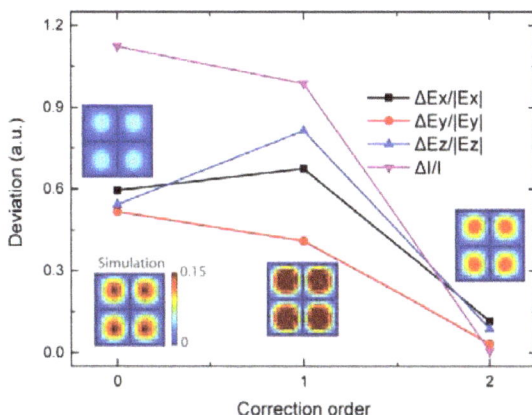

Figure 4. Average deviation of the restructured electric field at different order corrections. Insets: simulated and restructured electric field distributions $|E_x|$ of a periodic optical antenna array.

3.2. Diffractive Efficiency Optimization

The deflector based on the metasurface design is similar to the optical wedge and grating. One optical wedge can change the direction of the light without energy loss, which has a continuous linear phase profile to tune the wavefront. However, it is difficult to design and fabricate for a large deflection angle. The thickness of the high deflection angle wedge becomes large. Secondly, in diffractive optics, the grating can act as a deflector. Its diffractive efficiency is related to the number of phase levels. For a two-level grating, the DE is only about 40%, while the DE can achieve more than 90% for an eight-level grating. On the other hand, higher levels correspond to more complex procedures in fabrication.

Metasurface can provide a perfect solution theoretically. For an ideal case, the optical antenna can be designed with the scale much smaller than the wavelength. The designed metasurface with multiple phase levels can achieve a high DE. However, the optical antenna cannot be designed in a deep-subwavelength scale, which is limited by the optical properties of the materials in the visible and ultraviolet spectra. Therefore, the DE of the metasurface is not high enough due to the discontinuous phase profile. How to use the discrete optical antennas to control the continuous phase profile is a challenge in optical design.

To achieve a continuous phase profile, detail electric field distribution needs to be considered in the metasurface design. With this coupling decomposition method, it can be restructured via calculation after building a database of the detail radiated and coupled field distribution of optical antennas. The nanorod antenna is the most common phase modulator in the metasurface design due to its geometric phase effect to the cross-polarized light. The phase shift is twice the rotation angle without considering the coupling. To avoid time-consuming computation, the first-order coupling is chosen for diffractive efficiency optimization. The wavelength is changed to 400 nm, while the other parameters are the same. As shown in the simulation results in Figure 3c,d, the first-order coupling can significantly reduce the deviation. The electrical fields of optical antennas are calculated at a rotation step of 30 degrees. The electrical field distribution of the optical antenna with a more accurate rotation angle is calculated through linear interpolation. Only coupling from neighbored antennas is considered.

For deflector-based three-level optical antennas, the rotation angles can be simply set as $0°$, $60°$, and $120°$ (Figure 5a), corresponding to the phases of 0, $2\pi/3$, and $4\pi/3$, respectively. The designed phases are discrete. Its DE can achieve 73%, which is higher than ~60% for the theoretical limits of three-level phase gratings. From the simulated phase distribution shown in Figure 5c, it is found that the phase profile of the out plane of the metasurface is more continuous than that of a three-level grating. The optical antenna can operate not only as a simple phase shift element, but also as an element to generate a more detailed phase profile. The particle swarm optimization (PSO) algorithm is applied. The numbers of particles and cycles are set as 20 and 1000, respectively. The diffractive efficiency can be optimized to 90% with the first coupling compensation (92% without the coupling compensation) with the design of optical antennas at rotation angles of $10.6°$, $68.6°$, and $130.0°$ (Figure 5b). Via the full metasurface simulation, the diffractive efficiency of the optimized three-phase meta-deflector is 86%. Although there is a small deviation (4%) by the coupling compensation method, the optimization tendencies are the same. Meanwhile, it is certain that the calculated DE is more accurate after the compensation. Furthermore, Figure 5d indicates that the gradient phase distribution is smoother than the distribution without the optimization.

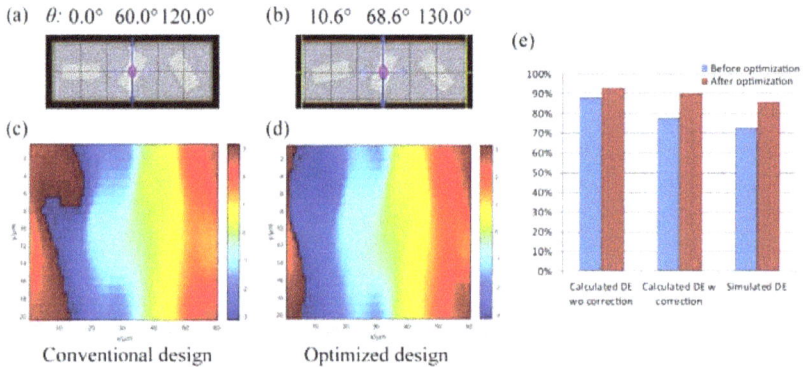

Figure 5. Schematics and simulated phase distributions of the *Ex* of (**a,c**) the conventional design and (**b,d**) optimized design of a meta-deflector. (**e**) The diffraction efficiency improvement at the calculation with and without the correction and simulation.

4. Conclusions

In this paper, a new computational method is proposed to analyze the electromagnetic coupling of metasurface by coupling decomposition into different orders. The relationship between coupling factor and design parameters, including the period, length of antenna, wavelength, and polarization are analyzed. These parameters affect the intensity of the coupling, but the calculated field distribution is always convergent. It takes a long time to build a database of radiated and coupled field distribution for different parameters. To get the detailed field information with a small grid size, the simulation time becomes longer. On the other hand, this method fully supports the parallel computation, which can greatly increase the simulation speed. Because the decomposed radiated and coupled fields are independent. Furthermore, after building a database, the light distribution of the whole metasurface can be restructured rapidly and accurately. By PSO with the first-order coupling correction, the DE of a deflector based on the three-element metasurface can be optimized to 86%, which is much higher than theoretical DE for the three-phase deflector (~60%). It indicates that the light manipulation in metasurface design is not confined by the lattice of arrangement. The more detailed field distribution can also be controlled by proper design. This method can be expanded to arbitrary linear optical systems. If the processor, memory, and storage space of the computation workstation is large enough, the accuracy of the field restruction can be further improved. It has the potential to be applied to reduce the noise in the meta-hologram and optimize the metasurface with the relatively low refractive index dielectric material, such as glass, and metal metasurface with smaller feature sizes.

Author Contributions: Conceptualization, Y.L. and M.H.; Methodology, Y.L.; Software, Y.L.; Investigation, Y.L.; Resources, M.H.; Data Curation, Y.L.; Writing-Original Draft Preparation, Y.L.; Writing-Review & Editing, M.H. and Y.L.; Project Administration, M.H.; Funding Acquisition, M.H.

Funding: This research was funded by RIE2020 Advanced Manufacturing and Engineering (AME) Individual Research Grant (IRG) (Grant No. A1883c0010) and Open Project of the State Key Laboratory of Luminescence and Applications (SKLA-2018-09).

Conflicts of Interest: The authors declare no conflict of interest.

References

1. Jackson, J.D. *Classical Electrodynamics*, 3rd ed.; John Wiley & Sons: Hoboken, NJ, USA, 2012.
2. Luo, X.G. Principles of electromagnetic waves in metasurfaces. *Sci. China Physics, Mech. Astron.* **2015**, *58*, 1–18. [CrossRef]
3. Plum, E.; Fedotov, V.A.; Schwanecke, A.S.; Zheludev, N.I.; Chen, Y. Giant optical gyrotropy due to electromagnetic coupling. *Appl. Phys. Lett.* **2007**, *90*, 223113. [CrossRef]
4. Khorasaninejad, M.; Chen, W.T.; Devlin, R.C.; Oh, J.; Zhu, A.Y.; Capasso, F. Metalenses at visible wavelengths: Diffraction-limited focusing and subwavelength resolution imaging. *Science* **2016**, *352*, 1190–1194. [CrossRef] [PubMed]
5. Huang, K.; Dong, Z.; Mei, S.; Zhang, L.; Liu, Y.; Liu, H.; Zhu, H.; Teng, J.; Luk'yanchuk, B.; Yang, J.K.W.; et al. Silicon multi-meta-holograms for the broadband visible light. *Laser Photonics Rev.* **2016**, *509*, 500–509. [CrossRef]
6. Devlin, R.C.; Khorasaninejad, M.; Chen, W.T.; Oh, J.; Capasso, F. Broadband high-efficiency dielectric metasurfaces for the visible spectrum. *Proc. Natl. Acad. Sci.* **2016**, *113*, 10473–10478. [CrossRef] [PubMed]
7. Zhan, A.; Colburn, S.; Trivedi, R.; Fryett, T.K.; Dodson, C.M.; Majumdar, A. Low-Contrast Dielectric Metasurface Optics. *ACS Photonics* **2016**, *3*, 209–214. [CrossRef]
8. Zhang, F.; Pu, M.; Li, X.; Gao, P.; Ma, X.; Luo, J.; Yu, H.; Luo, X. All-Dielectric Metasurfaces for Simultaneous Giant Circular Asymmetric Transmission and Wavefront Shaping Based on Asymmetric Photonic Spin–Orbit Interactions. *Adv. Funct. Mater.* **2017**, *27*, 1–7. [CrossRef]
9. Li, Y.; Li, X.; Chen, L.; Pu, M.; Jin, J.; Hong, M.; Luo, X. Orbital Angular Momentum Multiplexing and Demultiplexing by a Single Metasurface. *Adv. Opt. Mater.* **2016**, *1*, 1–5. [CrossRef]
10. Li, X.; Chen, L.; Li, Y.; Zhang, X.; Pu, M.; Zhao, Z.; Ma, X.; Wang, Y.; Hong, M.; Luo, X. Multicolor 3D meta-holography by broadband plasmonic modulation. *Sci. Adv.* **2016**, *2*, 1–7. [CrossRef] [PubMed]
11. Zheng, G.; Mühlenbernd, H.; Kenney, M.; Li, G.; Zentgraf, T.; Zhang, S. Metasurface holograms reaching 80% efficiency. *Nat. Nanotechnol.* **2015**, *10*, 308–312. [CrossRef] [PubMed]
12. Li, Y.; Li, Y.; Chen, L.; Hong, M. Reflection tuning via destructive interference in metasurface. *Opto-Electronic Eng.* **2017**, *44*, 313–318.
13. Born, M.; Wolf, E. *Principles of Optics: Electromagnetic Theory of Propagation, Interference and Diffraction of Light*, 7th ed.; Press of University of Cambridge: Cambridge, UK, 1999.
14. Raut, H.K.; Ganesh, V.A.; Nair, A.S.; Ramakrishna, S. Anti-reflective coatings: A critical, in-depth review. *Energy Environ. Sci.* **2011**, *4*, 3779–3804. [CrossRef]
15. Genet, C.; Ebbesen, T.W. Light in tiny holes. *Nature* **2007**, *445*, 39–46. [CrossRef] [PubMed]
16. Xu, T.; Wu, Y. K.; Luo, X.; Guo, L.J. Plasmonic nanoresonators for high-resolution colour filtering and spectral imaging. *Nat. Commun.* **2010**, *1*, 1–5. [CrossRef] [PubMed]
17. Wu, C.; Khanikaev, A.B.; Shvets, G. Broadband slow light metamaterial based on a double-continuum Fano resonance. *Phys. Rev. Lett.* **2011**, *106*, 107403. [CrossRef]
18. Burgos, S.P.; de Waele, R.; Polman, A.; Atwater, H.A. A single-layer wide-angle negative-index metamaterial at visible frequencies. *Nat. Mater.* **2010**, *9*, 407–412. [CrossRef]
19. Wang, Y.; Ma, X.; Li, X.; Pu, M.; Luo, X. Perfect electromagnetic and sound absorption via subwavelength holes array. *Opto-Electronic Adv.* **2018**, *1*, 18001301–18001306. [CrossRef]

20. Nemati, A.; Wang, Q.; Hong, M.; Teng, J. Tunable and reconfigurable metasurfaces and metadevices. *Opto-Electronic Adv.* **2018**, *1*, 18000901–18000925. [CrossRef]

21. Huang, L.; Chen, X.; Mühlenbernd, H.; Zhang, H.; Chen, S.; Bai, B.; Tan, Q.; Jin, G.; Cheah, K.-W.; Qiu, C.-W.; et al. Three-dimensional optical holography using a plasmonic metasurface. *Nat. Commun.* **2013**, *4*, 1–8. [CrossRef]

![materials logo] *materials*

MDPI

Article

Hybrid Metal-Dielectric Nano-Aperture Antenna for Surface Enhanced Fluorescence

Guowei Lu [1,2,*] , Jianning Xu [1], Te Wen [1], Weidong Zhang [1], Jingyi Zhao [1], Aiqin Hu [1], Grégory Barbillon [3] and Qihuang Gong [1,2]

[1]   State Key Laboratory for Mesoscopic Physics & Collaborative Innovation Center of Quantum Matter, School of Physics, Peking University, Beijing 100871, China; jnxu@pku.edu.cn (J.X.); wente@pku.edu.cn (T.W.); weidongzhang@pku.edu.cn (W.Z.); jingyi.zhao@pku.edu.cn (J.Z.); aiqinhu@pku.edu.cn (A.H.); qhgong@pku.edu.cn (Q.G.)
[2]   Collaborative Innovation Center of Extreme Optics, Shanxi University, Taiyuan 030006, China
[3]   EPF—École d'ingénieurs, 3 bis rue Lakanal, 92330 Sceaux, France; gregory.barbillon@epf.fr
*   Correspondence: guowei.lu@pku.edu.cn

Received: 26 June 2018; Accepted: 13 August 2018; Published: 14 August 2018

Abstract: A hybrid metal-dielectric nano-aperture antenna is proposed for surface-enhanced fluorescence applications. The nano-apertures that formed in the composite thin film consist of silicon and gold layers. These were numerically investigated in detail. The hybrid nano-aperture shows a more uniform field distribution within the apertures and a higher antenna quantum yield than pure gold nano-apertures. The spectral features of the hybrid nano-apertures are independent of the aperture size. This shows a high enhancement effect in the near-infrared region. The nano-apertures with a dielectric gap were then demonstrated theoretically for larger enhancement effects. The hybrid nano-aperture is fully adaptable to large-scale availability and reproducible fabrication. The hybrid antenna will improve the effectiveness of surface-enhanced fluorescence for applications, including sensitive biosensing and fluorescence analysis.

Keywords: plasmonics; nano-aperture; surface-enhanced fluorescence; antenna; hybrid

1. Introduction

Metallic nanostructures can be used as optical nanoantennas and they have attracted increasing attention for plasmon-enhanced spectroscopy and sensitive molecular fluorescence detection [1–3]. Nanoantennas can operate beyond the light diffraction limit and have been successfully implemented for single molecule analytical approaches, both in vitro and in vivo. One such design—nano-apertures formed in a metallic film also known as zero-mode waveguides—is an intuitive way to demonstrate the advantages of optical antennae. They offer localized enhancement of excitation light, modification of the fluorescence signal, and suppression of emission from species located outside the apertures [4–7]. Biophotonic applications of nanoantennas require the efficient enhancement of molecule fluorescence [3,7]. Unfortunately, metallic nanostructures, including nano-apertures antenna, usually suffer from serious absorption and scattering losses that would lead to low quantum yields and even fluorescence quenching [8–10]. For instance, due to localized surface plasmon (LSP), a typical nano-aperture can enable sub-wavelength confinement of the optical field at the side corner of holes, resulting in a highly localized excitation light field. The emission quantum efficiency is very low due to metal loss. This can ultimately quench the molecule fluorescence signal at the side corner [4,11].

Photonic structures, such optical micro-cavity or dielectric nanoparticles, have a high quantum yield due to low material loss, but the excitation field enhancement does not compare with the metallic nanostructures [12–14]. Meanwhile, hybrid plasmonic-photonic structures can dramatically increase the Purcell factor or achieve efficient waveguides over the uncoupled photonic and plasmonic components [15–18]. These hybrid systems have been reported in various systems and utilize both the highly localized plasmons and the low-loss photonic modes for Fano resonances, strong coupling, Raman scattering, and spontaneous emission [19–22]. The concept of hybrid photonic-plasmonic structures has not yet been reported for antennae with a nano-aperture geometry. This aperture antenna geometry is a powerful technology for studying single-molecule real-time dynamics of biological systems, and it is necessary to explore nano-apertures with hybrid configurations to optimize molecular fluorescence.

Here, we report on a hybrid metal-dielectric nano-aperture antenna for surface enhanced fluorescence applications. The nano-apertures formed in a composite film consisting of silicon and gold layers on a glass substrate. They have a more uniform field distribution and a higher antenna quantum yield than pure gold nano-apertures. We also investigated the dependence of surface-enhanced fluorescence on the aperture size and the thickness of the dielectric layer. Furthermore, nano-apertures with a dielectric gap are proposed and shown theoretically to offer better enhancement. The hybrid dielectric plasmonic nano-aperture antenna has better performance. In addition, we discuss several ways to further improve the surface enhancement effects, such as combining this scheme with periodic gratings or the use of lossless high-index dielectric materials.

2. Materials and Methods

To investigate the surface-enhanced fluorescence performances of the nano-aperture antennas, finite-difference time-domain (FDTD) was used to calculate the electromagnetic features of the nanostructures, including electromagnetic field distribution, Purcell factor, and antenna quantum efficiency [23–25]. The FDTD is a mature method that has been extensively employed to study both the near- and far-field electromagnetic responses of the nanostructures with different arbitrary shapes [26,27]. This method permits the computation of: (i) the electromagnetic field distribution of the nanostructure surroundings and (ii) the electromagnetic flux of a dipole source near metallic nanostructures. In calculations, the optical dielectric function of the gold and silicon materials are modeled while using a Drude-Lorentz dispersion function [28,29]. In all of the calculations, the refractive index of the surrounding media is taken to be 1.33 for water and 1.49 for silica glass. The geometry origin point is set at the center of the water-glass interface.

The scheme of the nano-aperture antenna is shown in Figure 1. It is formed in a thin film composed of Au and Si layers on a silica substrate. A single point dipole orienting along the z-direction is placed within the nano-aperture. During calculations, the horizontal dipole is usually positioned at the central point along the x-direction and a position 10 nm above the silica glass along the z-direction, i.e., position (0, 0, and 10 nm). The emission of a single point dipole source was referred to as a single quantum emitter. The optical antenna quantum yield is the ratio of P_{rad}, represents the energy that reaches the far field, to P_{tot} the total power dissipated by the emitter. P_{rad} was obtained by integrating the Poynting vector over closed surfaces that contain the nanoantenna and dipolar source, while P_{tot} was obtained over closed surfaces containing the dipolar source only. We note that the antenna quantum yield is actually the efficiency of radiation of an emitter interacting with an antenna, and the radiation efficiency is dependent on the emitter's distance and orientation with respect to the nanostructures. During the bottom collection scheme, i.e., $P_{rad\text{-}glass}$, a collecting planar monitor under a 10-nm water-glass interface with a region of 2 μm × 2 μm in the xy-plane is applied to integrate far field radiation. We compute these quantities by considering the power that is emitted by a classical oscillating dipole and normalize them with respect to the case without any structures [24,30]. Moreover, the FDTD method can calculate the electromagnetic field distribution under plane wave illumination of classical light from the bottom to estimate the field enhancement and thus compare the fluorescence enhancement effects.

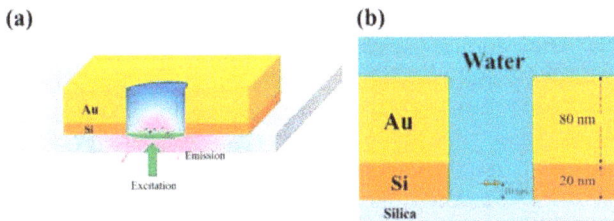

Figure 1. (a) Three-dimensional illustration of Au/Si hybrid nano-aperture used as antenna; (b) cross-sectional scheme of the hybrid structure consisting of an 80-nm Au layer on the glass separated by a 20-nm Si layer. The diameter of the nano-aperture is 40 nm, and the dipole is 10 nm above the glass surface. The background is water with a refractive index of 1.33.

3. Results and Discussion

For the nano-aperture structures that are studied here, we first calculate the electromagnetic field distributions and plot the cross-sections of the electromagnetic field distribution in both the xz-plane and the xy-plane (Figure 2a,b, respectively). In the gold aperture, it is clear that the maximum field enhancement occurs at the corners between the Au layer and the glass. The field intensity is low at the center. Figure 2b shows that the electrical field distribution of the Au/Si hybrid nano-aperture is more uniform than that shown in Figure 2a. Meanwhile, the field of the hybrid aperture presents a larger enhancement than the gold aperture in the center region. The antenna quantum yields of different apertures are also calculated for comparison.

We found that the antenna quantum yield is dependent on the emitter's position within the nano-aperture structures. Figure 2c,d illustrate the emitter's position and the relative intensity of the antenna quantum yields for an emitter at 660 nm in the Au nano-aperture or for an emitter at 814 nm in the Au/Si nano-aperture, respectively. Figure 2e shows that we fix the emitter at 10 nm above the glass surface (i.e., 10 nm in the z-axis), but at different positions along the x-direction. The antenna quantum yield of the gold nano-aperture decreases when the dipole is closer to the gold surface. The quantum yield becomes very low when the emitter is close to the gold surface at the side corner, although the field enhancement is larger for the gold nano-aperture. Hence, the fluorescence intensity is weak at the aperture side, and the highly efficient fluorescent enhancement occurs mostly at the central region for the pure gold nano-aperture.

In contrast to the Au nano-aperture, the antenna quantum yield of the Au/Si hybrid antenna remains high in a large region at the same x-position—even for the emitters that are located close to the side corners. We note that for the pure silicon aperture, the antenna quantum yield is higher due to low loss, but the field enhancement factor is low (Figure S5). These conclusions are similar to previous reports [24,31]. In addition, the relative photoluminescence (PL) enhancement is estimated approximately based on the product of excitation field enhancement and the antenna quantum yield. The calculated PL enhancements at the positions that are mentioned above are plotted in Figure 2g,h. These results imply that the Au/Si hybrid antenna has better surface-enhanced fluorescence performance than the Au nano-aperture antenna.

The antenna quantum yields for both the gold and hybrid aperture antennas show similar vertical variation of the emitter position, i.e., they decrease with increasing positions along the z-direction. In particular, signal collection only from the bottom side shows that the relative quantum yield is more sensitive to the dipole position in the z-direction (Figures S3 and S4). Hence, the high efficiency of the fluorescent emission occurs mainly within the shallow layer that is close to the water-glass interface when both the excitation and collection are from the bottom side. Finally, by considering the field enhancement and quantum yield together as above, an efficient PL enhancement occurs in a shallow region that is close to the water-glass interface, which is helpful for reducing the observation volume to beyond the diffraction limit.

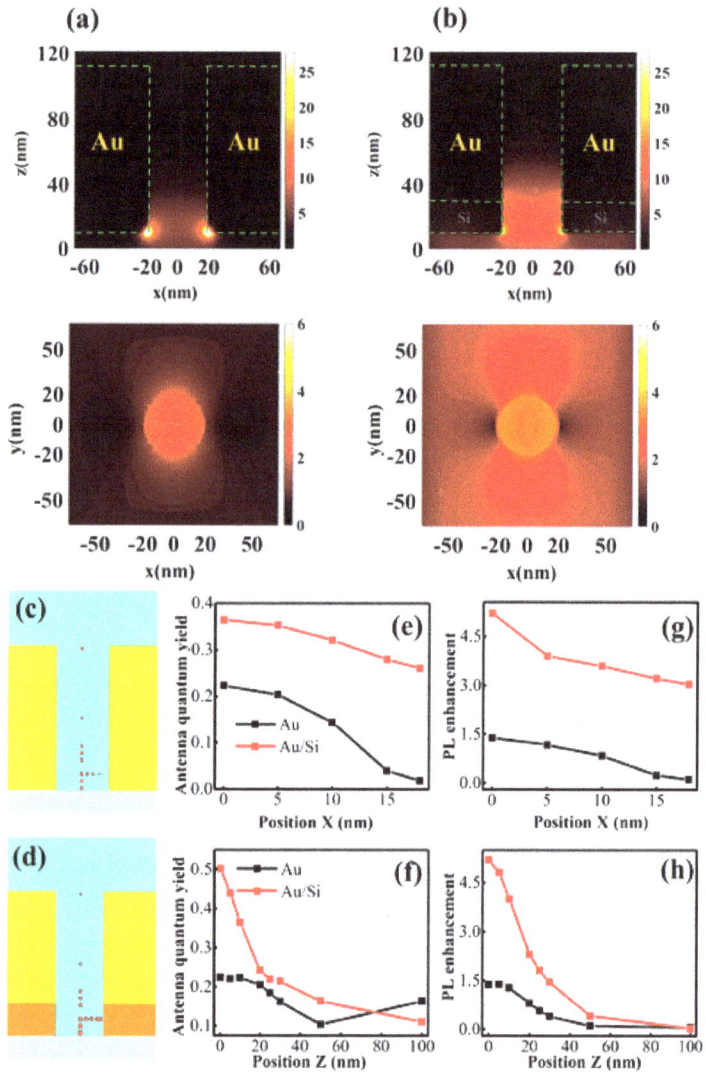

Figure 2. Electric field distribution $|E/E_0|^2$ in the *xz*-plane and *xy*-plane for (**a**) Au nano-aperture antenna at 573 nm and (**b**) Au/Si hybrid nano-aperture antenna at 785 nm illuminated by a plan wave with *x*-polarization from the bottom. The position-dependent quantum yields (the sizes of red disks represent the relative value of quantum yield at the positions correspondingly) (**c**) at 660 nm for the Au nano-aperture antenna; (**d**) at 814 nm for Au/Si hybrid nano-aperture antenna; (**e**,**f**) show quantum yields at different *x*- and *z*-positions indicated in (**c**,**d**), correspondingly. Panels (**g**,**h**) show photoluminescence (PL) enhancements at different *x*- and *z*-positions indicated in (**c**,**d**), correspondingly.

Having preliminarily demonstrated the advantages of the hybrid nano-aperture antenna, we next asked how the aperture size or layer thickness affects the enhancement effects. We calculated the far-field radiative rate enhancement and the antenna quantum yields as a function of the aperture diameter size. All of the calculations were executed for an emitter that was located at the center of a nano-aperture 10 nm above the silica surface, as above. The dipole orientates along the *x*-direction. We noted that the fluorescence enhancement is dependent on the dipole orientation [32]. For example, the antenna quantum yield of the z-orientated dipole is lower than the in-plane dipoles, and the z-component of electric filed is very weak under illumination of in-plane polarized light (the representative calculations are shown in supplemental materials). These factors result in much less signal from the z-orientated dipole. Hence, we focus our attention on the *x*-orientated dipole due to the aperture symmetry.

Figure 3 shows that the hybrid aperture antenna has different enhancement features versus pure gold nano-aperture as the aperture size varies from 20 to 120 nm. For the Au nano-aperture that is shown in Figure 3c, the far-field radiative rate enhancement factor generally increases with the increasing aperture diameter. It then decreases when the diameter reaches or is over 80 nm. Moreover, the maximum radiative rate always redshifts, and the half width at half maximum become broader. Concurrently, the antenna quantum yield increases monotonously with increasing diameter, due to increasing separation between the emitter and metal surface.

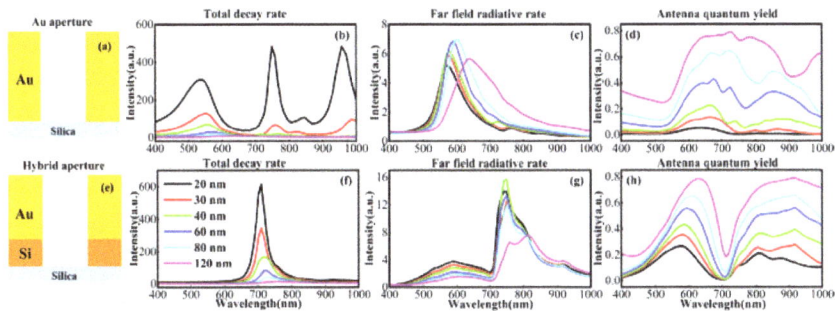

Figure 3. The enhancement effects are dependent on the aperture size of the antenna: (**b**) total decay rate enhancements; (**c**) far-field radiative rate enhancements; and (**d**) antenna quantum yields of (**a**) the Au nano-aperture antenna with diameters from 20 to 120 nm. Panels (**f–h**) show the same corresponding plots for (**e**) the Au/Si hybrid nano-aperture antenna with diameters from 20 to 120 nm, respectively. The thickness of the Au layer is 100 nm for gold aperture, and the hybrid structure consists of an 80 nm Au layer and a 20 nm Si layer.

Although smaller gold apertures have higher localized field enhancement, the antenna quantum yield is lower. Hence, the high enhancement effect occurs for the gold apertures with diameters of about 120 nm. (Larger apertures have low total fluorescent enhancement due to much lower field enhancement; data are not shown here). This simulation result agrees nicely with previous experimental studies [33].

The Au/Si hybrid apertures (Figure 3g) have enhanced far-field radiative rates with the increasing aperture diameter; this is independent of wavelength. Moreover, the radiative enhancement is higher than the gold apertures, which is consistent with the field enhancement calculations that are shown in Figure 2. The quantum yield also increases with the increasing aperture diameter (Figure 3h). The hybrid aperture still has a considerable antenna quantum yield, even down to 20 nm. This implies that the hybrid aperture has better performance beyond the diffraction limit for single molecule analysis, even at higher concentrations. The hybrid nano-aperture can also detect molecular fluorescence at the near-infrared region with a small size aperture. The dip in quantum yield at 700 nm is attributed to the surface mode between the metal and dielectric layers. Further calculations demonstrate that the hybrid nano-apertures with Si layer thickness from 20 to 60 nm have good surface enhancement effect (data is not shown here) that allows for a wide tolerance to experimentally fabricate such hybrid antennae.

In surface-enhanced spectroscopy, an important strategy to optimizing the enhancement effect is to construct plasmonic gap configurations [34]. The plasmon gap mode was first introduced into nano-aperture antenna by Lu et al. via a bowtie structure for molecule fluorescence analysis [35]. Later, the nano-aperture with plasmonic gap for molecule fluorescence analysis was further improved greatly via an antenna-in-box device by Punj et al. [36] Following this strategy, we introduce the dielectric nanogap structures into the Au/Si hybrid nano-aperture to further optimize the enhancement effect.

Figure 4 shows four kinds of nano-apertures simulated and compared to demonstrate the nanogap features. Two Si nanogaps are proposed to collaborate with a 120-nm diameter aperture. The first is a nanogap composed of two silicon disks. The thickness of the silicon disk is 20 nm, and the gap distance is 10 nm (Figure 4c). The second is a nanogap consisting of two silicon triangles—the thickness is 20 nm and the gap distance is 10 nm (Figure 4d). We calculate the electromagnetic field distribution in both the xz-plane and the xy-plane (Figure 4). Our simulations show that the field enhancement in the disk dimers nanogap is ~4-fold higher in magnitude than that of the pure gold aperture. The field enhancement could be optimized by adjusting the nanostructure parameters, e.g., smaller gap separation or sharper apex. The high field enhancement is very localized around the Si nanogap. This is useful for single molecule analysis at high concentrations.

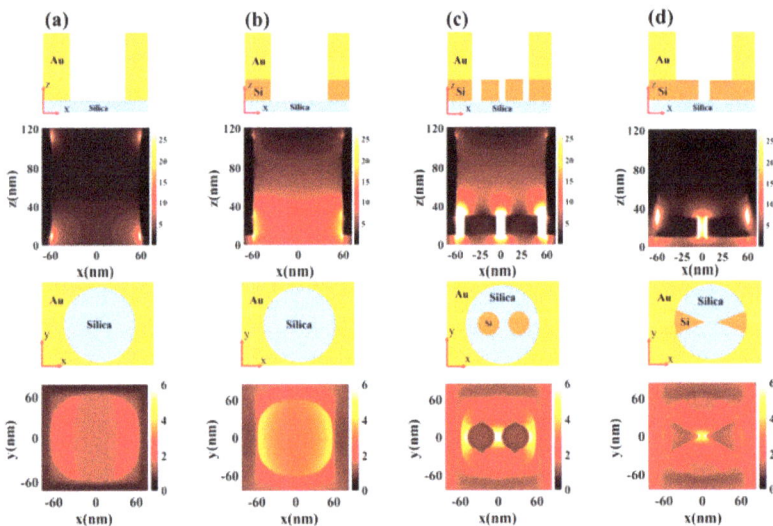

Figure 4. Electric field distribution $|E/E_0|^2$ and corresponding schematics in the xz-plane and the xy-plane for (**a**) Au nano-aperture at 573 nm; (**b**) Au/Si hybrid nano-aperture at 785 nm; (**c**) Au/Si hybrid nano-aperture with a silicon disk dimer at 758 nm; and (**d**) Au/Si hybrid nano-aperture with a silicon triangle gap at 758 nm. The diameters of all of the apertures are 120 nm and the silicon thickness is 20 nm. The disks' diameter in (**c**) is 40 nm, and the silicon gaps are 10 nm both in (**c**,**d**).

We also calculated the antenna quantum yields as a function of wavelength (Figure 5); here, an x-orientated dipole is placed at the center of the structures for all of the calculations. The antenna quantum yield of the nano-apertures with Si nanogaps is comparable to that of bare nano-apertures. Therefore, the hybrid apertures with a Si nanogap can greatly enhance the molecular fluorescence. There is always a dip near 700 nm (Figure 5) due to the surface mode between the metal and the dielectric layers. This intrinsic spectral feature could offer dual color analysis [7].

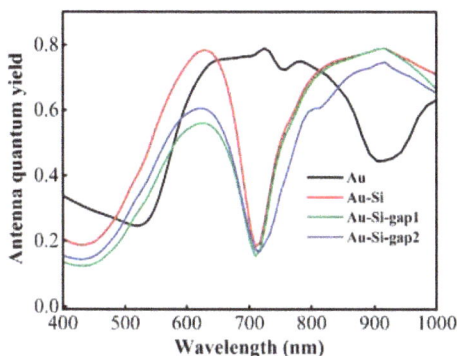

Figure 5. Antenna quantum yield as a function of wavelength from 400 nm to 1000 nm for four nano-aperture antennas shown in Figure 4. All of the calculations are performed for a dipole at the position (0, 0, and 10 nm). The Au-Si-gap1 and Au-Si-gap2 refer to the structures in Figure 4c,d, respectively.

Another important strategy to optimize the surface enhancement effect is to control the emission direction of the molecule's fluorescence, as first shown experimentally by Aouani et al. for a gold nano-aperture antenna [37]. The periodic grating structures usually surround the nano-aperture to both enhance the excitation rate and converge the emission for high collection efficiency [38,39]. Meanwhile, the silicon layer can be replaced with a lower loss dielectric material in the spectral region (e.g., GaP [40] in visible region). This would increase the antenna quantum yield. Experimentally, this hybrid nano-aperture antenna with or without Si nanogap is achievable via nanofabrication; experimental works are underway.

4. Conclusions

In summary, hybrid metal-dielectric nano-aperture antennas were investigated and compared theoretically by employing the FDTD method. The hybrid nano-apertures show better surface enhancement effect, rather than pure metal nano-apertures. The hybrid nano-apertures show a uniform field distribution within the apertures and higher antenna quantum yield than pure gold nano-apertures. The spectral feature of hybrid nano-apertures is independent from the aperture size and it shows a high enhancement effect in the near-infrared region even with small size aperture. Moreover, the hybrid nano-apertures show two high enhancement bands. This intrinsic feature benefits dual color analysis. Furthermore, the hybrid nano-apertures with dielectric gaps are useful for larger enhancement effects. The hybrid nano-aperture is fully adaptable to large-scale availability, and a Si layer thickness of 20 to 60 nm offers a good enhancement effect that facilitates a wide fabrication tolerance. The hybrid antennae will significantly improve the efficiency of surface-enhanced fluorescence for sensitive biosensing and molecular fluorescence applications.

Supplementary Materials: The following are available online at http://www.mdpi.com/1996-1944/11/8/1435/s1, Figure S1: Representative antenna quantum yields of the horizontal dipole along the z-, y-direction and the vertical dipole along the z-direction within (a) the Au aperture and (b) the Au/Si hybrid aperture at the position (0, 0, 10 nm; the origin is at the center of water-glass interface), correspondingly, Figure S2: Field distribution (a) $|Ex|2$, (b) $|Ey|2$ and (c) $|Ez|2$ in the xz-plane and xy-plane for Au nano-aperture antenna at a wavelength of 573 nm, Au/Si hybrid nano-aperture antenna at a wavelength of 785 nm illuminated by a plane wave with x-polarization from the bottom, Figure S3: Radiative rate of bottom collection (Prad-glass) and full integrated (Prad) for (a) the Au nano-aperture and (b) the Au/Si hybrid nano-aperture, Figure S4: Position-dependent quantum yields (the sizes of red disks represent the relative value of quantum yield at the positions correspondingly) (a,c) at a wavelength of 660 nm for Au nano-aperture antenna, (b,d) at a wavelength of 814 nm for Au/Si hybrid nano-aperture antenna; (e,f) show quantum yields with different x- and z-positions in the way of (a–b) full integrated and (c,d) bottom collection, correspondingly, Figure S5: Field distribution $|E/E_0|2$ in the xz-plane and xy-plane for (a) Au nano-aperture antenna at a wavelength of 573 nm, (b) Au/Si hybrid nano-aperture antenna at a wavelength of 785 nm, (c) Si nano-aperture antenna at a wavelength of 725 nm illuminated by a plane wave with *x*-polarization from the bottom. (d) shows the antenna quantum yields of structures above which all are calculated for a dipole located at position (0, 0, 10 nm), Figure S6: Aperture size of the antenna-dependent enhancement effects. (b) Total decay rate enhancements, (c) Far-field radiative rate enhancements, (d) Non-radiative rate, (e) Antenna quantum yields of (a) the Au nano-aperture antenna with diameters from 20 to 120 nm. (g,h) show the same corresponding plots for (f) the Au/Si hybrid nano-aperture antenna with diameters from 20 to 120 nm, respectively. The thickness of the Au layer is 100 nm for gold aperture, and the hybrid structure consists of an 80 nm Au layer and a 20 nm Si layer.

Author Contributions: G.L. conceived the research. J.X. and G.L. performed the numerical calculations. All of the authors participated in the data analysis and wrote the paper.

Funding: This work was supported by the National Natural Science Foundation of China (grant nos. 61422502, 11374026, 61521004, and 11527901). This research received no external funding.

Conflicts of Interest: The authors declare no conflict of interest.

References

1. Anker, J.N.; Hall, W.P.; Lyandres, O.; Shah, N.C.; Zhao, J.; Van Duyne, R.P. Biosensing with plasmonic nanosensors. *Nat. Mater.* **2008**, *7*, 442–453. [CrossRef] [PubMed]
2. Pelton, M. Modified spontaneous emission in nanophotonic structures. *Nat. Photonics* **2015**, *9*, 427–435. [CrossRef]
3. Holzmeister, P.; Acuna, G.P.; Grohmann, D.; Tinnefeld, P. Breaking the concentration limit of optical single-molecule detection. *Chem. Soc. Rev.* **2014**, *43*, 1014–1028. [CrossRef] [PubMed]
4. Levene, M.J.; Korlach, J.; Turner, S.W.; Foquet, M.; Craighead, H.G.; Webb, W.W. Zero-mode waveguides for single-molecule analysis at high concentrations. *Science* **2003**, *299*, 682–686. [CrossRef] [PubMed]
5. Eid, J.; Fehr, A.; Gray, J.; Luong, K.; Lyle, J.; Otto, G.; Peluso, P.; Rank, D.; Baybayan, P.; Bettman, B.; et al. Real-time DNA sequencing from single polymerase molecules. *Science* **2009**, *323*, 133–138. [CrossRef] [PubMed]
6. Rigneault, H.; Capoulade, J.; Dintinger, J.; Wenger, J.; Bonod, N.; Popov, E.; Ebbesen, T.W.; Lenne, P.F. Enhancement of single-molecule fluorescence detection in subwavelength apertures. *Phys. Rev. Lett.* **2005**, *95*, 117401. [CrossRef] [PubMed]
7. Zhu, P.; Craighead, H.G. Zero-mode waveguides for single-molecule analysis. *Annu. Rev. Biophys.* **2012**, *41*, 269–293. [CrossRef] [PubMed]
8. Khurgin, J.B. How to deal with the loss in plasmonics and metamaterials. *Nat. Nanotechnol.* **2015**, *10*, 2–6. [CrossRef] [PubMed]
9. LÜ, G.; Shen, H.; Cheng, Y.; Gong, Q. Advances in localized surface plasmon enhanced fluorescence. *Chin. Sci. Bull.* **2015**, *60*, 3169–3179.
10. Novotny, L.; Hecht, B. *Principles of Nano-Optics*; Cambridge University Press: Cambridge, UK, 2006.
11. Shen, H.M.; Lu, G.W.; Zhang, T.Y.; Liu, J.; Gong, Q.H. Enhanced Single-Molecule Spontaneous Emission in an Optimized Nanoantenna with Plasmonic Gratings. *Plasmonics* **2013**, *8*, 869–875. [CrossRef]
12. Caldarola, M.; Albella, P.; Cortes, E.; Rahmani, M.; Roschuk, T.; Grinblat, G.; Oulton, R.F.; Bragas, A.V.; Maier, S.A. Non-plasmonic nanoantennas for surface enhanced spectroscopies with ultra-low heat conversion. *Nat. Commun.* **2015**, *6*, 7915. [CrossRef] [PubMed]
13. Regmi, R.; Berthelot, J.; Winkler, P.M.; Mivelle, M.; Proust, J.; Bedu, F.; Ozerov, I.; Begou, T.; Lumeau, J.; Rigneault, H.; et al. All-Dielectric Silicon Nanogap Antennas to Enhance the Fluorescence of Single Molecules. *Nano Lett.* **2016**, *16*, 5143–5151. [CrossRef] [PubMed]
14. Barreda, A.I.; Saleh, H.; Litman, A.; Gonzalez, F.; Geffrin, J.M.; Moreno, F. Electromagnetic polarization-controlled perfect switching effect with high-refractive-index dimers and the beam-splitter configuration. *Nat. Commun.* **2017**, *8*, 13910. [CrossRef] [PubMed]
15. Barth, M.; Schietinger, S.; Fischer, S.; Becker, J.; Nusse, N.; Aichele, T.; Lochel, B.; Sonnichsen, C.; Benson, O. Nanoassembled plasmonic-photonic hybrid cavity for tailored light-matter coupling. *Nano Lett.* **2010**, *10*, 891–895. [CrossRef] [PubMed]
16. Chen, X.W.; Agio, M.; Sandoghdar, V. Metallodielectric hybrid antennas for ultrastrong enhancement of spontaneous emission. *Phys. Rev. Lett.* **2012**, *108*, 233001. [CrossRef] [PubMed]

17. Peng, P.; Liu, Y.C.; Xu, D.; Cao, Q.T.; Lu, G.; Gong, Q.; Xiao, Y.F. Enhancing Coherent Light-Matter Interactions through Microcavity-Engineered Plasmonic Resonances. *Phys. Rev. Lett.* **2017**, *119*, 233901. [CrossRef] [PubMed]

18. Nielsen, M.P.; Lafone, L.; Rakovich, A.; Sidiropoulos, T.P.H.; Rahmani, M.; Maier, S.A.; Oulton, R.F. Adiabatic Nanofocusing in Hybrid Gap Plasmon Waveguides on the Silicon-on-Insulator Platform. *Nano Lett.* **2016**, *16*, 1410–1414. [CrossRef] [PubMed]

19. Gurlek, B.; Sandoghdar, V.; Martín-Cano, D. Manipulation of Quenching in Nanoantenna–Emitter Systems Enabled by External Detuned Cavities: A Path to Enhance Strong-Coupling. *ACS Photonics* **2017**, *5*, 456–461. [CrossRef]

20. Kang, T.Y.; Lee, W.; Ahn, H.; Shin, D.M.; Kim, C.S.; Oh, J.W.; Kim, D.; Kim, K. Plasmon-Coupled Whispering Gallery Modes on Nanodisk Arrays for Signal Enhancements. *Sci. Rep.* **2017**, *7*, 11737. [CrossRef] [PubMed]

21. Thakkar, N.; Rea, M.T.; Smith, K.C.; Heylman, K.D.; Quillin, S.C.; Knapper, K.A.; Horak, E.H.; Masiello, D.J.; Goldsmith, R.H. Sculpting Fano Resonances to Control Photonic-Plasmonic Hybridization. *Nano Lett.* **2017**, *17*, 6927–6934. [CrossRef] [PubMed]

22. Milichko, V.A.; Zuev, D.A.; Baranov, D.G.; Zograf, G.P.; Volodina, K.; Krasilin, A.A.; Mukhin, I.S.; Dmitriev, P.A.; Vinogradov, V.V.; Makarov, S.V.; et al. Metal-Dielectric Nanocavity for Real-Time Tracing Molecular Events with Temperature Feedback. *Laser Photonics Rev.* **2018**, *12*, 1700227. [CrossRef]

23. Oskooi, A.F.; Roundy, D.; Ibanescu, M.; Bermel, P.; Joannopoulos, J.D.; Johnson, S.G. MEEP: A flexible free-software package for electromagnetic simulations by the FDTD method. *Comput. Phys. Commun.* **2010**, *181*, 687–702. [CrossRef]

24. Lu, G.; Zhang, T.; Li, W.; Hou, L.; Liu, J.; Gong, Q. Single-Molecule Spontaneous Emission in the Vicinity of an Individual Gold Nanorod. *J. Phys. Chem. C* **2011**, *115*, 15822–15828. [CrossRef]

25. Cheng, Y.; Lu, G.; Shen, H.; Wang, Y.; He, Y.; Chou, R.Y.; Gong, Q. Highly enhanced spontaneous emission with nanoshell-based metallodielectric hybrid antennas. *Opt. Commun.* **2015**, *350*, 40–46. [CrossRef]

26. Wang, Y.; Shen, H.; He, Y.; Cheng, Y.; Perriat, P.; Martini, M.; Tillement, O.; Gong, Q.; Lu, G. Amorphous nanoshell formed through random growth and related plasmonic behaviors. *Chem. Phys. Lett.* **2014**, *610–611*, 278–283. [CrossRef]

27. Shen, H.; Lu, G.; Zhang, T.; Liu, J.; Gu, Y.; Perriat, P.; Martini, M.; Tillement, O.; Gong, Q. Shape effect on a single-nanoparticle-based plasmonic nanosensor. *Nanotechnology* **2013**, *24*, 285502. [CrossRef] [PubMed]

28. Johnson, P.B.; Christy, R.W. Optical Constants of the Noble Metals. *Phys. Rev. B* **1972**, *6*, 4370. [CrossRef]

29. Green, M.A.; Keevers, M.J. Optical-Properties of Intrinsic Silicon at 300 K. *Prog. Photovolt.* **1995**, *3*, 189–192. [CrossRef]

30. Zhang, T.Y.; Lu, G.W.; Liu, J.; Shen, H.M.; Perriat, P.; Martini, M.; Tillement, O.; Gong, Q.H. Strong two-photon fluorescence enhanced jointly by dipolar and quadrupolar modes of a single plasmonic nanostructure. *Appl. Phys. Lett.* **2012**, *101*, 051109. [CrossRef]

31. Zyuzin, M.V.; Baranov, D.G.; Escudero, A.; Chakraborty, I.; Tsypkin, A.; Ushakova, E.V.; Kraus, F.; Parak, W.J.; Makarov, S.V. Photoluminescence quenching of dye molecules near a resonant silicon nanoparticle. *Sci. Rep.* **2018**, *8*, 6107. [CrossRef] [PubMed]

32. Chou, R.Y.Y.; Lu, G.W.; Shen, H.M.; He, Y.B.; Cheng, Y.Q.; Perriat, P.; Martini, M.; Tillement, O.; Gong, Q.H. A hybrid nanoantenna for highly enhanced directional spontaneous emission. *J. Appl. Phys.* **2014**, *115*, 244310. [CrossRef]

33. Gérard, D.; Wenger, J.; Bonod, N.; Popov, E.; Rigneault, H.; Mahdavi, F.; Blair, S.; Dintinger, J.; Ebbesen, T.W. Nanoaperture-enhanced fluorescence: Towards higher detection rates with plasmonic metals. *Phys. Rev. B* **2008**, *77*, 045413.

34. Cao, Z.M.; He, Y.B.; Cheng, Y.Q.; Zhao, J.Y.; Li, G.T.; Gong, Q.H.; Lu, G.W. Nano-gap between a gold tip and nanorod for polarization dependent surface enhanced Raman scattering. *Appl. Phys. Lett.* **2016**, *109*, 233103. [CrossRef]

35. Lu, G.W.; Li, W.Q.; Zhang, T.Y.; Yue, S.; Liu, J.; Hou, L.; Li, Z.; Gong, Q.H. Plasmonic-Enhanced Molecular Fluorescence within Isolated Bowtie Nano-Apertures. *ACS Nano* **2012**, *6*, 1438–1448. [CrossRef] [PubMed]

36. Punj, D.; Mivelle, M.; Moparthi, S.B.; van Zanten, T.S.; Rigneault, H.; van Hulst, N.F.; Garcia-Parajo, M.F.; Wenger, J. A plasmonic 'antenna-in-box' platform for enhanced single-molecule analysis at micromolar concentrations. *Nat. Nanotechnol.* **2013**, *8*, 512–516. [CrossRef] [PubMed]

37. Aouani, H.; Mahboub, O.; Bonod, N.; Devaux, E.; Popov, E.; Rigneault, H.; Ebbesen, T.W.; Wenger, J. Bright unidirectional fluorescence emission of molecules in a nanoaperture with plasmonic corrugations. *Nano Lett.* **2011**, *11*, 637–644. [CrossRef] [PubMed]
38. Chou, R.Y.; Li, G.; Cheng, Y.; He, Y.; Zhao, J.; Cao, Z.; Gong, Q.; Lu, G. Surface enhanced fluorescence by metallic nano-apertures associated with stair-gratings. *Opt. Express* **2016**, *24*, 19567–19573. [CrossRef] [PubMed]
39. Shen, H.; Lu, G.; Zhang, T.; Liu, J.; He, Y.; Wang, Y.; Gong, Q. Molecule fluorescence modified by a slit-based nanoantenna with dual gratings. *J. Opt. Soc. Am. B* **2013**, *30*, 2420–2426. [CrossRef]
40. Cambiasso, J.; Grinblat, G.; Li, Y.; Rakovich, A.; Cortes, E.; Maier, S.A. Bridging the Gap between Dielectric Nanophotonics and the Visible Regime with Effectively Lossless Gallium Phosphide Antennas. *Nano Lett.* **2017**, *17*, 1219–1225. [CrossRef] [PubMed]

materials

MDPI

Article

Al/Si Nanopillars as Very Sensitive SERS Substrates

Giovanni Magno [1], Benoit Bélier [1] and Grégory Barbillon [2],[*]

[1] Centre de Nanosciences et de Nanotechnologies, CNRS, University Paris Sud, Université Paris-Saclay, C2N-Orsay, CEDEX, 91405 Orsay, France; giovanni.magno@c2n.upsaclay.fr (G.M.); benoit.belier@c2n.upsaclay.fr (B.B.)

[2] EPF-Ecole d'Ingénieurs, 3 Bis Rue Lakanal, 92330 Sceaux, France

[*] Correspondence: gregory.barbillon@epf.fr

Received: 20 July 2018 ; Accepted: 23 August 2018; Published: 26 August 2018

Abstract: In this paper, we present a fast fabrication of Al/Si nanopillars for an ultrasensitive SERS detection of chemical molecules. The fabrication process is only composed of two steps: use of a native oxide layer as a physical etch mask followed by evaporation of an aluminum layer. A random arrangement of well-defined Al/Si nanopillars is obtained on a large-area wafer of Si. A good uniformity of SERS signal is achieved on the whole wafer. Finally, we investigated experimentally the sensitivity of these Al/Si nanopillars for SERS sensing, and analytical enhancement factors in the range of $1.5 \times 10^7 - 2.5 \times 10^7$ were found for the detection of thiophenol molecules. Additionally, 3D FDTD simulations were used to better understand optical properties of Al/Si nanopillars as well as the Raman enhancement.

Keywords: SERS; sensors; aluminum; silicon

1. Introduction

During this last decade, Surface Enhanced Raman Scattering (SERS) is mainly employed as a powerful technique for detection of biological/chemical molecules. The fabrication of SERS substrates having high enhancement factors (EF) is the key point for the improvement of this biological/chemical sensing. Several groups investigated a great number of novel SERS substrates, which demonstrated a large Raman enhancement, such as colloidal metallic nanoparticles [1–3] and metallic nanostructures on different surfaces fabricated by various lithographic techniques [4–11]. Indeed, this large Raman enhancement is mainly due to the presence of hotspots in these different SERS substrates. The mechanism of this enhancement due to the hotspots is well-described in References [12,13], and the development of this type of SERS substrates with high densities of hotspots is demonstrated in References [8,9,14–17]. However, certain fabrication techniques cited previously are technologically demanding in terms of time and expensive for a mass production destined to industrial applications. Besides, Nanoimprint Lithography (NIL) [18–20] and Nanosphere Lithography (NSL) [21–23] allows fabricating these SERS substrates with a lower cost. Nevertheless, they can be plagued by poor definition of nanostructures obtained on large surfaces, which are required for practical/industrial applications. Another way for obtaining higher EFs is to use silicon nanowires (SiNW) coupled to metallic nanoparticles. This type of nanostructures allows obtaining a better detection limit [16,24,25]. Moreover, disordered SiNWs can be fabricated by large-surface techniques. Although all these SERS substrates have great potential for a very sensitive detection of chemical or biological molecules, most of the applications are hampered by the non-uniformity of the SERS signals. Several groups have already addressed this non-uniformity issue of the SERS signal, and they have demonstrated a good uniformity of the latter [26,27].

In this paper, the aim is to present a simple and fast process to produce very sensitive SERS substrates composed of Al/Si nanopillars at the large-area wafer-scale, which will have a good uniformity of the SERS signal. In our case, aluminum was chosen as a plasmonic material for its attractive properties including low cost, high natural abundance, compatibility with CMOS technology and optoelectronic devices, and plasmonic resonances in the spectral domains of UV and visible [28,29]. Moreover, aluminum plasmonics can be applied to a wide range of applications such as SERS in ultraviolet [30] and visible domains [31–33], SEIRA (Surface-Enhanced Infrared Absorption Spectroscopy) [31,34], photocatalysis [35], and metal-enhanced fluorescence [36]. Although several groups have already worked on Al plasmonics nanostructures for SERS sensing [30,31,34], there was little consideration on the synergy between silicon and aluminum in order to improve the performance of SERS sensors. Here, the ability of Al/Si nanopillars to be very sensitive SERS sensors is investigated and evaluated using thiophenol solutions. To further deepen the comprehension of the SERS signal enhancement, 3D FDTD simulations are made.

2. Experimental Details

2.1. Two-Step Fabrication of Al/Si Nanopillars

The fabrication process of large area of Al/Si nanopillars (NPs) is composed of two steps (see Figure 1): (i) etching through the mask obtained by the native oxide layer; and (ii) depositing of titanium and aluminum layers under vacuum by Electron Beam Evaporation (EBE). In this fabrication process, no pre-patterning of the Si surface is required. Indeed, we only use the native oxide layer of Si wafer as a physical etch mask. Then, an anisotropic RIE (Reactive Ion Etching) process consisting in sixty cycles of passivation and etching steps is realized on the Si wafer through the native oxide layer by using ICP-SPTS (Inductively Coupled Plasma-Sumitomo Precision Products Process Technology Systems) equipment. Gases involved in this protocol are SF_6 (300 sccm), C_4F_8 (180 sccm) and O_2 (200 sccm). This anisotropic RIE process is a switched process in which fluorine from SF_6 etches the Si while C_4F_8 passivates the surface, and it starts with a cycle of passivation. During this first and short passivation cycle, only certain nanoscale zones of the native oxide layer are randomly passivated, which will then serve as etch mask and thus produce Si nanopillars at the end of these zones. The organization of the obtained nanopillars is completely random. The pressure and power used in this process are 20 mTorr and 25 W, respectively. By modifying the process parameters such as cycle times, number of cycles, platen and coil power, and substrate temperature, the size distribution, depth and density can be controlled. To finish, a 2 nm titanium layer used as adhesion layer, and an aluminum layer of 50 nm are deposited by EBE under normal incidence. The evaporation rate used in this process are 0.05 nm/s and 0.3 nm/s for Ti and Al layers, respectively.

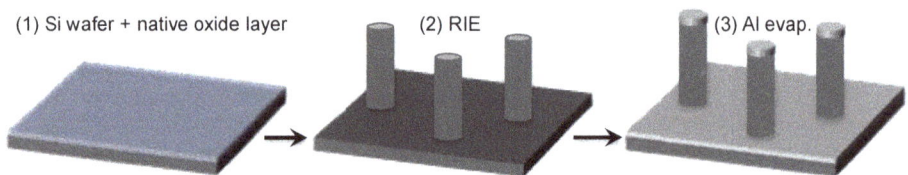

Figure 1. Principle scheme of the Al/Si nanopillar fabrication. The metal evaporation step is made under normal incidence.

2.2. Thiophenol Deposition on Al/Si Nanopillars

For our SERS investigations, thiophenol molecules were used to test the sensitivity of these Al/Si nanopillars because they are excellent model molecules. The deposition protocol is as follows: (i) preparation of a 1 µM solution of thiophenol in ethanol; (ii) dipping the SERS sample in the solution for 3 h; and (iii) the SERS sample was allowed to nitrogen dry in a specific box. For our reference experiment, the deposition protocol is: (i) preparation of a 1 M solution of thiophenol in ethanol;

(ii) dipping the reference sample (Si substrate with nanopillars without metal) in the solution for 3 h; and (iii) the reference sample was allowed to nitrogen dry in a specific box.

2.3. Raman Characterization

For all the Raman measurements, we employed a Labram spectrophotometer from Horiba Scientific, which has a spectral resolution of 1 cm^{-1}. The excitation wavelength (λ_{exc} = 633 nm) and an acquisition time of 10 s were used for all the SERS and Raman (reference) experiments. For these characterizations, the laser was focused on the substrate using a microscope objective (100×, N.A. = 0.9). The Raman signal from the SERS substrates (or reference experiment) was collected by the same objective in a backscattering configuration, and the used laser power was 1 mW. The average of SERS intensities and relative standard deviations (RSD) were calculated on the basis of 25 SERS spectra.

2.4. FDTD Simulations of SERS Substrates

To calculate the extinction spectrum of the SERS substrates, 3D Finite-Difference Time-Domain (FDTD) method was used. For these FDTD simulations, we considered an isolated Al/Si nanopillar, which corresponds to the experimental case for Al/Si nanopillars (see Figures 2 and 3). The nanopillar diameter (D) is 150 nm, its height (h_{pillar}) is 1450 nm, and the Al layer thickness (h_{Al}) is 50 nm on the top of nanopillar and on Si substrate. The top corners of the Al/Si nanopillar are not rounded. Both materials used for this study have been modelled by fitting the real and imaginary parts of the permittivities reported in the reference [37]. The nanopillar on substrate, centred in a computational cell of 3 × 3 × 5 µm^3, is surrounded by Perfectly Matched Layers (PML) in order to absorb radiation leaving the calculation region. For providing an excellent resolution of the fields, a uniform mesh of 2 × 2 × 2 nm^3 was used for discretising the computational cell. Finally, the extinction spectrum has been calculated by exciting the structure with a broadband plane wave source (spectral range from 400 nm to 800 nm) impinging from above the pillar and by collecting the reflected (R) and transmitted (T) powers. This simulated extinction spectrum does not consider the thiophenol layer. Thus, these simulations shed light on the optical properties of Al/Si nanopillars.

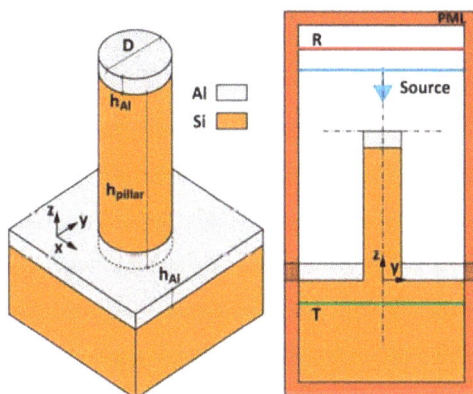

Figure 2. Nanopillar morphology used for the FDTD simulations, the diameter (D) is 150 nm, the height (h_{pillar}) is 1450 nm, and the Al layer thickness (h_{Al}) is 50 nm on the top of nanopillar and on Si substrate. On the right, the broadband plane wave source and monitors (R and T) for calculating the extinction spectrum are displayed.

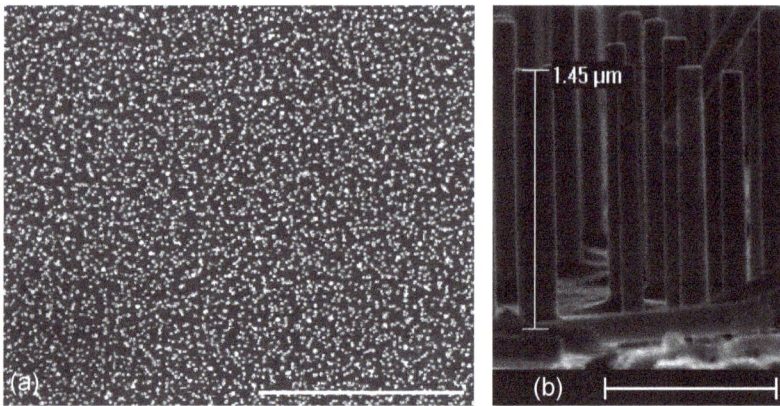

Figure 3. SEM images of Al/Si nanopillars obtained with our fabrication technique: (**a**) on a large zone (scale bar = 20 µm); and (**b**) cross-section view of the nanopillars (scale bar = 1 µm).

3. Results and Discussion

Firstly, Al/Si nanopillars were fabricated with the process in Section 2.1. Figure 3 displays SEM images of these Al/Si nanopillars. The diameter and the height of the Al/Si NPs were determined to be 150 ± 40 nm, and 1450 ± 50 nm, respectively. The homogeneity of Al/Si nanopillars is correct in terms of dimensions.

Next, thiophenol molecules (see molecular scheme in Figure 4) were deposited on Al/Si NPs directly after their fabrication with the protocol in Section 2.2, and then characterized directly by Raman measurements. Figure 4 reveals the SERS spectra of thiophenol on Al/Si nanopillars recorded at the excitation wavelength of 633 nm. On all SERS spectra, we observed Raman shifts, which are characteristic of thiophenol molecules [38–40] as those at 1000 cm^{-1} corresponding to the C-C stretching mode (named: $\nu(CC)$, see References [40–42]); at 1025 cm^{-1} corresponding to the combination of the following modes: C-C stretching and C-H in-plane bending (named: $\nu(CC)$ and $\delta(CH)$, respectively, see References [40–42]); at 1075 cm^{-1} corresponding to the combination of the following modes: C-C stretching, C-H in-plane bending and C-S stretching (named: $\nu(CC)$, $\delta(CH)$ and $\nu(CS)$, respectively, see References [40–42]); and at 1575 cm^{-1} corresponding to the C-C stretching (named: $\nu(CC)$, see References [40–42]). Besides, some multi-phonon peaks of Si in the range of 900–980 cm^{-1} are observed [43,44]. In the inset of Figure 4, a reference Raman spectrum of thiophenol obtained with only Si nanopillars (without metal) is displayed. No significant Raman shift studied here is visible, because they are very weak.

To evaluate the sensitivity of Al/Si nanopillars, the analytical enhancement factor (AEF) is calculated for the 4 characteristic Raman peaks of the thiophenol molecules previously cited. *AEF* is given by the following formula:

$$AEF = \frac{I_{SERS}}{I_{Raman}} \times \frac{C_{Raman}}{C_{SERS}} \tag{1}$$

where I_{SERS}, I_{Raman} represent the SERS and Raman intensities, respectively (see Table 1). C_{SERS} (1 µm), C_{Raman} (1 M) are the concentrations of thiophenol for SERS and reference Raman experiments, respectively.

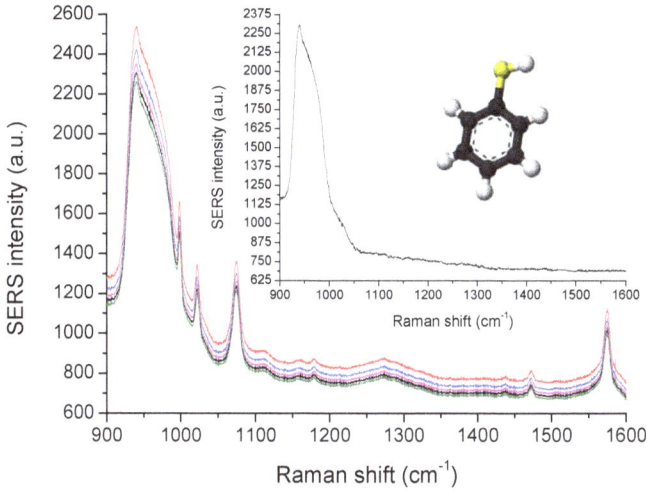

Figure 4. Five SERS spectra of thiophenol molecules recorded randomly on the whole substrate composed of Al/Si nanopillars. The inset depicts the Raman spectrum of thiophenol obtained with only Si nanopillars (without metal). Moreover, the molecular scheme of the thiophenol molecule is also displayed.

From the results in Table 1, the largest AEF value, which was found for the Al/Si nanopillars, is 2.4×10^7 the Raman shift of 1575 cm^{-1}. In addition, some groups have obtained excellent AEFs of $\sim 10^6$ with Ag nanoparticles on Si/ZnO nanotrees [45] and around 2×10^6 with Au nanostructured electrodes [46]. Furthermore, other groups have demonstrated good EF with similar SERS substrates such as Ag nanoparticles on Si nanowires (for [17]: EF$\sim 4 \times 10^6$; for [16]: EF $= 10^7 - 2.3 \times 10^8$; and for [24]: EF $= 10^8 - 10^{10}$), and Si nanopillars covered on the nanopillar top by Ag lumps (EF $\sim 5 \times 10^6$) [25]. By comparison, our Al/Si nanopillars are faster to fabricate and a better sensitivity is achieved for all the Raman peaks studied here ($1.5 \times 10^7 <$ AEF $< 2.5 \times 10^7$) except for Ag nanoparticles on Si nanowires of References [16,24] concerning to the sensitivity. Besides, the relative standard deviation (RSD) is calculated for all the four peaks of our investigation in order to quantify the uniformity. To do that, 25 SERS spectra of thiophenol molecules were recorded from several randomly chosen zones on the whole wafer under same experimental conditions. In Table 1, a good uniformity (RSD $< 7\%$) of SERS signal is obtained for each Raman peak on the large-area wafer of the Al/Si nanopillars.

Table 1. For the excitation wavelength of 633 nm and four Raman peaks (RS) studied here, λ_{Raman} associated to RS, the intensities I_{Raman} and I_{SERS}, RSDs associated to I_{SERS} values, analytical enhancement factors (AEF) and EF values (in arbitrary unit) obtained with the E^4 model are presented.

Name	RS (cm^{-1})	λ_{Raman} (nm)	I_{Raman}	I_{SERS}	RSD (%)	AEF	EF (a.u.)
1	1000	676	16	271	6.6	1.7×10^7	0.220
2	1025	677	11	209	4.8	1.9×10^7	0.222
3	1075	679	20	409	4.4	2.1×10^7	0.224
4	1575	703	14	334	3.6	2.4×10^7	0.238

To understand these experimental results, we calculated the extinction spectrum of the SERS substrate (see Figure 5). From this, we easily observe the positions of different resonances observed for these Al/Si nanopillars compared to the positions of the excitation wavelength and Raman wavelengths associated to the Raman shifts measured experimentally. Moreover, λ_{Raman} is the Raman scattering wavelength corresponding the studied Raman shift, which is determined with the following formula:

$$\Delta\omega = 10^7 \left(\frac{1}{\lambda_{exc}} - \frac{1}{\lambda_{Raman}} \right)$$

(2)

where $\Delta\omega$ is the studied Raman shift (in cm^{-1}), λ_{exc} is the excitation wavelength used in the experiments (in nm), and λ_{Raman} is the Raman scattering wavelength to be determined (in nm, see Table 1).

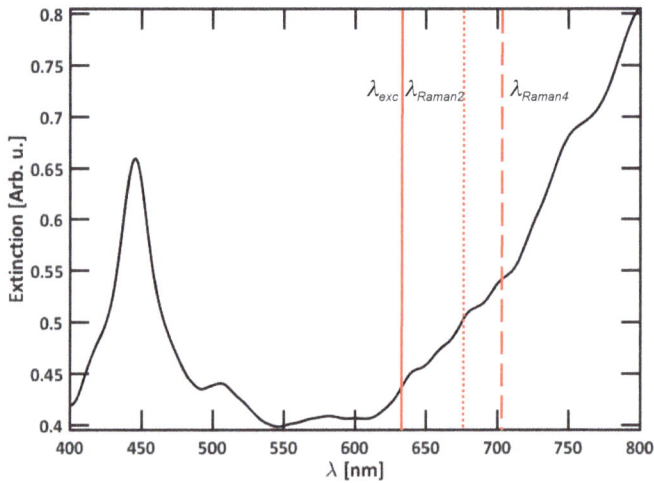

Figure 5. Calculated extinction spectrum of Al/Si nanopillars. λ_{exc} corresponds to the excitation wavelength (λ_{exc} = 633 nm, continuous red line). λ_{Raman2} and λ_{Raman4} correspond to the Raman scattering wavelengths for the Raman shifts of 1025 cm^{-1} and 1575 cm^{-1} (λ_{Raman2} = 677 nm, dotted red line, and λ_{Raman4} = 703 nm, dashed red line), respectively. For the sake of readability, only λ_{Raman2} and λ_{Raman4} are displayed, since λ_{Raman1} and λ_{Raman3} are very close to λ_{Raman2}.

Finally, we can qualitatively analyze the SERS enhancement, which can be obtained by using the E^4 model, assuming that enhancement factor is proportional to the extinction intensities (Q_e) at λ_{exc} and λ_{Raman}, i.e., EF $\sim Q_e(\lambda_{exc}) \times Q_e(\lambda_{Raman})$ [47]. In Figure 5 and Table 1, we observe that EF$_4$ is the highest value, and the EF values increased when λ_{Raman} also increased, i.e., $Q_e(\lambda_{Raman})$ increased with λ_{Raman}. The different EF values correspond to enhancement factors for the couples (λ_{exc}, λ_{Raman1}), (λ_{exc}, λ_{Raman2}), (λ_{exc}, λ_{Raman3}) and (λ_{exc}, λ_{Raman4}), respectively. These FDTD results suggest that the AEF values observed experimentally (see Table 1) have the same behavior as the EF values obtained with the E^4 model.

4. Conclusions

In this paper, we demonstrate the fast fabrication of very sensitive SERS substrates composed of Al/Si nanopillars for chemical detection. The key point of this fabrication process is the use of a native oxide layer as a physical etch mask. This fabrication allowed obtaining well-defined nanopillars at the large-area wafer-scale. The sensitivity of these Al/Si nanopillars was investigated and compared to the results obtained for gold nanostructured electrodes [46], Ag nanoparticles on Si/ZnO nanotrees [45], Ag nanoparticles on Si nanowires [16,17,24], and Si nanopillars covered on the nanopillar top by Ag lumps [25]. The AEF values achieved with our Al/Si nanopillars ($1.5 \times 10^7 <$ AEF $< 2.5 \times 10^7$) is better than the SERS substrates cited previously, except for Ag nanoparticles on Si nanowires of References [16,24]. Moreover, an excellent uniformity of SERS signal (RSD < 7%) was achieved on the whole wafer, which is a key point for industrial applications. Thus, such Al/Si nanopillars could be integrated on a lab-on-chip for label-free chemical/biological detection processes.

Author Contributions: G.B. conceived the research. G.M. performed the FDTD simulations. G.M., B.B. and G.B. performed the experiments and wrote the paper.

Funding: This research received no external funding.

Conflicts of Interest: The authors declare no conflict of interest.

References

1. Rodriguez-Fernandez, D.; Langer, J.; Henriksen-Lacey, M.; Liz-Marzan, L.M. Hybrid Au-SiO$_2$ core-satellite colloids as switchable SERS tags. *Chem. Mater.* **2015**, *27*, 2540–2545. [CrossRef]
2. La Porta, A.; Sanchez-Iglesias, A.; Altantzis, T.; Bals, S.; Grzelczak, M.; Liz-Marzan, L.M. Multifunctional self-assembled composite colloids and their application to SERS detection. *Nanoscale* **2015**, *7*, 10377–10381. [CrossRef] [PubMed]
3. De Jimenez Aberasturi, D.; Serano-Montes, A.B.; Langer, J.; Henriksen-Lacey, M.; Parak, W.J.; Liz-Marzan, L.M. Surface enhanced Raman scattering encoded gold nanostars for multiplexed cell discrimination. *Chem. Mater.* **2016**, *28*, 6779–6790. [CrossRef]
4. Yu, Q.; Guan, P.; Qin, D.; Golden, G.; Wallace, P.M. Inverted size-dependence of surface-enhanced Raman scattering on gold nanohole and nanodisk arrays. *Nano Lett.* **2008**, *8*, 1923–1928. [CrossRef] [PubMed]
5. Faure, A.C.; Barbillon, G.; Ou, M.; Ledoux, G.; Tillement, O.; Roux, S.; Fabregue, D.; Descamps, A.; Bijeon, J.-L.; Marquette, C.A.; et al. Core/shell nanoparticles for multiple biological detection with enhanced sensitivity and kinetics. *Nanotechnology* **2008**, *19*, 485103. [CrossRef] [PubMed]
6. Vo-Dinh, T.; Dhawan, A.; Norton, S.J.; Khoury, C.G.; Wang, H.N.; Misra, V.; Gerhold, M.D. Plasmonic nanoparticles and nanowires: design, fabrication and application in sensing. *J. Phys. Chem. C* **2010**, *114*, 7480–7488. [CrossRef] [PubMed]
7. Dhawan, A.; Duval, A.; Nakkach, M.; Barbillon, G.; Moreau, J.; Canva, M.; Vo-Dinh, T. Deep UV nano-microstructuring of substrates for surface plasmon resonance imaging. *Nanotechnology* **2011**, *22*, 165301. [CrossRef] [PubMed]
8. Pérez-Mayen, L.; Olivat, J.; Torres-Castro, A.; De la Rosa, E. SERS substrates fabricated with star-like gold nanoparticles for zeptomole detection of analytes. *Nanoscale* **2015**, *7*, 10249–10258. [CrossRef] [PubMed]
9. Bryche, J.-F.; Gillibert, R.; Barbillon, G.; Sarkar, M.; Coutrot, A.-L.; Hamouda, F.; Aassime, A.; Moreau, J.; Lamy de la Chapelle, M.; et al. Density effect of gold nanodisks on the SERS intensity for a highly sensitive detection of chemical molecules. *J. Mater. Sci.* **2015**, *50*, 6601–6607. [CrossRef]
10. Bryche, J.-F.; Gillibert, R.; Barbillon, G.; Gogol, P.; Moreau, J.; Lamy de la Chapelle, M.; Bartenlian, B.; Canva, M. Plasmonic enhancement by a continuous gold underlayer: Application to SERS sensing. *Plasmonics* **2016**, *11*, 601–608. [CrossRef]
11. Gillibert, R.; Sarkar, M.; Bryche, J.-F.; Yasukuni, R.; Moreau, J.; Besbes, M.; Barbillon, G.; Bartenlian, B.; Canva, M.; Lamy de la Chapelle, M. Directional surface enhanced Raman scattering on gold nano-gratings. *Nanotechnology* **2016**, *27*, 115202. [CrossRef] [PubMed]
12. Itoh, T.; Yamamoto, Y.S.; Ozaki, Y. Plasmon-enhanced spectroscopy of absorption and spontaneous emissions explained using cavity quantum optics. *Chem. Soc. Rev.* **2017**, *49*, 3904–3921. [CrossRef] [PubMed]
13. Ding, S.-Y.; You, E.-M.; Tian, Z.-Q.; Moskovits, M. Electromagnetic theories of surface-enhanced Raman spectroscopy. *Chem. Soc. Rev.* **2017**, *46*, 4042–4076. [CrossRef] [PubMed]
14. Yamamoto, Y.S.; Hasegawa, K.; Hasegawa, Y.; Takahashi, N.; Kitahama, Y.; Fukuoka, S.; Mursase, N.; Baba, Y.; Ozaki, Y.; Itoh, T. Direct conversion of silver complexes to nanoscale hexagonal columns on a copper alloy for plasmonic applications. *Phys. Chem. Chem. Phys.* **2013**, *15*, 14611–14615. [CrossRef] [PubMed]

15. Barbillon, G.; Sandana, V.E.; Humbert, C.; Bélier, B.; Rogers, D.J.; Teherani, F.H.; Bove, P.; McClintock, R.; Razeghi, M. Study of Au coated ZnO nanoarrays for surface enhanced Raman scattering chemical sensing. *J. Mater. Chem. C* **2017**, *5*, 3528–3535. [CrossRef]

16. Galopin, E.; Barbillat, J.; Coffinier, Y.; Szunerits, S.; Patriarche, G.; Boukherroub, R. Silicon nanowires coated with silver nanostructures as ultrasensitive interfaces for surface-enhanced Raman spectroscopy. *ACS Appl. Mater. Interfaces* **2009**, *1*, 1396–1403. [CrossRef] [PubMed]

17. Akin, M.S.; Yilmaz, M.; Babur, E.; Ozdemir, B.; Erdogan, H.; Tamer, U.; Demirel, G. Large area uniform deposition of silver nanoparticles through bio-inspired polydopamine coating on silicon nanowire arrays for pratical SERS applications. *J. Mater. Chem. B* **2014**, *2*, 4894–4900. [CrossRef]

18. Barbillon, G.; Hamouda, F.; Held, S.; Gogol, P.; Bartenlian, B. Gold nanoparticles by soft UV nanoimprint lithography coupled to a lift-off process for plasmonic sensing of antibodies. *Microelectron. Eng.* **2010**, *87*, 1001–1004. [CrossRef]

19. Hamouda, F.; Sahaf, H.; Held, S.; Barbillon, G.; Gogol, P.; Moyen, E.; Aassime, A.; Moreau, J.; Canva, M.; Lourtioz, J.-M.; et al. Large area nanopatterning by combined anodic aluminum oxide and soft UV-NIL technologies for applications in biology. *Microelectron. Eng.* **2011**, *88*, 2444–2446. [CrossRef]

20. Cottat, M.; Lidgi-Guigui, N.; Tijunelyte, I.; Barbillon, G.; Hamouda, F.; Gogol, P.; Aassime, A.; Lourtioz, J.-M.; Bartenlian, B.; de la Lamy Chapelle, M. Soft UV nanoimprint lithography-designed highly sensitive substrates for SERS detection. *Nanoscale Res. Lett.* **2014**, *9*, 623. [CrossRef] [PubMed]

21. Masson, J.F.; Gibson, K.F.; Provencher-Girard, A. Surface-enhanced Raman spectroscopy amplification with film over etched nanospheres. *J. Phys. Chem. C* **2010**, *114*, 22406–22412. [CrossRef]

22. Bryche, J.-F.; Tsigara, A.; Bélier, B.; Lamy de la Chapelle, M.; Canva, M.; Bartenlian, B.; Barbillon, G. Surface enhanced Raman scattering improvement of gold triangular nanoprisms by a gold reflective underlayer for chemical sensing. *Sens. Actuator B Chem.* **2016**, *228*, 31–35. [CrossRef]

23. Barbillon, G.; Noblet, T.; Busson, B.; Tadjeddine, A.; Humbert, C. Localised detection of thiophenol with gold nanotriangles highly structured as honeycombs by nonlinear sum frequency generation spectroscopy. *J. Mater. Sci.* **2018**, *53*, 4554–4562. [CrossRef]

24. Zhang, M.L.; Fan, X.; Zhou, H.W.; Shao, M.W.; Zapien, J.A.; Wong, N.B.; Lee, S.T. A high-efficiency surface-enhanced Raman scattering substrate based on silicon nanowires array decorated with silver nanoparticles. *J. Phys. Chem. C* **2010**, *114*, 1969–1975. [CrossRef]

25. Schmidt, M.S.; Hübner, J.; Boisen, A. Large area fabrication of leaning silicon nanopillars for surface enhanced Raman spectroscopy. *Adv. Mater.* **2012**, *24*, OP11–OP18. [CrossRef] [PubMed]

26. Gómez-Graña, S.; Fernández-López, C.; Polavarapu, L.; Salmon, J.-B.; Leng, J.; Pastoriza-Santos, I.; Pérez-Juste, J. Gold nanooctahedra with tunable size and microfluidic-induced 3D assembly for highly uniform SERS-active supercrystals. *Chem. Mater.* **2015**, *27*, 8310–8317.

27. Rodal-Cedeira, S.; Montes-Garcia, V.; Polavarapu, L.; Solis, D.M.; Heidari, H.; La Porta, A.; Angiola, M.; Martucci, A.; Taboada, J.M.; Obelleiro, F.; et al. Plasmonic Au@Pd nanorods with boosted refractive index susceptibility and SERS efficiency: A multifunctional platform for hydrogen sensing and monitoring of catalytic reactions. *Chem. Mater.* **2016**, *28*, 9169–9180. [CrossRef]

28. Knight, M.W.; King, N.S.; Liu, L.; Everitt, H.O.; Nordlander, P.; Halas, N.J. Aluminum for Plasmonics. *ACS Nano* **2014**, *8*, 834–840. [CrossRef] [PubMed]

29. Martin, J.; Plain, J. Fabrication of aluminum nanostructures for plasmonics. *J. Phys. D Appl. Phys.* **2015**, *48*, 184002. [CrossRef]

30. Jha, S.K.; Ahmed, Z.; Agio, M.; Ekinci, Y.; Löffler, J.F. Deep-UV surface-enhanced resonance Raman scattering of Adenine on aluminum nanoparticle arrays. *J. Am. Chem. Soc.* **2012**, *134*, 1966–1969. [CrossRef] [PubMed]

31. Ayas, S.; Topal, A.E.; Cupallari, A.; Güner, H.; Bakan, G.; Dana, A. Exploiting Native Al_2O_3 for Multispectral Aluminum Plasmonics. *ACS Photonics* **2014**, *1*, 1313–1321. [CrossRef]

32. Mogensen, K.B.; Gühlke, M.; Kneipp, J.; Kadkhodazadeh, S.; Wagner, J.B.; Palanco, M.E.; Kneipp, H.; Kneipp, K. Surface-enhanced Raman scattering on aluminum using near infrared and visible excitation. *Chem. Commun.* **2014**, *50*, 3744–3746. [CrossRef] [PubMed]

33. Lay, C.L.; Koh, C.S.L.; Wang, J.; Lee, Y.H.; Jiang, R.B.; Yang, Y.J.; Yang, Z.; Phang, I.Y.; Ling, X.Y. Aluminum nanostructures with strong visible-range SERS activity for versatile micropatterning of molecular security labels. *Nanoscale* **2018**, *10*, 575–581. [CrossRef] [PubMed]

Materials **2018**, *11*, 1534

34. Cerjan, B.; Yang, X.; Nordlander, P.; Halas, N.J. Asymmetric Aluminum Antennas for Self-Calibrating Surface-Enhanced Infrared Absorption Spectroscopy. *ACS Photonics* **2016**, *3*, 354–360. [CrossRef]

35. Honda, M.; Kumamoto, Y.; Taguchi, A.; Saito, Y.; Kawata, S. Plasmon-enhanced UV photocatalysis. *Appl. Phys. Lett.* **2014**, *104*, 061108. [CrossRef]

36. Chowdhury, M.H.; Ray, K.; Gray, S.K.; Pond, J.; Lakowicz, J.R. Aluminum Nanoparticles as Substrates for Metal-Enhanced Fluorescence in the Ultraviolet for the Label-Free Detection of Biomolecules. *Anal. Chem.* **2009**, *81*, 1397–1403. [CrossRef] [PubMed]

37. Palik, E.D. *Handbook of Optical Constants of Solids*, 1st ed.; Academic Press: San Diego, CA, USA, 1998; pp. 1–3224, ISBN 978-0-12-544415-6.

38. Carron, K.T.; Gayle Hurley, L. Axial and azimuthal angle determination with surface-enhanced Raman spectroscopy: Thiophenol on copper, silver, and gold metal surfaces. *J. Phys. Chem.* **1991**, *95*, 9979–9984. [CrossRef]

39. Humbert, C.; Pluchery, O.; Lacaze, E.; Tadjeddine, A.; Busson, B. A multiscale description of molecular adsorption on gold nanoparticles by nonlinear optical spectroscopy. *Phys. Chem. Chem. Phys.* **2012**, *14*, 280–289. [CrossRef] [PubMed]

40. Tetsassi Feugmo, C.G.; Liégeois, V. Analyzing the vibrational signatures of thiophenol adsorbed on small gold clusters by DFT calculations. *ChemPhysChem* **2013**, *14*, 1633–1645. [CrossRef] [PubMed]

41. Li, S.; Wu, D.; Xu, X.; Gu, R. Theoretical and experimental studies on the adsorption behavior of thiophenol on gold nanoparticles. *Appl. Spectrosc.* **2007**, *38*, 1436–1443. [CrossRef]

42. Lin-Vien, D.; Colthup, N.; Fateley, W.; Grasselli, J. *The Handbook of Infrared and Raman Characteristic Frequencies of Organic Molecules*, 1st ed.; Academic Press: New York, NY, USA, 1991; Chapter 17, pp. 277–306, ISBN 978-0124511606.

43. Temple, P.A.; Hathaway, C.E. Multiphonon Raman Spectrum of Silicon. *Phys. Rev. B* **1973**, *7*, 3685–3697. [CrossRef]

44. Khorasaninejad, M.; Walia, J.; Saini, S.S. Enhanced first-order Raman scattering from arrays of vertical silicon nanowires. *Nanotechnology* **2012**, *23*, 275706. [CrossRef] [PubMed]

45. Cheng, C.W.; Yan, B.; Wong, S.M.; Li, X.L.; Zhou, W.W.; Yu, T.; Shen, Z.X.; Yu, H.Y.; Fan, H.J. Fabrication and SERS performance of silver-nanoparticle-decorated Si/ZnO nanotrees in ordered arrays. *ACS Appl. Mater. Interfaces* **2010**, *2*, 1824–1828. [CrossRef] [PubMed]

46. Zong, X.L.; Zhu, R.; Guo, X.L. Nanostructured gold microelectrodes for SERS and EIS measurements by incorporating ZnO nanorod growth with electroplating. *Sci. Rep.* **2015**, *5*, 16454. [CrossRef] [PubMed]

47. Etchegoin, P.G.; Le Ru, E.C. Basic Electromagnetic Theory of SERS. In *Surface Enhanced Raman Spectroscopy: Analytical, Biophysical and Life Science Applications*; Schlücker, S., Ed.; Wiley-VCH: Weinheim, Germany, 2011; pp. 1–34, ISBN 978-3-527-32567-2.

![materials logo] *materials*

MDPI

Review

Light Concentration by Metal-Dielectric Micro-Resonators for SERS Sensing

Andrey K. Sarychev [1], Andrey Ivanov [1,*], Andrey Lagarkov [1] and Grégory Barbillon [2]

[1] Institute for Theoretical and Applied Electrodynamics, Russian Academy of Sciences, 125412 Moscow, Russia; sarychev_andrey@yahoo.com (A.K.S.); lag@dol.ru (A.L.)

[2] EPF-Ecole d'Ingenieurs, 3 bis rue Lakanal, 92330 Sceaux, France; gregory.barbillon@epf.fr

* Correspondence: av.ivanov@physics.msu.ru

Received: 28 October 2018; Accepted: 27 December 2018; Published: 29 December 2018

Abstract: Metal-dielectric micro/nano-composites have surface plasmon resonances in visible and near-infrared domains. Excitation of coupled metal-dielectric resonances is also important. These different resonances can allow enhancement of the electromagnetic field at a subwavelength scale. Hybrid plasmonic structures act as optical antennae by concentrating large electromagnetic energy in micro- and nano-scales. Plasmonic structures are proposed for various applications such as optical filters, investigation of quantum electrodynamics effects, solar energy concentration, magnetic recording, nanolasing, medical imaging and biodetection, surface-enhanced Raman scattering (SERS), and optical super-resolution microscopy. We present the review of recent achievements in experimental and theoretical studies of metal-dielectric micro and nano antennae that are important for fundamental and applied research. The main impact is application of metal-dielectric optical antennae for the efficient SERS sensing.

Keywords: metal-dielectric resonance; plasmon; metasurface; nanoparticles; sensing; surface-enhanced Raman scattering (SERS)

---

1. Introduction

In the short review, we present recent results in plasmonics of metal-dielectric composites and metasurfaces. Optical properties of metal nanoparticles are intensively studied for more than one hundred years [1–3]. Surface plasmons can confine the electromagnetic (EM) field to a nanoscale (hotspots), which can enhance greatly this EM field. The modern technology allows for designing and producing metal nanostructures of different shapes and sizes. The specially designed metal nanostructures serve as optical antennae, which opens exciting opportunities in fundamental physics studies, but also in plasmonic applications such as optical signal processing on a nanoscale, medical imaging and biodetection, optical super-resolution microscopy [4], magnetic recording assisted by heat [5,6], quantum electrodynamics studies [7], nanolasing, and solar energy concentrators [8]. A strongly amplified electromagnetic field can be generated by disordered metal-dielectric composites in a broad spectral range [9]. Periodically ordered nanostructures enhance local EM field at selected frequencies [10–23]. Plasmon modes propagating in a chain of metallic nanoparticles, where the particle radius a is relatively equal to the distance δ between particles, are precisely investigated in [24]. Concerning these modes, the near-field interaction allows the jump of the dipolar excitation between each particle. EM field is confined in the chain vicinity. Guided modes of the nanoparticle chain, which propagate in the domain $a \gg \lambda$, were studied in [25–27]. These modes are similar to dipoles modes propagating around an optically thin cylinder. Scattering/diffracting experiments of an EM wave with nanorod periodic arrays were showed in [28,29]. Recently, the wave propagation along metallic nanorods is an attractive issue. Indeed, the negative refraction can be realized with these types of systems [30–33]. Fascinating optical effects such as Doppler shift, Cherenkov-Vavilov radiation, light pressure and Magnus effects are anomalous/inverse in negative refractive materials [34–37]. Stacked nanorods are proposed for the microwave and optical superresolution imaging [38–44]. The EM field mappings for the metal nanoparticles and nanoshells in a close-packed configuration were demonstrated in [45] and [46,47], respectively. In [11], an metallic nanocylinder array of which the cylinders are very close is investigated. At the resonance, the EM field is significantly improved in the gaps between metallic nanocylinders. This enhancement is due to the excitation surface plasmons (SPs) in the gap between almost touching ("kissing") metal cylinders. Electromagnetic resonators can be formed by dielectric optical microcavities. The optical microcavities can confine the light at the micro/nanoscale. Optical resonators can be applied to different domains such as resonators enabling data transmission by using optical fibers. Furthermore, they can be used for obtaining narrow spot-size laser beams for the reading and writing of CD/DVD. Microcavities can force an atom or quantum dot to spontaneously producing a photon in a given direction. In quantum optical devices, dissipation can be overcome in order to potentially obtain a quantum entanglement of the matter and radiation (see Refs. [7,48,49]). In addition, the optical behavior of metals is strongly damped due to important losses as interband transitions and intraband transitions due to impurity scattering in solids and additional surface scattering. The losses result in the heat degradation of metal nanoparticles. The degradation reaches the highest value in maxima of the local electric field. Another issue for metalic nanoparticles is the chemical instability. Plasmonic nanostructures of gold are well-known for being the most chemically stable. Unfortunately, large losses of gold due to interband transitions occur in the visible domain for the wavelengths $\lambda < 600$ nm [50]. The blue loss in gold gives the yellow color. The optical properties of metals described previously have a negative effect on the sensor efficiency. For all the low loss dielectrics, electromagnetic resonances can be obtained by light excitation. However, the quality factor Q varies with the type of the EM modes and, for some of them, they have losses even for huge resonators [51]. For instance, the whispering-gallery modes (WGM) excited in dielectric resonators made of silica, CaF_2, MgF_2, GaN or GaAs have large Q-factors. The shape of the WGM resonators is usually a torus, disk or sphere. The values of the Q-factor for a WGM resonator can achieve 10^7 and 10^9 [52–57]. These WGM resonators can allow the realization of filters, modulators, sensors or lasers. Due to a long lifetime of a WGM, a single molecule or virus can be detected on the cavity surface [58,59]. In this last decade, the concentration of electric and magnetic fields in the dielectric micro-structures has had a great amount of attention [60,61]. For two resonant dielectric spheres, the electric field located within the gap of these spheres was enhanced and demonstrated in [62,63]. A Yagi–Uda antenna can be composed of a chain of six dielectric nanoparticles [64]. This latter can significantly enhance the radiation of a dipole placed between these particles. Therefore, the antenna can confine the incident light in this same location. The enhancement of light was obtained with a ring of plasmonic nanoparticles coupled to a dielectric micro-resonator [65]. The surface plasmon resonance of the nanoparticle ring enables the EM field enhancement close to the dielectric micro-resonator surface. On the contrary, the electric field is significantly less important in the case of a dielectric resonator and when the metallic nanoparticles are spatially separated [66]. The light propagation in dielectric metamaterials is discussed in the review paper [67]. For instance, the EM wave can be confined within a nanoscopic volume with a dielectric waveguide having an anisotropic cladding [68]. The electric field confinement between dielectric rectangular resonators was demonstrated by [69]. The super resolution of resonant microstructures can be achieved by a dielectric microsphere (see [70] and references therein).

The optical nonlinearity can be enhanced by using a magnetic resonance obtained with four closely packed dielectric disks [71]. Kim et al. have showed an effective absorption of the EM field in periodic semiconductor metafilms for solar cells (see [72]). Some groups have demonstrated a strong electric field and SERS enhancement for periodic metafilms made of rectangular dielectric bars [12,16]. Sharp minima in the reflectance of the dielectric metasurface can be achieved in the microwave and optical domains. Distributed dielectric resonances in randomly cracked ceria metafilms were also considered [13]. Dielectric metamaterials can be used for biosensing (see Refs. [73–75]). All-dielectric and metal-dielectric 2D and 3D light concentrators were investigated in [16,17]. It was shown that plasmon and dielectric resonances can be independently managed, i.e., the frequencies of the resonances can be independently tuned by varying of the shape, arrangement and the nature of the metals and dielectrics. Thus, the enhancement of SERS signal is additionally increased by combining plasmonic and dielectric resonators.

2. Plasmon Resonance and Field Enhancement

Plasmon resonance can be explained in terms of L-C-R circuit [45,76]. Knowing the negativity of the metal permittivity in the optical frequency range, the optical electric current inside a metal nanoparticle is opposite to the direction of the displacement current outside the particle. Therefore, the metallic particle can be modeled as an inductance L. The excitation of the L-C-R contour models the interaction between the plasmonic nanoparticle and the EM field (Figure 1a). Here, the plasmonic nanoparticle is thus represented by the inductance L of which small losses are modeled by resistance R, and the surrounding medium is modeled by the capacitance C. Thus, the resonance in the L-C-R circuit models the plasmon resonance of a single metallic nanoparticle. An array of L-C-R circuits' models the EM coupling between two adjoined plasmonic nanoparticles. It is quite evident from the lumped circuit for the almost touching particles (Figure 1b) that there exists a longitudinal resonance when all the gap capacitances operate at the same phase. In addition, there are transverse modes propagating along a L-C chain, which represents the interparticle gap [11]. A narrow gap between metallic nanoparticles can be considered as a metal-dielectric waveguide, where the standing plasmon waves are excited (see, e.g., Figure 2).

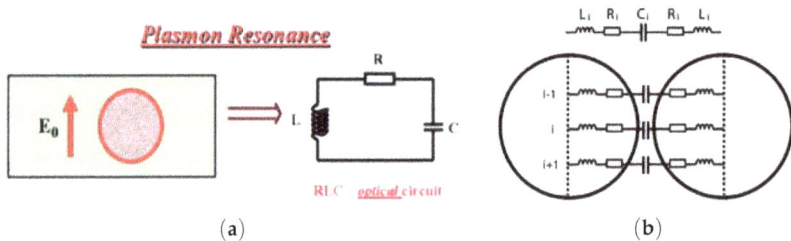

(a)

(b)

Figure 1. (a) L-C-R contour mapping of a plasmonic nanoparticle resonance. Plasmonic nanoparticle is represented by the Inductance (L) and resistance (R), and the surrounding medium by the capacitor (C); (b) plasmonic response depending on the frequency described as an L-C-R circuit array, reprinted with permission from [45], Copyright 2004 American Chemical Society, and from [76], Copyright 2007 World Scientific Publishing Co. Pte. Ltd.

The L-C-R model (Figure 1b) shows that the frequencies of the collective plasmon resonances for nanoparticle arrays decrease (corresponding to high values of L and C) with increasing of diameter-spacing ratio a/δ. In addition, the model agrees with the broadening of the plasmon bandwidth that occurs with red-shifting of resonance frequencies. The L-C-R model gives a description of the EM field enhancement depending on a, δ and ε_m. The electric field mapping and field enhancement in a two-dimensional (2D) periodic array of infinite metallic cylinders are shown in Figure 2. The analytical expression for the EM field enhancement in the gap between the nanorods gives the results that are very similar to computer simulations [11]. The EM field mapping shows excitation of multiple plasmon resonances. In a system of nanorods, SPs are strongly localized between the rods, and a large enhancement of the local EM field is achieved. The resonance frequencies are given by this simple equation:

$$\mathrm{Re}[\varepsilon_m(w_q)] = -\varepsilon_d \frac{\gamma^{2q}+1}{\gamma^{2q}-1}, \tag{1}$$

where $q = 1, 2, 3, \ldots$, ε_d is the permittivity of the host medium, the parameters are $\gamma = l/a + \sqrt{1 + (l/a)^2}$, $l = \sqrt{\delta(a + \delta/4)}$, and δ is the distance between cylinders of the radius a. For the closely packed cylinders, when $\delta \ll a$ the characteristic scale $l \approx \sqrt{a\delta}$ and the parameter $\gamma \approx 1 + l/a$ is close to unity. The position of the resonances can be controlled by adjusting the diameter-spacing ratio. The above defined characteristic length l is those of the effective plasmonic waveguide between nanocylinders. The electric field between the cylinders can be found by conform mapping and by using new coordinates $u + iv = \ln[(il + z)/(il - z)]$ instead of the original coordinates $z = x + iy$ to solve the Laplace equation. Thus, the field enhancement in the middle point between the cylinders $(x = y = 0)$ is given by the Equation:

$$|E/E_0|^2 = \left|1 + 8\sum_{q=1}^{\infty}(-1)^q \frac{q\,\alpha_m}{\gamma^{2q} + \alpha_m}\right|^2, \qquad (2)$$

where E_0 corresponds to the incident field amplitude and $\alpha_m = (\epsilon_d - \epsilon_m)/(\epsilon_d + \epsilon_m)$ is proportional to the polarizability of a metallic nanocylinder. The electric field is the sum of the resonance terms. Recall that the permittivity of metals is almost negative in the optical spectral domain of which the imaginary part is small $\varepsilon''_m \ll |\varepsilon'_m|$. The denominators in the sum in Equation (2) almost vanishes at the resonance frequencies given by Equation (1). The maximum of the electric field at the q-th resonance estimates as

$$|E_q/E_0|^2 \simeq 256\,q^2 \frac{\gamma^{4q}\epsilon_d^2}{\epsilon''_m(\omega_q)^2\,(\gamma^{2q} - 1)^4}, \qquad (3)$$

where E_q is the q-th resonance electric field in the middlepoint between the cylinders, $\varepsilon''_m(\omega_q)$ is the value of the imaginary part of the metal permittivity at the resonance frequency ω_q. The resonances are well-seen in Figure 2. The field enhancement $|E(x,y)/E_0|^2$ in the system of the adjoined nanocylinders can be as large as 10^5 or even larger [11]. We note below that SERS is proportional to $|E(x,y)/E_0|^4$, and, therefore, is indeed huge in the system of the metal nanocylinders.

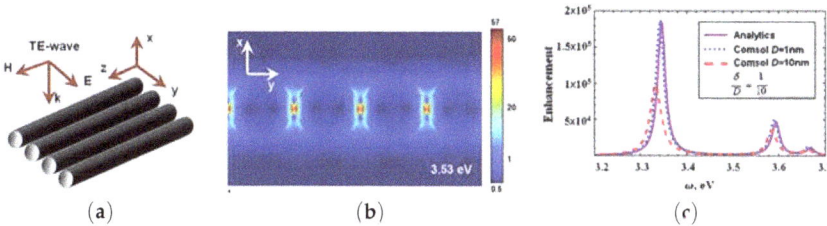

Figure 2. (a) scheme of the TE-wave propagation for an array of closely packed nanocylinders; (b) electric field mapping of surface plasmon resonances in an array of Ag cylinders with the following parameters: $\omega = 3.53$ eV, cylinder radius $a = 5$ nm, interparticle distance $\delta = 1$ nm; (c) comparison of analytical (purple line) and numerical (blue and red dashed lines) enhancements $|E_m/E_0|^2$ of the electromagnetic (EM) field in the middle of dimer. The ratio δ/D is equal to 0.1, the nanorod diameter for COMSOL simulations is $D = 2a = 10$ nm or 1 nm, and E_0 is amplitude of incident light, reprinted by permission from Springer Nature: Springer Nature, Applied Physics A: Materials Science and Processing [11], Copyright 2012.

Giant electric field fluctuations and the related enhancement of nonlinear optical phenomena in semicontinuous metallic films are an area of active studies. Random metal-dielectric films are generally carried out on a glass substrate (insulating substrate) with several evaporation techniques such as thermal evaporation or sputtering of metallic layer. The static conductivity of the gold/glass composite is decreased when the percent of metal is decreased. At a critical percent p_c called percolation threshold, the composite undergoes a composition-dependent metal-insulator transition. The composite behaves as a dielectric below this threshold (Figure 3a). In a series of works, it was shown that at the percolation threshold fluctuations of the EM field reach the enormous values (Figures 3b and 4) [76–80]. The local electric field strongly fluctuates in any nanocomposite, where the local permittivity fluctuates between negative and positive values (see Figure 4).

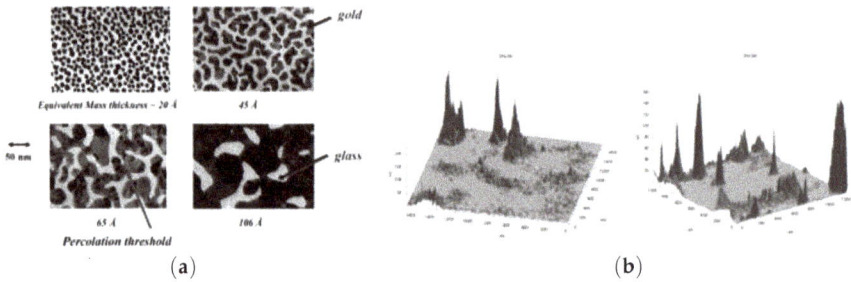

Figure 3. (a) gold/glass composite for four metal concentrations p. At percolation threshold $p = p_c$, continuous gold channel spans the system; (b) experimental images of the localized optical excitation in a gold/glass composite at percolation threshold ($p = p_c$). These images are collected with a SNOM using an excitation wavelength of $\lambda = 780$ nm, reprinted with permission from [76], Copyright 2007, World Scientific Publishing Co. Pte. Ltd.

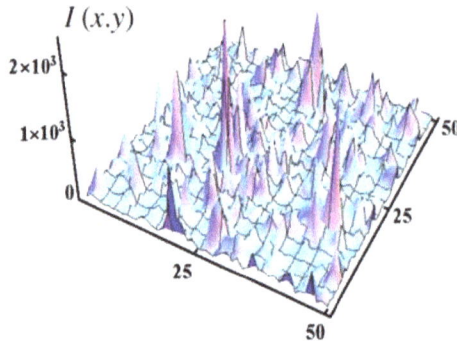

Figure 4. Collective volume plasmons in manganite $La_{0.7}Ca_{0.3}MnO_3$ with nanoscale phase separation; computer simulation of infrared electric field $I = |E(x,y)/E_0|^2$ at volume concentration $p = 0.28 > p_c$ of conducting phase; local permittivity $\varepsilon(r)$ is positive in the dielectric phase $Re[\varepsilon(r)] > 0$ and it is negative in conducting phase $Re[\varepsilon(r)] < 0$, reprinted figure with permission from [81]; Copyright 2006 by the American Physical Society.

3. Surface Enhanced Raman Scattering (SERS)

The resonances present in dielectric or metal-dielectric micro/nanostructures allow the confinement of the electromagnetic field at the nanoscale, and thus they can potentially improve the enhancement of the Raman scattering [82–84]. The Raman scattering is a phenomena of inelastic scattering, in which the vibration modes of the bonds of a molecule modulate an incident optical field at a higher or lower frequency and thereby get imprinted onto it. Spectrally, it observes a central peak corresponding to the central carrier frequency and other peaks with higher frequencies corresponding to anti-Stokes shifts and lower frequencies corresponding to Stokes shifts. Therefore, Raman spectroscopy can probe the structural information of the given molecules thanks to the vibrational modes of the bonds which compose them. The main interests of the Raman spectroscopy are its working mode in visible domain instead of an infrared one, its utility for molecular sensing due to the abundance of both high intensity coherent radiation sources, as well as of sensitive detectors operating in the visible domain, but also a possible extraction of the molecular structure informations at a high spatial resolution as a result of a substantially lower diffraction limit at visible wavelengths [85]. Surface enhanced Raman scattering (SERS) is one major physical phenomena of the last quarter of the twentieth century. Nevertheless, Raman scattering has a very weak scattering cross section. Moreover, the Raman signal can be potentially hampered by the background luminescence [85–97].

The generation of several plasmon modes on a metallic surface and further Raman scattering of the plasmons by analyte molecules are the basis of the SERS sensing (Figure 5). The molecules excited by the incident light and the plasmons can generate secondary plasmons which can be significantly enhanced. The radiation coming from these secondary plasmons produces a SERS signal [76,77]. Thus, the intensity of SERS signal is depending on the fourth power of local enhancement of the incident electric field. Nano- and micro-structures act as antennae efficiently amplifying the Raman signal. Several groups have already demonstrated enhancement factors from 10^4 to 10^9 for SERS substrates composed of clusters of gold or silver nanoparticles encapsulated in a dielectric matrix [98–109]. Moreover, a record enhancement up to 10^{12} of the Raman signal was reported by some authors [110]. It should be noted that a main contribution to SERS is the electromagnetic enhancement. The chemical enhancement strongly depends on local electronic structures of the molecules and the substrate it interacts with as each of their wavefunctions begin to overlap [111–113]. Some groups have demonstrated a chemical enhancement of SERS but with significant difficulties. Moreover, its influence is significantly weaker than electromagnetic enhancement. Indeed, the magnitude order of the chemical enhancement is only 10^2 [114,115]. Thus, SERS effect (electromagnetic enhancement) allows detecting a weak concentration of biological and chemical molecules. However, in spite of all the efforts by many bright researchers, there is no self-consistent theory of the SERS effect. For instance, contemporary theories do not explain why the enhancement is so different for various Raman spectral lines as it is clearly seen in the next section. The authors heard the following opinion from the people dealing with SERS: "The surface enhances what ever it wants". More recently, new highly sensitive SERS substrates have been carried out by using semiconducting materials (silicon or zinc oxide) with a metal (Au, Ag, Al). A couple of groups already demonstrated higher SERS enhancements obtained with Si nanowires coupled to metallic nanoparticles [116–118] and Si nanopillars coupled to a metallic layer [119–122]. Furthermore, other groups also demonstrated the same thing with ZnO nanowires/nanopillars coupled to a metallic layer [20,123,124] or metallic nanoparticles [125–127].

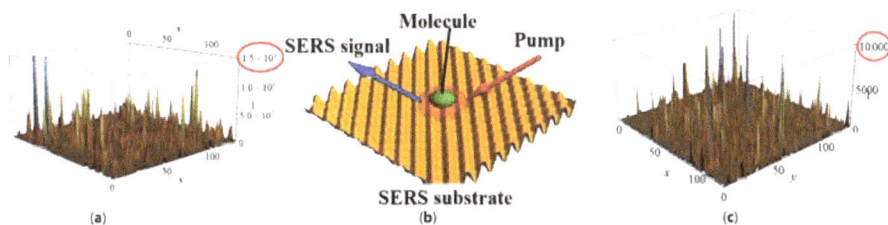

Figure 5. Schematic picture of SERS effect, see (**b**); incident light excites the collective plasmon field in semicontinuous metal film (see **c**, primary field); Raman active molecules, being pumped by the primary electric field, excite secondary EM field at the Stokes shifted frequency, see (**a**).

SERS phenomenon is also used in clinical diagnostics, which include the cancer detection and imaging, cancer therapy and drug delivery [128–157]. Important application of the SERS sensing is quantitative control of glycated proteins for diabetes detection [75,158–161]. Another very important SERS application is detection of the cardiovascular biomarkers for early diagnosis of acute myocardial infarction [162–164]. SERS detection of hormone Estradiol E2 is used for the clinical diagnosis of precocious puberty [165]. Environmental and food safety issues can be resolved using SERS real-time monitoring of pathogenic bacteria, pesticides and toxic molecules [166–173]. Ultra-low quantities of nerve gases, explosive substances and other hazard substances are also detected [174,175].

4. Field Enhancement in Dielectric Resonators

To illustrate SERS effect, we consider the EM field enhancement in the dielectric, transparent layer that is placed on a metallic substrate. The EM field for the incident light excitation, which propagates along a "z" axis normal to the layer, can be written as $E = E_0 \exp(ikz)$, where $k = \omega/c = 2\pi/\lambda$ is the wave-vector, λ is the wavelength. The electric field on the surface of the dielectric layer equals:

$$E_{sur} = 2E_0 \exp(-idk)/\left[1 + in\cot(ndk)\right],\tag{4}$$

where n denotes the refractive index of the dielectric layer. The surface field achieves the maximum $E_s = 2E_0$, where E_0 when the layer thickness $d = (2m+1)\lambda/4n$, $m = 1, 2, \ldots$. The surface EM field amplitude is larger than the incident field by factor 2. The enhancement of the Raman scattering is given by:

$$G = \frac{\langle |E_\omega(\mathbf{r})|^2 |E_{\omega-\Delta\omega}(\mathbf{r})|^2 \rangle}{|E_0|^4} \simeq \frac{\langle |E_\omega(\mathbf{r})|^4 \rangle}{|E_0|^4},\tag{5}$$

where ω is the excitation frequency, $\Delta\omega$ corresponds to the Stokes shift of the frequency; the second equation holds for $\Delta\omega \ll \omega$ (see Refs. [76,81,176]). From Equations (4) and (5), the enhancement achieves its maximum value of $G = 16$. Therefore, a simple dielectric layer with a thickness of a few hundreds of nanometers increases the Raman signal more than an order of magnitude.

The excitation of plasmon and dielectric resonances can enhance the Raman signal. From the hypothesis that the resonance frequency equals ω_m and the resonance width is higher than the Stokes shift of $\Delta\omega$. Moreover, from Equation (5), the effective SERS is obtained when the frequency of the excitation laser ω is within the interval is $\omega_m - \Delta\omega < \omega < \omega_m$. McFarland et al. were the first to obtain experimentally these types of results [98]. For all the dielectric nanoparticles, both EM resonances can be excited, for instance, with ceria, silica, and other dielectric materials [60,61]. Explosive molecules are detected thanks to a semiconducting resonator that is part of a plasmonic laser [177]. The EM field enhancement for the WGM resonators can be employed for the detection of biological and chemical molecules [178]. The latter allowed for detecting different particles with several sizes [179], batteries or viruses [59], and single molecules [4]. The EM field for a dielectric resonator can be confined in a hotspot. For example, a simple dipolar Mie resonance excited in the spherical ceria particle is displayed in Figure 6. Recently, several investigations have been made for obtaining the refractive index of ceria [180–187]. The refractive index n of ceria is depending on the wavelength, structure of the films [182,183,186], temperature [181,185], and the RF power employed for the magnetron sputtering of the film deposition [187]. The refractive index is higher with the denser ceria films. Thus, for the monocrystal particles, the refractive index of ceria is $n \simeq 2.3$ is discussed below.

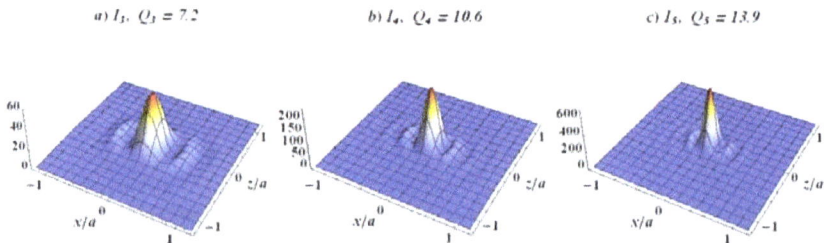

Figure 6. Electric field distribution $I_m = |E/E_0|^2$ and Q_m factor for a ceria (CeO_2) spherical particle ($a = 400$ nm). (**a**) $m = 3$, $\lambda_3 = 716$ nm; (**b**) $m = 4$, $\lambda_4 = 507$ nm; (**c**) $m = 5$, $\lambda_5 = 393$ nm. All the parameters are available in [16], reprinted with permission from [16], the Optical Society (OSA).

When the dipolar mode is excited, the EM field is enhanced at the center (see Figure 6). The highest value of the EM field is estimated in the following manner. The energy outflow from an eigenmode is approximated as $S_e \sim a^2 |E_0|^2$, where a corresponds to the dielectric sphere radius, and E_0 corresponds to the surface electric field. The field intensity $I_m = |E_m|^2$ concentrates in the center of the particle for the dipole eigenmode (Figure 6). The radius of the field maximum is estimated as $r_m \sim a/m$, where m in the order of the dipole resonance. The energy outflow S_m from the maximum is $S_m \sim r_m^2 I_m$. By equating the energy flows S_e and S_m, we obtain $I_m \sim E_0^2 m^2$. The highest value of the electric field is m^2 times elevated than the outside field as displayed in Figure 6. This result holds when the dissipate loss is much smaller than the radiation loss as discussed in [60,188]. The eigenfrequency, as well as electric field mapping in the electric dipolar resonance of Mie for an excited dielectric sphere, can be found analytically (see, e.g., [189]). When calculating the eigenmode, the radiation boundary conditions are imposed. That is, the outgoing wave propagating outside the sphere is assumed. The EM field is decaying due to the radiation loss; therefore, the eigenfrequency of any EM mode is a complex value $\omega_m = \omega_m' - i\omega_m''$. We obtain the quality factor for the dipole resonances:

$$Q_m = \frac{\omega_m'}{2\omega_m''} \approx \frac{\pi\left(2\pi^2 m^3 - \pi^2 m^2 - 2m - 1\right)}{4\left[(\pi^2 m^2 + 1)\operatorname{arccoth} n + n\right]} \simeq \frac{\pi}{4}(2m - 1)\,n, \tag{6}$$

where m is the radial number (see Figure 6) and n is the refractive index [16]. This simple equation holds for $n > 1$ and $m > n$. Thus, a significant confinement can be achieved by using a dielectric structure. Now, the molecules to be studied are placed in the slit as depicted in Figure 7. Next, the Raman enhancement G proportional to $\langle |E|^4 \rangle$ can be evaluated as $G \sim I_m^2 E_0^{-4} \sim Q^4 \sim (mn)^4 \gg 1$. Thus, the SERS enhancement for dielectric structures could be even important than this obtained for plasmonic structures.

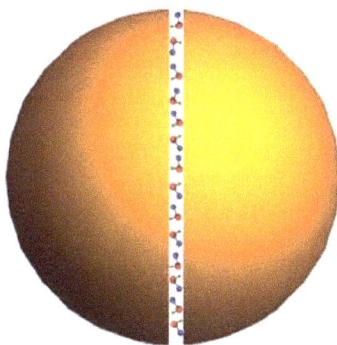

Figure 7. Slotted spherical dielectric resonator, reprinted with permission from [16], the Optical Society (OSA).

In addition, the electric field can be enhanced for dielectric structures for certain characteristic frequencies as depicted in Figure 6. For the planar dielectric metamaterials, discussed in the next sections, the EM field can also be confined in gaps between each dielectric structure (cf., Figures 2 and 7). The Raman signal has few characteristic Stokes shifts $\Delta\omega_i \ll \omega$ [190]. When the set of dielectric resonances coincides with the set of Raman spectral lines, the dielectric resonator can be used as a sensor for the particular molecule. Both closely packed dielectric resonances corresponding to resonance splitting can be organized as follows: two dielectric spheres separating a d center-to-center distance are considered. There are three independent dipole modes p_x, p_y and p_z. The interaction energy of dipoles is estimated as $\Delta U_m \sim (p_m \cdot p_m)/d^3 \sim p_m^2/d^3$, where p_m corresponds to the dipolar moment of the m eigenmode. The dipolar moment square estimates as $p_m^2 \sim \varepsilon^2 I_m r_m^6$, where I_m and $r_m \sim a/m$ are the intensity and the radius of the m-th mode, respectively (see Figure 6). The ratio of interaction energy ΔU_m to the eigenmode energy $U_m = \varepsilon I_m r_m^3 \sim \varepsilon I_m (a/m)^3$ gives the frequency split $\Delta\omega_m/\omega_m \sim \Delta U_m/U_m \sim \varepsilon(a/dm)^3$. Mainly, the Stokes shifts are weaker than the excitation frequency of laser. Therefore, it is enough to have the gap between the spheres $(2a - d) \sim a$ and excite the second order dipole modes to obtain the proper resonance shifts. Then, the cluster of dielectric spheres can be used for the SERS sensing of particular molecules. EM resonances in a dielectric lamp, whose symmetry is less than spherical symmetry, may have quasi-continuous spectrum (see Ref. [18]). Small clusters of metal particles are considered in [191]. The metallic nanoparticles arranged in a regular shape of pentagon demonstrate two resonances with a low Q which are separated by $\Delta\lambda$ ($\Delta\lambda > 100$ nm or $\Delta\omega > 5 \times 10^3$ cm^{-1}). Thus, this type of plasmonic structure can be employed with difficulty for the application of SERS sensors.

In [13], the authors have demonstrated the combination of plasmon resonances obtained with gold nanoparticles and localized EM resonances obtained with cerium dioxide films for a very sensitive detection of chemical and biological molecules by SERS. Morphology of cerium dioxide films is displayed in Figure 8. Nanostructures can form large clusters with a size of several hundred nanometers which are arranged in order to form structures with facets. The facet perimeter can correspond to a shape of curb. Each facet is spaced by small cracks of several tens of nanometers. Figure 9 depicts an optical image and a Raman intensity mapping for the Raman shift of 456 cm^{-1} of cerium dioxide. From these images, an irregular mapping of the signal on the film surface is demonstrated. The highest intensity of signal is generally located on the borders of facets. Toppgraphy of cerium dioxide films after deposition of gold nanoparticles (AuNP) is shown in Figure 10a. One to eight percent of the (CeO$_2$) surface are occupied by the immobilized AuNPs. The non-regular mapping of the Raman signal on the (CeO$_2$) film could indicate that a supplementary signal increasing is occurred after the immobilization of SERS tags on the facet surface. Raman signal was investigated from the conjugate of the DTNB and AuNP. The DTNB molecules (5,5'-Dithiobis-(2-nitrobenzoic acid)), also known as Elleman's Reagent, bounded to the AuNP surface are the source of the Raman signal. Figure 10b depicts the SERS signal depending on the thickness of the (CeO$_2$) film. The magnitude of the signal oscillates as a function of the film thickness. The additional enhancement for the SERS signal can be evaluated by normalizing this latter by the signal magnitude obtained with the film of 2800 nm thickness (minimal signal). For the Raman shift of 1060 cm^{-1}, this additional enhancement is of a factor 200 times more important than with for the film of 2400 nm thickness. The SERS enhancement depends on the Raman shift studied. In a similar way, this means that the films with different thicknesses have a different selectivity relative to the vibrational modes having different frequencies. This important experimental result is in an obvious contradiction with the electromagnetic theory of SERS. All of the vibration modes of a molecule are excited by the same surface enhanced electric field. Therefore, the relative value of the Raman signals of various frequencies, i.e., various Stokes shifts depend on the Raman polarizability. The Raman polarizability is a molecule property and it should not depend on the ceria film thickness. Therefore, the relative value of the Raman signals with different Stokes shifts should not depend on the film thickness. In reality, we see in Figure 10b that the Raman signal from DTNB molecules with Stokes shift of 1558 cm^{-1} is larger than the signal with the Stokes shift of 1060 cm^{-1} in the film with the thickness from 1400 to 2100 nm. With further increasing of the thickness from 2100 nm to 2800 nm, the signal with Stokes shift of 1558 cm^{-1} becomes *smaller* than the signal with Stokes shift of 1060 cm^{-1}. In addition, note that the Raman signal with Stokes shift of 1060 cm^{-1} is almost eight times smaller than this with a shift of 1338 cm^{-1} for the film thickness of 2000 nm. It is enough to slightly increase the film thickness from 2000 to 2400 nm and the ratio of these signals decreases from eight to two. It is not clear where such behavior of the SERS in the ceria facet film comes from. The experimental results cannot be explained using contemporary theories of SERS.

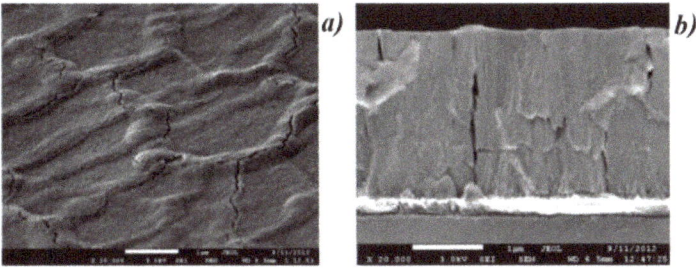

Figure 8. General (**a**) and cross-section (**b**) SEM views of CeO_2 film that have facet structure; the facet structure is clearly seen: the facets are separated by cracks whose thickness is about $\simeq 50\,nm$.

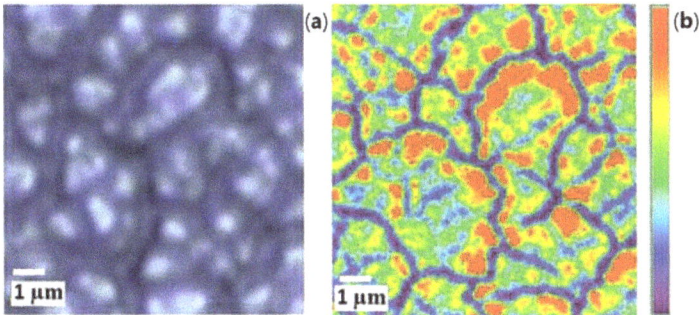

Figure 9. On (**a**), an optical image of 2400 nm thick CeO_2 surface is displayed, and on (**b**), an intensity distribution map for the main Raman peak, whose Stokes shift is of $1338\,cm^{-1}$ from the laser frequency 12.738×10^3 cm^{-1}, i.e., wavelength $\lambda = 785\,nm$ (CCD: red = 1700 counts, violet = 300 counts).

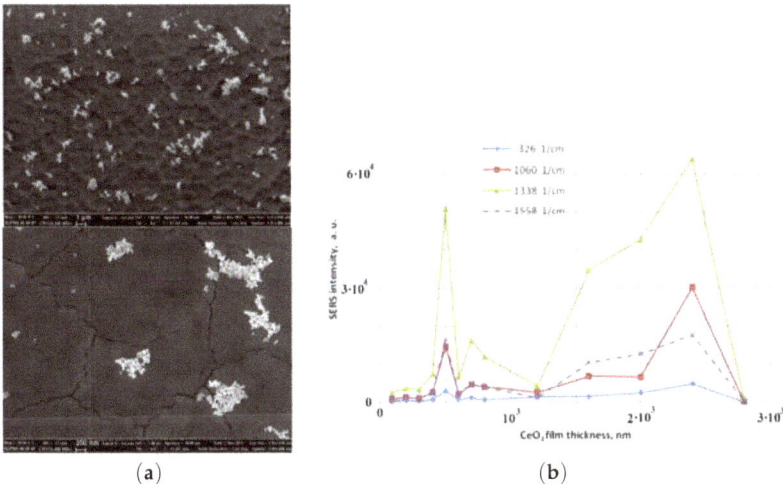

Figure 10. (**a**) SEM images of facet CeO_2 films with gold nanoparticles (AuNP); scale bars are 1 μm and 200 nm for the top and bottom images, respectively; (**b**) SERS intensity for four Raman shifts from conjugate of DTNB/AuNP on CeO_2 films as function of the film thickness (after normalization that is the Raman signal is divided into the number of AuNP in the spot, where the signal is collected from).

5. Metal-Dielectric Resonances

High EM field enhancements are mainly achieved with plasmonic resonators composed of gold (Au) or silver (Ag) nanoparticles. Metallic nanoparticles are well-known for having resonances thanks to the Faraday work (for recent references, see [1,2,76,192,193]). For a plasmonic sphere of which radius a is much smaller than the skin depth (i.e., $ak\sqrt{|\varepsilon_m|} \ll 1$), the electric field E behaves as $E_{in} \simeq 3E_0/(\varepsilon_m + 2)$, where ε_m is the metal permittivity. From Ref. [50], it was demonstrated that an isolated gold nanoparticle presents a plasmon resonance at $\lambda \simeq 500$ nm when this nano particle is excited in air. The resonance frequency can be tuned using a dielectric envelope. A hybrid (metal/dielectric) resonator which consists of a gold nanosphere with a radius a and a dielectric shell of thickness Δ is considered (see Figure 11).

The metal-dielectric resonator was considered using the hybridization approach (see Refs. [194,195] and references therein). The result of computational simulations of enhancement factor obtained with a hybrid resonator composed of gold nanoparticles (core) and a dielectric shell is displayed in Figure 11b. When $\Delta > a$, the resonance of this resonator behaves almost a dipolar resonance of the dielectric sphere with a radius $b = a + \Delta$. By using the asymptotic solution of Equation (7) referenced in the paper [16], the resonance of a dielectric sphere with the following parameters ($a = 200$ nm and $n = 2.5$) occurs at a wavelength of $\lambda = 690$ nm. This result is in good agreement with Figure 11b. The dielectric layer increases the internal electric field due to the existence of a plasmon-dielectric resonance. Several groups have investigated the effect of the cascade field enhancement for hybrid opto-plasmonic systems [16,19,20,65,66,196,197]. Santiago et al. have showed the detection of proteins with WGM resonators having hybrid (photonic/plasmonic) modes [198]. The Q-factor enhancement for hybrid (metal-dielectric) resonator is explained in the following way. The Q-factor increases in metal-dielectric resonators by decreasing the radiative losses. In Ref. [50], it was demonstrated for a resonator that the conducting current present in the metallic core and the polarization current present in the dielectric move with the opposite directions when the metal permittivity is mainly negative. The radiation is comparable to the sum of the currents, and the radiative losses decrease for a resonator, when the permittivities of resonator materials have different signs.

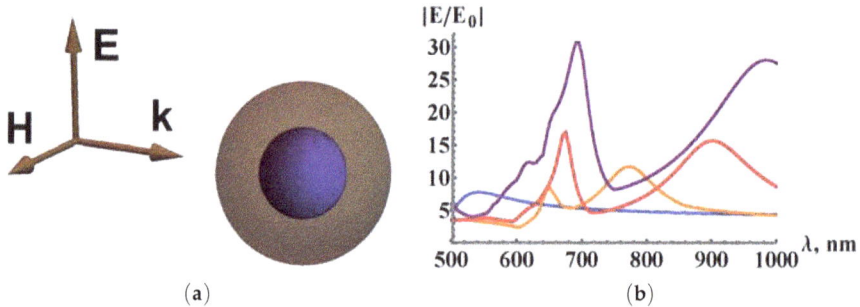

Figure 11. (**a**) scheme of the hybrid spherical resonator; (**b**) simulation electric field enhancement $|E_{\max}/E_0|$ in the resonator (gold core) for different thicknesses Δ of the dielectric shell. All the parameters are available in [18]. The blue line corresponds to isolated Au particle without dielectric shell and the orange, red and purple lines correspond to isolated Au particle with a dielectric shell of the thickness $\Delta = 50$ nm, $\Delta = 100$ nm and $\Delta = 150$ nm , respectively, reprinted with permission from [18], the Optical Society (OSA).

In order to quantitatively examine the dielectric screening, the eigenstate of a plasmonic nanosphere with a dielectric shell (radius $b = a + \Delta > a$, with a and Δ corresponding to the sphere radius and the dielectric thickness, respectively) is considered. Then, for simplicity, the electric dipole eigenstate is only studied. Matching the solution of the Maxwell equations in metallic and dielectric parts of the resonator with outgoing wave, we obtain a dispersion expression for an eigenfrequency $\omega = \omega' - i\omega''$, which have an imaginary part due to the radiative losses. The resonator quality factor $Q = \omega'/2\omega''$ can give the definition of the EM field enhancement in the resonator. A plasmon resonance in the gold nanoparticle of which the radius is a, is considered. For an approximate estimate, the Drude model can be used: $\varepsilon_m = \varepsilon_b - (\omega_p/\omega)^2/(1 + \omega_\tau/\omega)$, where $\varepsilon_b = 4.1$, $\omega_p = 8.7$ eV, $\omega_\tau = 0.11$ eV are chosen to correspond to the experiment [50] for $\omega < 2$ eV. The Q-factor values of the gold nanoparticles without additional materials or layers (with the following parameters: $a_1 = 50$ nm and $a_2 = 100$ nm) are $Q_1 \approx 12$ and $Q_2 \approx 1.4$, respectively. This large difference of the Q-factor is due to the radiation loss when the radius was doubled. Indeed, the radiation is comparable to $(ka)^3$. When the metallic nanoparticle is surrounded by the dielectric layer, the radius of the metal-dielectric resonator increases. However, radiation losses decrease and the Q-factor increases, as displayed in Figure 12. We think that the radiation screening obtained by the dielectric layer improves Q-factor.

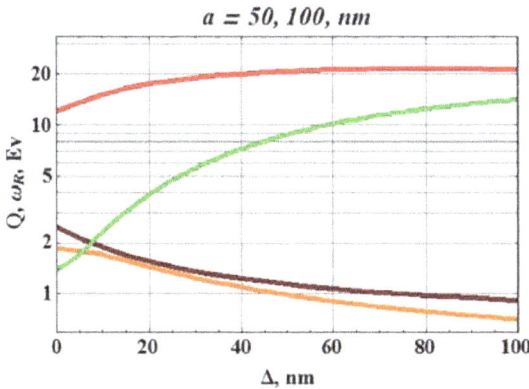

Figure 12. Resonance frequency ω_r (brown and orange curves) and Q-factor (red and green curves) of the gold-dielectric resonator, displayed in Figure 11, as function of the shell thickness Δ (refractive index of dielectric shell is $n = 4$). Brown/red and orange/green curves correspond to the gold core radius $a = 50$ nm, and $a = 100$ nm, respectively, reprinted with permission from [18], the Optical Society (OSA).

6. Metal-Dielectric SERS Metasurfaces

6.1. Periodic Bars

Refs [16,17] are devoted to the investigation of metamaterials of which the sharp resonances are separated by $\Delta\omega_i$ coming from metal-dielectric resonators. These types of metamaterials can be controlled in order to detect specific analytes with the signature $\Delta\omega_i$. A metal-dielectric composite metamaterial has been proposed. This metamaterial is composed of a silicon substrate on which is deposited a thick gold layer, then periodic bars of polymethylmethacrylate (PMMA) are fabricated (Figure 13). This system of periodic bars exhibits deep resonances in the wavelength range from 600 nm to 800 nm for *p*- and *s*-polarized waves, where the EM field enhancement is significantly high as depicted in Figure 14. The main peak of the field intensity $|E/E_0|^2$ is on top of the PMMA bars. It shifts to a shorter wavelength with increasing the angle of incidence α. Moreover, the resonances for *s*- and *p*-polarized waves take place at different frequencies. It opens a new opportunity to tune SERS substrates for a particular analyte.

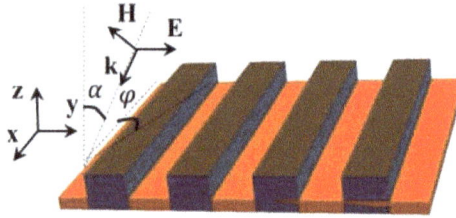

Figure 13. Scheme of the periodic polymethylmethacrylate (PMMA) bars on gold substrate, from [17].

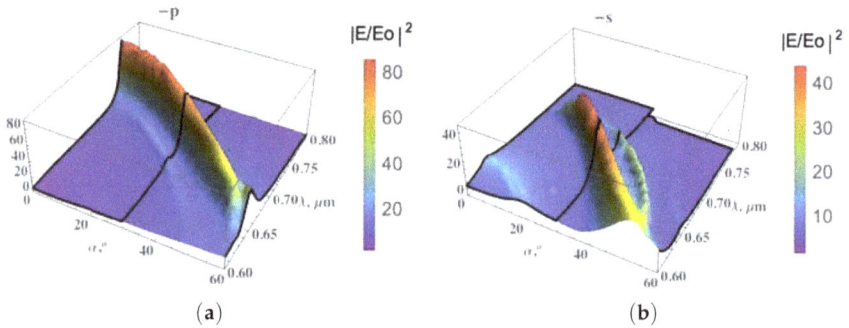

(a)

(b)

Figure 14. Intensity of electric field $|E/E_0|^2$ on top of PMMA bars for different wavelength λ and incidence angle α for (**a**) p-polarized wave and (**b**) s- polarized wave. The parameters are: an azimuthal angle of $\phi = 2°$, a period of $L = 635\,\text{nm}$, a bar width of $D = 313\,\text{nm}$, a bar height of $H_{\text{PMMA}} = 124\,\text{nm}$, a refractive index of PMMA: $n_{\text{PMMA}} = n_1 + i n_2$ with $n_1 = 1.5$ and $n_2 = 0.001$, and the thickness of the gold substrate is $H_{\text{Au}} = 100\,\text{nm}$ (from [17]).

The experimental realization of such SERS substrates based on PMMA was proposed in [16] to detect DTNB molecules immobilized on the surface as shown in Figures 15–17. The metal-dielectric metasurface as described previously (Figure 15) was realized. Firstly, the sample size of $10 \times 10 \times 0.3\,\text{mm}^3$ was cut from n-type phosphorus doped silicon wafer of which the resistivity is of $0.3\ \Omega{\cdot}\text{cm}$. After a chemical cleaning (in petroleum ether), the samples were cleaned with deionized water. Then, an adhesion layer of 4-nm Ti is evaporated followed by the deposition of a 40-nm gold film on Si substrate. Next, a PMMA thin film was spin-coated on the gold film in order to produce the PMMA film of thickness 600–1200 nm. An electron beam lithography (Raith 150) has been employed for fabricating the nanobars on the sample surface. The dimensions of fabricated PMMA bars are 350-nm-wide lines with a periodicity of 700 nm on a $100 \times 100\ \mu\text{m}^2$ surface area.

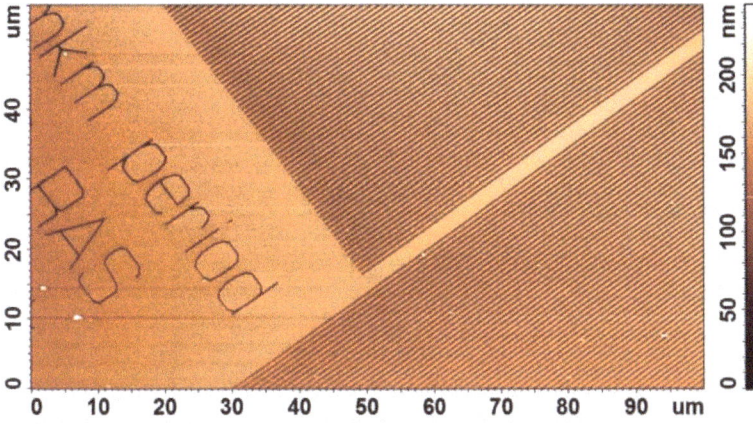

Figure 15. AFM of periodic dielectric structures based on Au and PMMA. The period equals 670–680 nm, gold thickness equals 40 nm, reprinted with permission from [16], the Optical Society (OSA).

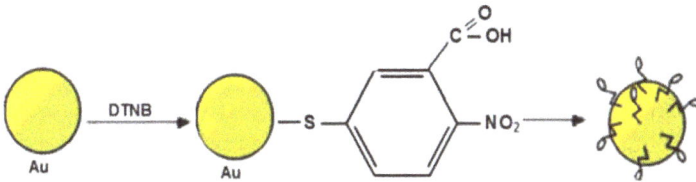

Figure 16. Principle scheme for preparation of AuNP-DTNB conjugate.

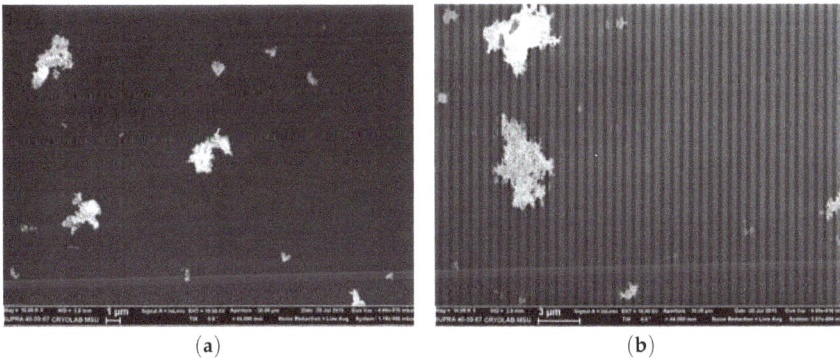

(**a**)   (**b**)

Figure 17. SEM images of AuNP-DTNB conjugates on: (**a**) flat and non-structured PMMA film; (**b**) PMMA bars; small bright spots are gold nanoparticles, some of them are organized in conglomerates, reprinted with permission from [16], the Optical Society (OSA).

To prepare SERS active particles, AuNPs of average size of 56 ± 1 nm were modified by DTNB molecules as depicted in Figure 16. The conjugates of AuNPs and DTNB molecules were adsorbed to the bar-shaped metamaterial after a deposition of polyelectrolyte (poly(diallyldimethylammonium chloride)) with the process reported in the paper [199] (see Figure 17). The conjugate of AuNP with DTNB (AuNP-DTNB) has well-known Raman peaks and can serve as SERS indicator [200] (Figure 18).

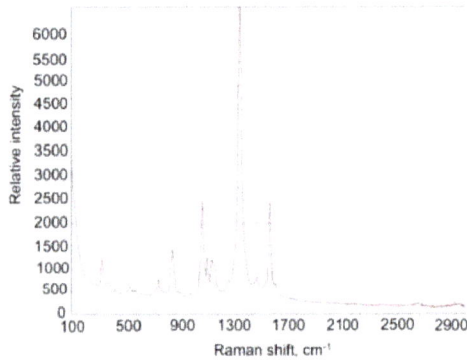

Figure 18. SERS intensity of AuNP-DTNB conjugates placed on PMMA bars (after normalization to the amount of gold nanoparticles as discussed in Figure 10), reprinted with permission from [16], the Optical Society (OSA).

The relative SERS intensity of AuNP-DTNB for the Raman shifts of 1338 and 1558 cm^{-1} was enhanced of a factor 5 for the bar-structured area compared to the flat PMMA layer. It should be noted that the Raman signal can be hindered by the background luminescence. The Raman/luminescence ratio (signal/noise) is most important for an effective SERS (for example, see Refs. [17,20,201]). We speculate that SERS substrates made of silicon look rather promising. Silicon has the advantage for SERS of having no luminescence background.

6.2. Periodic Blocks

We believe the main drawback of the existing SERS substrates is an insufficient selectivity. In this section, we discuss high selective SERS substrates based on anisotropic periodic dielectric structures. We consider periodic dielectric metasurfaces and double periodic metal-dielectric metasurfaces, fabricated from dielectric cuboids made of ceria dioxide (CeO$_2$, n = 2.3). The cuboids are placed on silver and gold substrates (see Figure 19). Multiple plasmon resonances are excited near the metal-dielectric boundary as shown in Figure 19c. Figure 19b demonstrates that the Raman signal enhancement G has two maxima for the structure with different periods D_x and D_y. The enhancement in the SERS substrate, shown in Figure 19a, has two peaks $\omega_1 = 12.77 \times 10^3$ cm^{-1} and $\omega_2 = 13.95 \times 10^3$ cm^{-1} with the widths of 200 cm^{-1} and 360 cm^{-1}, correspondingly. The difference $\Delta\omega = \omega_2 - \omega_1 = 1180$ cm^{-1} corresponds to the characteristic Stokes shift in the trinitrotoluene (TNT). This result was achieved by optimizing the dimensions of D_x, D_y of the unit-cell and dielectric cuboid dimensions of d_x, d_y, and height h. The fabrication method of plasmon nanostructures consisting of dielectric blocks and highly sensitive SERS sensors was discussed in the previous subsection [13,202].

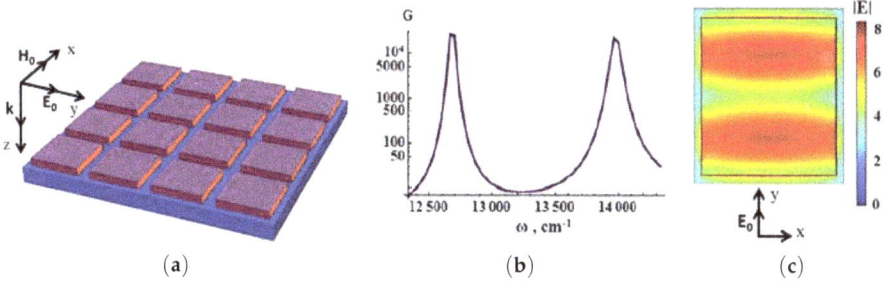

Figure 19. (a) 16 dielectric cuboids (purple) placed on the silver substrate (blue); (b) enhancement G as function of wavenumber for the dielectric cuboids placed on silver substrate, unit-cell dimensions of $D_x = 650$ nm, $D_y = 550$ nm, dielectric cuboid dimensions of $dx = 585$ nm, $dy = 495$ nm, and height $h = 55$ nm. The incident laser beam is normal to metasurface, electric field in the beam is directed along the diagonal of the dielectric blocks. The enhancement G is obtained at the surface of dielectric cuboid; (c) local electric field mapping on the top of dielectric cuboids of the first resonance shown in (b) (from [17]).

To further increase the Raman enhancement G, a regular lattice of thin metal nanodisks was deposited on the top of dielectric cuboids, as shown in Figure 20.

Thereby, the cascade enhancement of the electric field is obtained. The resonating metallic nanoparticles are excited by the enhanced field of the composite metal-dielectric substrate (see Figure 20). The optimal surface concentration of the nanodisks equals $p_m \approx 7\%$. Silver nanodisks allow to increase the enhancement factor up to $10^7 - 10^9$ (see Figure 20). Note that decreasing of the distance between nanodisks results in the increase of dipole-dipole interaction and the peaks are dualized [203,204], see the discussion in the paragraph next to Equation (6). The surface morphology of SERS substrates can be more complicated. We consider the periodic planar dielectric structures of silicon dioxide (SiO_2, $n = 1.46$) dielectric blocks which are placed on a metallic substrate as shown in Figure 21. The chosen lattice is anisotropic with elementary cell dimensions of $D_x \times D_y$. The dielectric cuboids inside the elementary cell have dimensions of d_{xi}, d_{yi} and the heights h_i. The resonance frequencies are tuned by the elementary cell design. Figure 21 demonstrates the high local electric field at the air-dielectric boundary. The SERS amplitude is given by Equation (5), where E_0 is the amplitude of the incident light. Figure 22 shows that the electric field enhancement has three maxima at three adjustment frequencies for the structure where the period $D_x > D_y$. The wavelength dependence of the local electric field intensity $|E/E_0|^2$ has three peaks, and can be tuned for two Raman spectral lines. The resonance frequencies are determined by independent variation of the periods D_x and D_y, for $(D_x - D_y) < D_x, D_y$. The difference between the peaks $\Delta\omega_1 = \omega_3 - \omega_1 = 1338\,\text{cm}^{-1}$, $\Delta\omega_2 = \omega_3 - \omega_2 = 326\,\text{cm}^{-1}$ ($\omega_1 = 11.40 \times 10^3\ \text{cm}^{-1}$, $\omega_2 - 12.412 \times 10^3\ \text{cm}^{-1}$, $\omega_3 = 12.738 \times 10^3\ \text{cm}^{-1}$) are tuned in order to correspond to the Stocks shifts of DTNB. Therefore, the discussed simple structure can be used for DTNB sensing. The enhancement of SERS signal can be additionally increased by combining plasmonic and dielectric resonators [16]. Thin metal nanodisks with diameter d_c and height h_c were inserted into the surface of the dielectric blocks (see Figure 22). We assume, for simplicity, that the nanodisks are made of the same material as the substrate. We vary the aspect ratio of the disks to tune the resonance frequency. As a result, the Raman enhancement $G \sim |E/E_0|^4$ reaches the value of 10^9 or even more for the substrate and nanodiscs made of silver (see Figure 22).

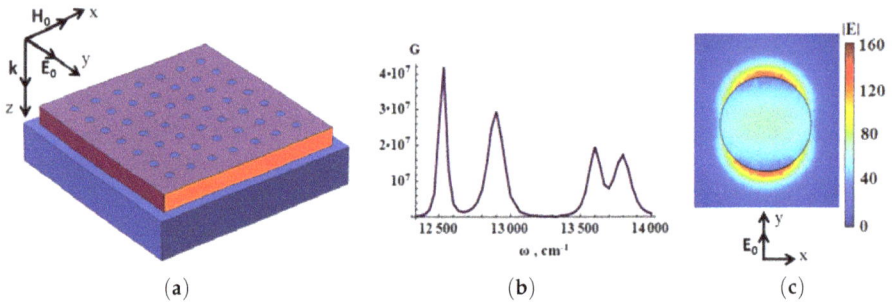

(a)   (b)   (c)

Figure 20. (a) silver nanodisks (small blue spots) deposited on dielectric (CeO$_2$, $n = 2.3$) cuboids (dark red) which are placed on silver substrate (blue); (b) Raman enhancement G as function of wavenumber; (c) distribution of local electric field $|E/E_0|$ around silver nanodisk with a wavelength of the incident light of $\lambda = 785$ nm. Geometric parameters: elementary sell sizes are $D_x = 650$ nm and $D_y = 550$ nm (light blue in (a); cuboid dimensions are $d_x = 585$ nm, $d_y = 495$ nm , and height $h = 47$ nm (dark red), diameter of silver nanodisk $d_c = 21$ nm and heght $h_c = 3$ nm , aspect ratio of the nanodisk $h_c/d_c = 1/7$, distance between nanodisks ≈ 60 nm (from [17]).

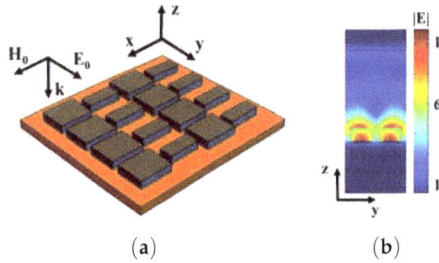

(a)   (b)

Figure 21. (a) 16 silicon dioxide blocks (eight elementary cells) placed on metallic substrate (orange). The metasurface has the following parameters: unit-cell dimensions of $D_x = 1634$ nm and $D_y = 698$ nm, dimensions of "large" dielectric blocks $d_{x1} = 704$ nm, $d_{y1} = 568$ nm, dimensions of "small" dielectric blocks $d_{y1} = 704$ nm, $d_{y2} = 284$ nm, height of the blocks $h_1 = h_2 = 148$ nm [16]; (b) electric field distribution $|E/E_0|$ when the incident light is normal to the metasurface at the frequency $\omega_1 = 11.4 \times 10^3$ cm$^{-1} = 1.41$ eV (from [17]).

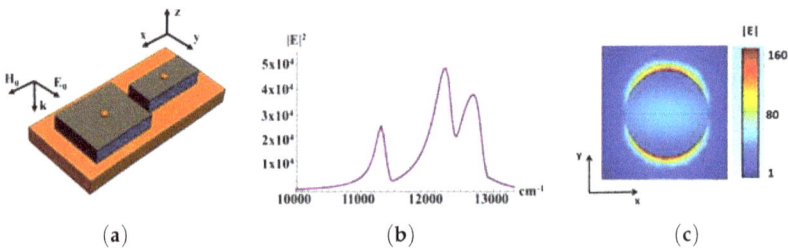

(a)   (b)   (c)

Figure 22. (a) single elementary cell of metal-dielectric substrate with gold nanodisks (orange color) placed on each of dielectric cuboids (gray), geometrical parameters are: lattice unit dimensions of $D_x = 1634$ nm, $D_y = 698$ nm, dimensions of dielectric resonators are $d_{x1} = d_{x2} = 704$ nm and $d_{y1} = 568$ nm, $d_{y2} = 284$ nm, height of the resonators is $h = 148$ nm. The parameters of the top gold disk are $d_c = 50$ nm, aspect ratio $h_c/d_c = (1/14)$; (b) electric field enhancement $|E/E_0|^2$ as function of wavenumber; gold disks are placed in the center of the top surface of dielectric cuboids. (c) electric field distribution $|E/E_0|$ over the gold disk. The metasurface is illuminated by light with an amplitude E_0 and a wavelength of 785 nm , the light is incident normal to the metasurface (from [17]).

7. 3D Dielectric Resonators for Surface Field Enhancement

7.1. WGM Resonators

Large ohmic losses occur in metals that induce a damping of their optical response. Novel materials are to be developed in order to obtain good performances for the application to optical devices [205–207]. A promising way of development of such materials is based on dielectric materials [97]. A great number of EM modes exist, and the whispering-gallery modes (WGM) have the advantage of having large Q-factors. WGM has been a well-known phenomena through light interaction with dielectric interfaces for 100 years [3]. It is known in architectural acoustics that the sound propagates with a relative preference along the concave surfaces. In addition, the light can suffer a total internal reflection at the interface between a dense medium and a less dense medium for a certain angle of incidence. Thus, WGM can be understood as waves of the total internal reflection (Figure 23). The resonators based on WGM modes can be constituted of silica, CaF_2, MgF_2, GaN, GaAs, and have different shapes such as disk, torus, sphere or cylinder. The values of the Q-factors with such resonators can achieve 10^7 and 10^9 [4,53–57,208,209].

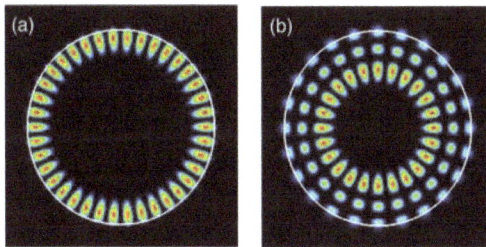

Figure 23. Whispering-gallery modes (WGMs) in a dielectric microdisk. The refractive index of GaAs is $n = 3.3$, and the polarization is TM. The different numbers are (for more details (see [61])): (a) $l = 1$ and $m = 19$; (b) $l = 3$ and $m = 12$, reprinted figure with permission from [61], Copyright (2015) by the American Physical Society.

Usually, the shape of the disk or sphere for resonators is employed. Some more complex 3D geometries exist and have supplementary degrees of freedom. Very interesting effects can be obtained with these complex geometries compared to simple ones. Sumetsky has demonstrated the light localization in a resonator with a shape of bottle named "whispering gallery bottle" [57,208,209]. Other shapes were investigated for dielectric resonators such as conical shapes [210–212]. In addition, a weak variation of the radius of an optical microcylinder implies strongly localized WGMs in the conical shape. The deformation of symmetry can lead to fascinating phenomena. In Ref. [213], an effect of unidirectional lasing with $In_{0.09}Ga_{0.91}N/In_{0.01}Ga_{0.99}N$ multiple-quantum-well spiral micropillars was demonstrated. The highest value of emission is achieved by the notch of spiral microcavity for an angle of about 40° from the notch normal (Figure 24).

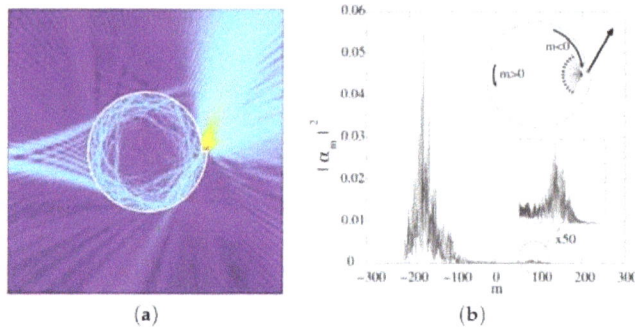

(a)   (b)

Figure 24. (a) real-space plot of the electric field modulus concerning to a calculated quasibound state at $nkr_0 \approx 200$ with an eccentricity deformation ($\epsilon = 0.10$); (b) angular momenta distribution for the resonance plotted in (a). The peak at negative m corresponds to clockwise rotation, and the weak peak at positive m corresponds to counterclockwise rotation; the counterclockwise modes are the diffracted waves emitting from the notch. Reproduced from [213], with the permission of AIP Publishing.

Thanks to their high Q-factor, WGM resonators can be employed for realizing filters, switches, lasers and sensors. In addition, several groups have reported the effect of quantum chaos for WGM resonators [61,214–216]. In addition, the nature and degree of the shape deformation can imply a change of whispering gallery orbits from regular shape to partially or fully chaotic. A long lifetime in a WGM resonator can enable the detection of single molecules or viruses onto the surface of this type of cavity [58,59] as mentioned above.

7.2. Cone-Shaped Resonator

In Ref. [18], Lagarkov et al. have studied the light interaction with a tip-shaped metasurfaces composed of silicon cones (see Figure 25).

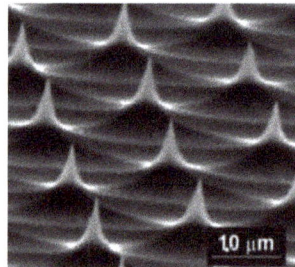

Figure 25. SEM of silicon tip-shaped metasurface, reprinted by permission from Springer Nature: Applied Physics A: Materials Science and Processing [217], Copyright 1998.

The geometrical parameters are: a square lattice period of $w = 2.1\,\mu m$, a height of 0.3–0.7 µm, the opening angle of the cone of $2\theta_0 \approx 30°$ a tip curvature radius of ≤ 10 nm, and a whole area of $2 \times 2\,mm^2$. This metasurface composed of cones can be seen as a diffraction grating due to the fact that the inter-cone distance d is larger than the excitation wavelength λ in the visible domain. Indeed, the condition for obtaining a positive interference is that the difference in optical paths must be equivalent to an integer number of vacuum wavelengths. For obtaining a higher diffraction order, the necessary condition is that $-1 < (m_1\lambda)/d < 1$, $-1 < (m_2\lambda)/d < 1$, where m_1 and m_2 are diffraction orders. For instance, a HeNe laser ($\lambda = 632.8$ nm) (see Figure 26) is used, and 37 diffracting modes occurred (the case where $m_1 = m_2 = 0$ corresponds to the reflected wave, and the cases $m_1 = m_2 = \pm 1, \pm 2, \pm 3$ correspond to the diffracted waves).

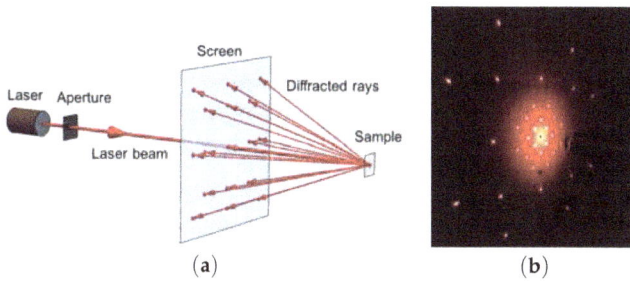

Figure 26. (**a**) principle scheme of the laser beam illuminating the metasurface for obtaining a diffraction pattern on the other side of the screen; (**b**) picture of this diffraction pattern obtained with a laser of wavelength $\lambda = 632.8$ nm. The other parameters are available in [18], reprinted with permission from [18], the Optical Society (OSA).

The traditional approach is used in the experiment. A HeNe laser ($\lambda = 632.8$ nm) illuminates the sample (metasurface) and reflects from this sample by producing a diffraction pattern on the other side of the screen as displayed in Figure 26. The parameters and settings are available in [18]. The diffraction pattern contains all the 37 modes having different radiances $I(m_1, m_2)$ as shown in Figure 26. The total reflection ($R = R_{00} + \sum R_{m_1 m_2}$) is about $\simeq 0.32$, where 0.26 corresponds to zero order reluctance and 0.06 to the diffraction. In another manner, 19% of the energy reflected by the metasurface (sample) is provided to the diffraction beams. Moreover, these Si tips occup only 8% of the total surface of the sample. Thus, the diffraction is very efficient that we could call an "extraordinary" optical diffraction. This experiment is also possible by using another wavelength as $\lambda = 405$ nm and the result also gave a bright diffraction pattern Figure 26, even with large losses in silicon for wavelengths ($\lambda < 500$ nm) (see [218]). This observation is in agreement with discussions realized in [18] on the fact that the enhancement of the electric field is independent optical losses. No change of the diffraction pattern occurs when the system is moved with respect to the laser beam. We can deduce that the periodicity and the cone shape are well-defined on the whole sample. The highest value of the enhancement is obtained from the resonance involving Si cone and metallic nanoparticles placed on its lateral surface as seen in Figure 27. Surface plasmons in metal nanoparticles interact with EM modes in the dielectric cone, which results in huge enhancement of the local electric field.

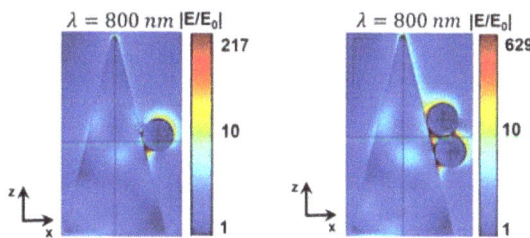

Figure 27. Electric field mapping $|E(z, x)/E_0|$ for Au nanoparticles (on the left, for one NP and, on the right, for two NPs, sphere diameter = 100 nm) placed on the lateral surface of a silicon cone for the resonant wavelength $\lambda = 800$ nm, reprinted with permission from [18], the Optical Society (OSA).

The EM field at the resonance is confined in a close vicinity to the cone. In addition, the interaction between each cone is negligible due to the distance between each cone of $w = 2.1$ μm, which is greatly longer than the cone size. Furthermore, the collective interaction can occur in the mid-IR frequency range ($\lambda \simeq w$). In this case, we think that collective surface modes can be excited. Lagarkov et al. have analyzed in a semi-quantitative manner the different resonances and the electric field enhancement for a dielectric resonator with a conical shape [18]. These dielectric resonators present several resonances in the visible and near-IR domains (see Figure 28). Due to an axial symmetry of the cone, the angular momentum is quantified and the modes with polar quantum number l, azimuthal quantum number m, and radial quantum number q are excited. The strongest value of the Q-factor corresponds to the WGM with the values of the numbers l and q which are minimal; however, the value of the number m is high (see Figure 29). The modes of which the values of polar and radial "quantum" numbers are weak propagate along the lateral surface of the cone. In addition, external modes also exist with weaker values of Q-factor where the electric field is partially located outside the cone. Moreover, the "leaky" region corresponds to these external modes with the needed condition that the total internal reflection is violated.

Figure 28. EM resonances spectrum for a conical resonator. The opening angle of cone is 30°, the refractive index is $n = 4$, and the height is $h = 595\,\text{nm}$. The red and blue lines correspond to frequencies of electric and magnetic resonances, respectively, reprinted with permission from [18], the Optical Society (OSA).

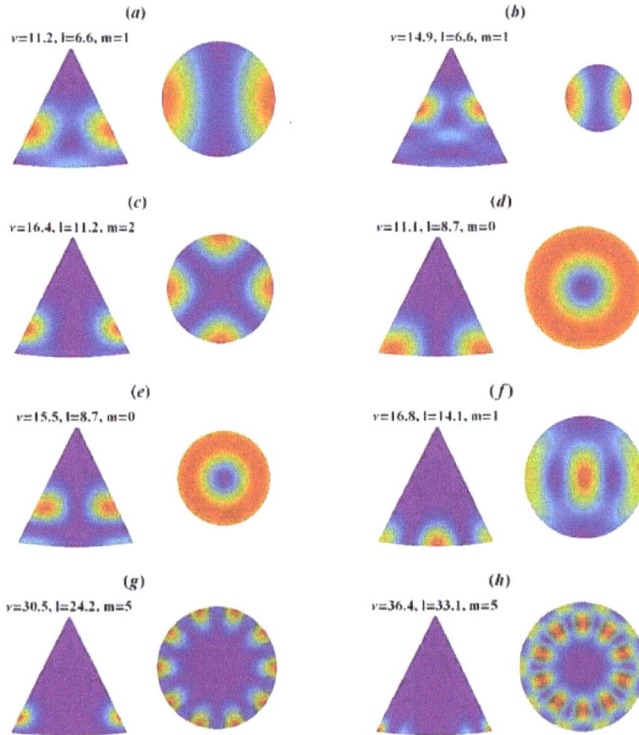

Figure 29. Electric field intensity $|E(r, \theta, \phi)|^2$ in a dielectric conical resonator having an opening angle of 30° and a refractive index of $n \gg 1$. "Electric" and "magnetic" resonances are depicted in Figures 28a,b,c,g and 28d,e,f,h, respectively. Vertical distribution of field is determined by dimensionless frequency $m = nkh$, where n, $k = \omega/c$, and h are refractive index, wavevector, and cone height, respectively. Each resonance is labeled by orbital "number" l and the azimuthal number m. The low symmetry of the resonator results in a non-integer l, reprinted with permission from [18], the Optical Society (OSA).

The metasurface composed of Si tips has been covered by SERS active tags consisting of Au nanoparticles of which the average size is 55 ± 5 nm, on which a DTNB monolayer has been deposited (see Figure 30). The DTNB molecules are grafted on gold nanopaticles thanks to their sulfate groups [13]. The Raman signal has been measured from AuNP–DTNB conjugate in order to have an approximate value of the EM field enhancement. We obtain a great enhancement of the intensity distribution for the metasurface compared to a flat region. This can be understood by excitation of the hybrid resonances (metal/dielectric) of Si cones covered by AuNPs. These conical Si tips serving as resonators enable converting the excitation light into longitudinal electric field. A detailed research has demonstrated that SERS intensity is depending on the position of gold nanoparticles on the surface (see Figure 31).

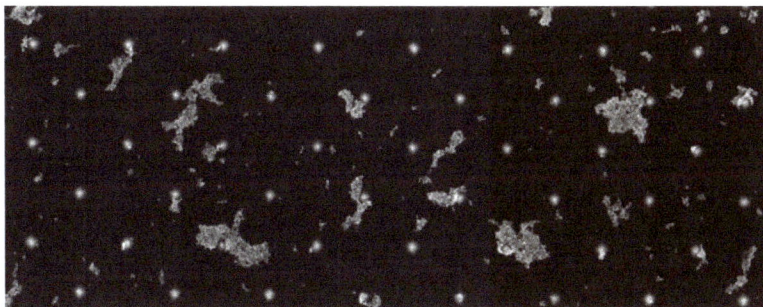

Figure 30. Aggregates of gold nanoparticles, which are seen as small bright white spots, are deposited on cone-shaped regular silicon metasurfaces; apexes of the cones are seen as a square lattice of 48 blurred specks, reprinted with permission from [18], the Optical Society (OSA).

A strong SERS enhancement occurs when the gold nanoparticles are placed in a suitable manner on the lateral face of the cone which will induce hotspots. Thus, an electric field enhancement of more than three orders of magnitude can be achieved as displayed in Table 1. The light localization in this type of resonators (metasurface) opens the way to new possibilities in R and D for the fabrication of highly sensitive SERS substrates applied to biological and chemical sensing.

Table 1. Raman intensity for the Raman shift of 1338 cm^{-1} of the conjugate AuNP-DTNB on the metasurface vs. flat plane (in a.u.). The signal is collected from red frames, shown in Figure 31; normalized signal is obtained by dividing by the number of Au nanoparticles in a frame.

Au-NP Localization	Signal (Counts)	Au-NP Number	Normalized Signal
Pyramid side	786	1	786
Between pyramids	553	30	18.4
Outside of grating	553	89	6.2

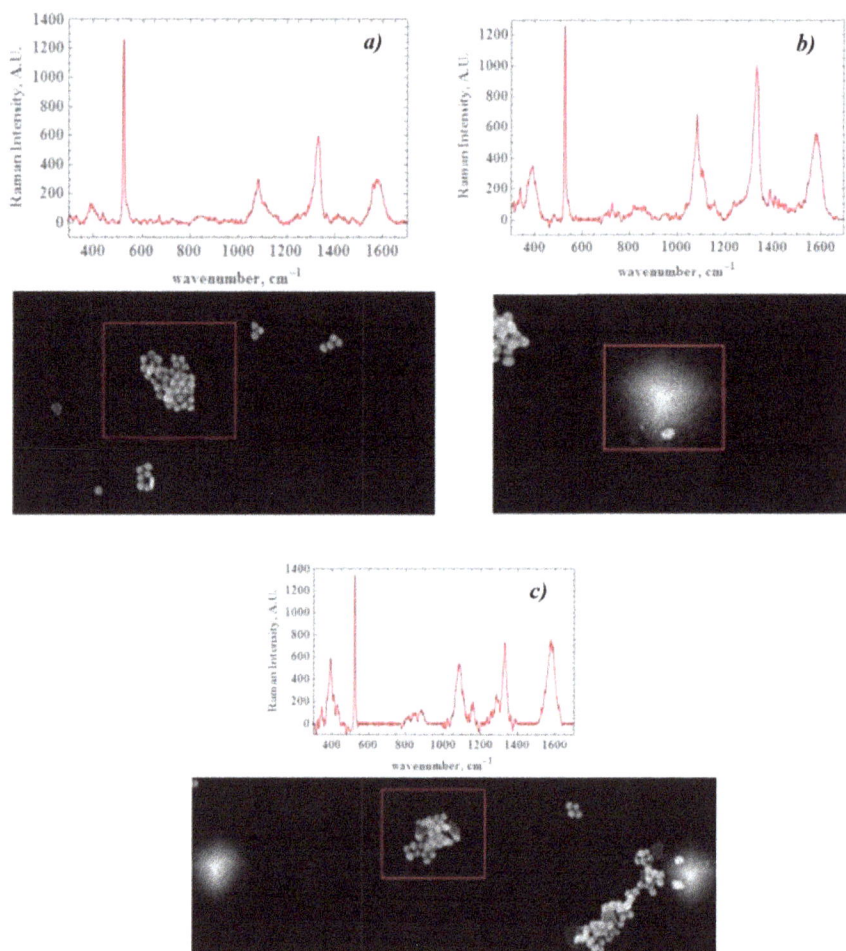

Figure 31. Raman intensity (SERS) obtained with AuNP-DTNB conjugates placed on different areas of cone-shaped metasurface; the areas where the Raman signal is collected are shown by red squares; apexes of the cones look like blurred specks, AuNP-DTNB conjugates look like bright spots. SEM images and SERS spectra recorded: (**a**) outside of grating, (**b**) on the pyramid side and (**c**) between pyramids. Reprinted with permission from [18], the Optical Society (OSA).

8. Local-Field Dielectric Transducer (LFDT)

Recent development of the plasmonics provides the possibility to concentrate the light onto a nano-area. Metal nanoantennae and subwavelength apertures [219] are used to better increase the local electric field in the subwavelength volume. The optical field enhancement and concentration are achieved by excitation of surface plasmons [220,221]. Nanosized metal particles are indispensable for most optical transducers and concentrators. Concentration of the huge electric field in a metal nanoparticle results in the fast degradation or even destruction of the particle [222]. To avoid negative effects of the large optic loss, it was proposed all-dielectric Local-Field Dielectric Transducer (LFDT), which enables confining of the light into a hotspot of nanometric size [189,223]. Negative thermal effects are nearly suppressed for LFDT allowing new possibilities in magnetic recording [224–226], optical sensing, and nanolaser pumping. LFDT is similar to waveguide gallery resonator with the notch, shown in Figure 24. The discussed LFDT is composed of a spherical dielectric resonator coupled to a dielectric nanostick, where the electric field is concentrated as shown in Figure 32. The EM field is stronger for the stick apex which is linked to the resonator. The field enhancement located at the stick apex was discussed by Novotny et al. [227]: the electric field aligned along the stick, periodically drives the bounded electrons. Electrons of an atom in the dielectric moves along the stick shaft with the same frequency as the excitation field. Around the apex, a large surface charge is present due to the small surface of apex and the uniform movement of bounded electrons. The accumulated charges gives the giant electric field (see Figure 32). With all the optical dense materials having small losses, an EM field enhancement can occur (see, e.g., [61,228]). The incident light is converted into a longitudinal electric field by means of the spherical resonator. Note that a simple waveguide cannot be used to effectively excite the longitudinal electric field [229]. Wang et al. have demonstrated the production of a longitudinal electric field with metallic LFDT composed of plasmonic lenses [230]. In the work [223], the dielectric stick is excited by using a waveguide connected to an optical resonator (Figure 32). This spherical resonator serves as an accumulator of the EM energy, and the energy stored in this latter excites the elliptical dielectric stick attached to the sphere. Note that the channelling of the whispering gallery modes was realized with an excellent efficiency in the connected waveguide with no supplementary loss (see Ref. [231]).

The Si waveguide with a cylindrical shape is connected to the resonator as depicted in Figure 32. As displayed in Figure 32, the stick is located at the interface between the resonator and the waveguide or opposite to the waveguide. The electric field is stronger for the inclined configuration of the stick. We think that for this configuration the field is more confined at the interface between the resonator and the waveguide. The field enhancement at the stick surface is shown in Figure 33. The field distribution was simulated by using three FePt nanospheres with a diameter of 2 nm located within in the apex's vicinity of the stick (see Figure 34). For magnetic recording assisted by heat, FePt nanoparticles are mainly employed [219,232].

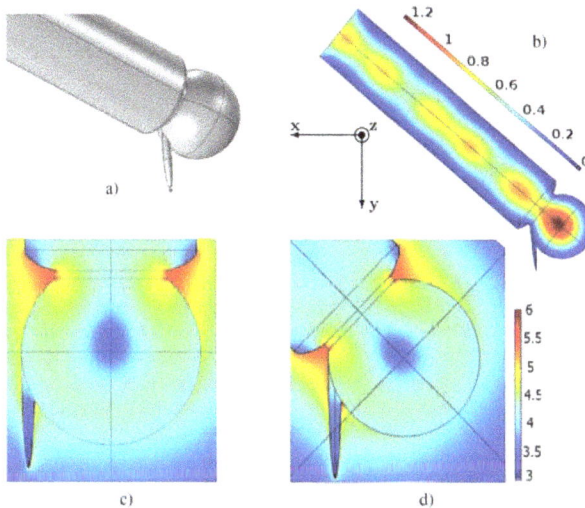

Figure 32. (**a**) scheme of a LFDT; (**b**) numerical simulations of the magnetic field $|H|$. Excitation of the stick by the electric field $|E|$ with the following orientation of the waveguide: (**c**) vertical, and (**d**) inclined. All the parameters are available in [189], reprinted figure with permission from [189], Copyright (2017) by the American Physical Society.

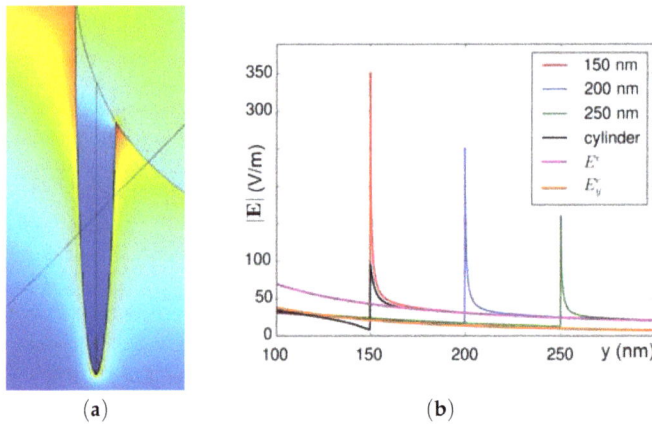

(a)   (b)

Figure 33. (a) scheme of a dielectric elliptical stick. The different parameters are available in [189]; **(b)** $|E^b(y)|$ corresponds to the field along the stick shaft for several lengths b. External E^e-field and its tangent projection E^e_y in the absence of the stick are also displayed. Field peaks shows that the electrical field is confined in the nanovolume near the apex of the stick, reprinted figure with permission from [189]. Copyright (2017) by the American Physical Society.

The LFDT can be used for local sensing of various chemical and biological objects. There is a fascinating possibility to "illuminate" the investigated object and then collect the Raman signal from the spot, whose size can be less than one nanometer.

To integrate the LFDT in the domain of electronics, the layer by layer growth can be more easily used, which is a well-known technique of the thin film technology. In order to decrease the number of the fabrication steps, a right cylindrical geometry of LFDT is adopted (in the direction of the growth which is perpendicular to the plane of Figure 35: z-direction).

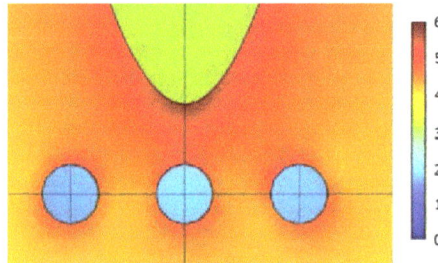

Figure 34. Tip heats magnetic nanoparticles (2 nm) made of FePt alloy [232]. For the magnetic particle placed perfectly under the tip, the heat production is 1.4 times greater than the other neighboring particles, reprinted figure with permission from [189]. Copyright (2017) by the American Physical Society.

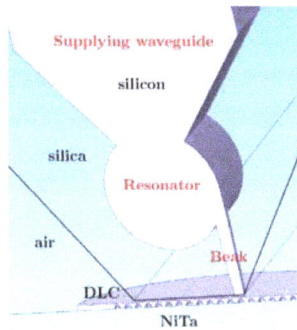

Figure 35. Scheme of a local-field dielectric transducer made of silicon. Diamond-like carbon (DLC) layer with equally spaced FePt grains is placed above NiTa substrate. FePt grains of size 14 nm are arranged in the square lattice with the period of 20 nm. The other parameters and informations can be obtained in Ref. [189]. Optical properties of FePt alloy and silicon were obtained from [232] and [233], respectively. The permittivity of NiTa alloy is calculated as the arithmetic mean of Ni and Ta permittivities. DLC permittivity is around $\varepsilon_{DLC} = 3.5$ coming from [234,235], reprinted figure with permission from [189]. Copyright (2017) by the American Physical Society.

The disk resonator is considered. The waveguide and the stick have a rectangular shape. Several groups have demonstrated EM field confinements between two rectangular dielectric waveguides (see Refs. [236–238]), and other groups have studied a great number of disk resonators for RF applications for a couple of decades [239]. More recently, optical magneto-dipole resonances in silicon disks were demonstrated in Refs. [60,240]. The numerical simulations of 2.5-dimensional silicon LFDT are shown in Figure 36.

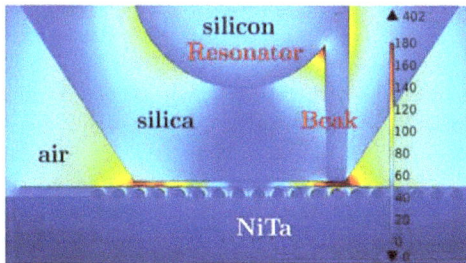

Figure 36. Electric field mapping below the LFDT, reprinted figure with permission from [189]. Copyright (2017) by the American Physical Society.

The silica cladding incorporates the plane waveguide, the disk resonator, and the rectangular stick. The disk resonator and the plane waveguide have the same height (the size in the direction normal to the figure plane). The 2.5D LFDT is a solid-state device that can have a large size (macroscopic scale) and can be incorporated in present electronics. LFDTs can be used for the local sensing including the collection of the Raman signal from a single molecule. To demonstrate efficiency of 2.5D LFDT, the nanostructured substrate is considered. The substrate is a carbon matrix, where FePt nanoparticles are distributed [241,242]. The computer simulations give the heating of the Fe-Pt under the action of the LFDT as displayed in Figure 37.

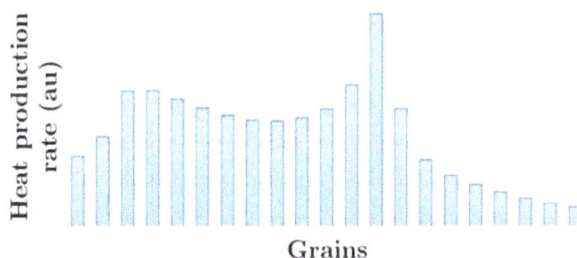

Figure 37. Heat production rate inside the grains in the central line. The highest heat production rate corresponds to the grain placed perfectly below the tip as displayed in Figure 36, reprinted figure with permission from [189]. Copyright (2017) by the American Physical Society.

When the tip is perfectly placed above a grain, the heat production is 1.6 times greater than in the case of neighboring grains. Thus, the 2.5D LFDT can be employed for the local detection of molecules with a spatial resolution of ~10 nm (Figure 37). We think that the supplementary optimization could allow a resolution of ~1 nm.

9. Conclusions

The low-loss and high-quality optical resonators are of great interest for the fundamental as well as applied research, as they are indefensible in optical filters, solar energy concentrator, medical imaging and biodetection including SERS, optical super-resolution microscopy, and magnetic recording assisted by heat, quantum electrodynamics and nanolasing. Optical resonators include regular or disordered plasmonic and dielectric micro- and nano-structures, which can serve optical antennae by concentrating large electromagnetic fields at micro- and nano-scales. The morphology of an optical resonator has an important key role in the control of its optical response. Thus, resonators with periodic structure allow for concentrating EM energy at any specified frequency. Such resonators can be used to reach additional SERS enhancements and increase the selectivity of SERS sensors. The approach based on "dielectric plasmonics" allows for fabricating high-Q light concentrators with low radiation losses. Dielectric hierarchical structures are designed in order to achieve a great field enhancement at the nanoscale. A great field confinement is obtained without losses. Thus, the EM field transducer is composed of a dielectric resonator and a stick having a sharp apex. The accumulation of the EM energy delivered by the waveguide is realized by the resonator. Therefore, the waveguide effectively pumps the resonator, which illuminates the stick and produces a strong electric field at the apex. Dielectric LFDTs arranged in an array can be employed as a SERS substrate without a luminescent background.

Author Contributions: A.K.S. wrote all the sections, edited the draft, prepared the original draft; A.I. wrote all the sections, edited the draft; A.L. wrote all the sections, edited the draft; G.B. wrote Section 3, prepared the original draft, edited of the draft.

Funding: This research is partially supported by the Russian Foundation for Basic Research (RFBR) (Grant Nos. 17-08-01448, 18-58-00048), the Russian Science Foundation (Grant No. 16-14-00209) and the Presidium of Russian Academy of Science (Program No. 56).

Acknowledgments: The authors acknowledge important discussions with Ilya A. Ryzhikov, Ilya N. Kurochkin, and Sergei S. Vergeles.

Conflicts of Interest: The authors declare no conflict of interest.

Abbreviations

The following abbreviations are used in this manuscript:

AFM	Atomic force microscopy
ASNOM	Apertureless scanning near-field optical microscopy
AuNP	Gold nanoparticle
DLC	Diamond like carbon
DTNB	5,5′-Dithiobis-(2-nitrobenzoic acid)
EM	Electromagnetic
HAMR	Heat-assisted magnetic recording
LFDT	Local field dielectric transducer
PMMA	Polymethylmethacrylate
QED	Quantum electrodynamics
R & D	Research and development
SEM	Scanning electron microscopy
SERS	Surface-enhanced Raman scattering
SNOM	Scanning near-field optical microscopy
SP	Surface plasmon
TERS	Tip-enhanced Raman scattering
TNT	Trinitrotoluene
WGM	Whispering gallery mode

References

1. Faraday, M. Experimental relations of gold (and other metals) to light. *Philos. Trans. R. Soc. Lond.* **1857**, *147*, 145–181.
2. Garnett, J.C.M. Colours in metal glasses and in metallic films. *Philos. Trans. R. Soc.* **1904**, *203*, 385, doi:10.1098/rsta.1904.0024.
3. Rayleigh, J.W.S. The problem of the whispering gallery. *Philos. Mag.* **1910**, *20*, 1001–1004, doi:10.1080/14786441008636993.
4. Armani, A.M.; Kulkarni, R.P.; Fraser, S.E.; Flagan, R.C.; Vahala, K.J. Label-free, single-molecule detection with optical microcavities. *Sci. Mag.* **2007**, *317*, 783–787, doi:10.1126/science.1145002.
5. Challener, W.; Peng, C.; Itagi, A.; Karns, D.; Peng, W.; Peng, Y.; Yang, X.; Zhu, X.; Gokemeijer, N.; Hsia, Y.; et al. Heat-assisted magnetic recording by a near-field transducer with efficient optical energy transfer. *Nat. Photonics* **2009**, *3*, 220–224.
6. Vedantam, S.; Lee, H.; Tang, J.; Conway, J.; Staffaroni, M.; Yablonovitch, E. A plasmonic dimple lens for nanoscale focusing of light. *Nano Lett.* **2009**, *9*, 3447–3452, doi:10.1021/nl9016368.
7. Aoki, T.; Dayan, B.; Wilcut, E.; Bowen, W.P.; Parkins, A.S.; Kippenberg, T.J.; Vahala, K.J.; Kimble, H.J. Observation of strong coupling between one atom and a monolithic micro-resonator. *Nature* **2006**, *443*, 671–674, doi:10.1038/nature05147.
8. Solano, M.; Faryad, M.; Hall, A.S.; Mallouk, T.E.; Monk, P.B.; Lakhtakia, A. Optimization of the absorption efficiency of an amorphous-silicon thin-film tandem solar cell backed by a metallic surface-relief grating. *Appl. Opt.* **2013**, *52*, 966–979.
9. Gresillon, S.; Aigouy, L.; Boccara, A.C.; Rivoal, J.C.; Quelin, X.; Desmarest, C.; Gadenne, P.; Shubin, V.A.; Sarychev, A.K.; Shalaev, V.M. Experimental observation of localized optical excitations in random metal dielectric films. *Phys. Rev. Lett.* **1999**, *82*, 4520–4523.
10. Ivanov, A.V.; Shalygin, A.N.; Saryhev, A.K. TE-wave propagation through 2D array of metal nanocylinders. *Solid State Phenom.* **2012**, *190*, 577–580, doi:10.4028/www.scientific.net/SSP.190.577.
11. Ivanov, A.; Shalygin, A.; Lebedev, V.; Vorobev, V.; Vergiles, S.; Sarychev, A.K. Plasmonic extraordinary transmittance in array of metal nanorods. *Appl. Phys. A* **2012**, *107*, 17–21, doi:10.1007/s00339-011-6731-3.
12. Ivanov, A.V.; Vaskin, A.V.; Lagarkov, A.N.; Sarychev, A.K. The field enhancement and optical sensing in the array of almost adjoining metal and dielectric nanorods. In *Plasmonics: Metallic Nanostructures and Their Optical Properties XII*; SPIE: Bellingham, WA, USA, 2014; p. 91633C, doi:10.1117/12.2063141.

13. Kurochkin, I.S.; Ryzhikov, I.A.; Sarychev, A.K.; Afanasiev, K.N.; Budashov, I.A.; Sedova, M.S.; Boginskaya, I.A.; Amitonov, S.; Lagarkov, A.N. New SERS-active junction based on cerium dioxide facet dielectric films for biosensing. *Adv. Electromagn.* **2014**, *3*, 57–60, doi:10.7716/aem.v3i1.266.

14. Afanasiev, K.N.; Boginskaya, I.A.; Budashov, I.A.; Ivanov, A.V.; Kurochkin, I.S.; Lagarkov, A.N.; Ryzhikov, I.A.; Sarychev, A.K. Giant field fluctuations in dielectric metamaterials and Raman sensors. In *Plasmonics: Metallic Nanostructures and Their Optical Properties XII*; SPIE: Bellingham, WA, USA, 2015; p. 95441Y1, doi:10.1117/12.2187134.

15. Ivanov, A.V.; Boginskaya, I.A.; Vaskin, A.V.; Afanas'ev, K.N.; Ryzhikov, I.A.; Lagarkov, A.N.; Sarychev, A. The field enhancement and optical sensing in the surface photonic chrystal. In Proceedings of the 2015 Days on Diffraction (DD), St. Petersburg, Russia, 25–29 May 2015; pp. 146–149, doi:10.1109/DD.2015.7354849.

16. Lagarkov, A.; Budashov, I.; Chistyaev, V.; Ezhov, A.; Fedyanin, A.; Ivanov, A.; Kurochkin, I.; Kosolobov, S.; Latyshev, A.; Nasimov, D.; et al. SERS-active dielectric metamaterials based on periodic nanostructures. *Opt. Express* **2016**, *24*, 7133–7150, doi:10.1364/OE.24.007133.

17. Sarychev, A.K.; Lagarkov, A.N.; Ivanov, A.V.; Boginskaya, I.A.; Bykov, I.V.; Ryzhikov, I.A.; Sedova, M.V.; Vaskin, A.V.; Kurochkin, I.N.; Rodionov, I.A.; et al. Metal-dielectric resonances in tip silicon metasurface and SERS based nanosensors. In *Plasmonics: Design, Materials, Fabrication, Characterization, and Applications XV*; SPIE: Bellingham, WA, USA, 2017; p. 103460C, doi:10.1117/12.2273097.

18. Lagarkov, A.; Boginkaya, I.; Bykov, I.; Budashov, I.; Ivanov, A.; Kurochkin, I.; Ryzhikov, I.; Rodionov, I.; Sedova, M.; Zverev, A.; et al. Light localization and SERS in tip-shaped silicon metasurface. *Opt. Express* **2017**, *25*, 17021–17038, doi:10.1364/OE.25.017021.

19. Bryche, J.F.; Gillibert, R.; Barbillon, G.; Gogol, P.; Moreau, J.; Lamy de la Chapelle, M.; Bartenlian, B.; Canva, M. Plasmonic enhancement by a continuous gold underlayer: Application to SERS sensing. *Plasmonics* **2016**, *11*, 601–608, doi:10.1007/s11468-015-0088-y.

20. Barbillon, G.; Sandana, V.E.; Humbert, C.; Belier, B.; Rogers, D.J.; Teherani, F.H.; Bove, P.; McClintock, R.; Razeghi, M. Study of Au coated ZnO nanoarrays for surface en-hanced Raman scattering chemical sensing. *J. Mater. Chem. C* **2017**, *5*, 3528–3535, doi:10.1039/C7TC00098G.

21. Barbillon, G.; Bijeon, J.L.; Lerondel, G.; Plain, J.; Royer, P. Detection of chemical molecules with integrated plasmonic glass nanotips. *Surf. Sci.* **2008**, *602*, L119–L122, doi:10.1016/j.susc.2008.06.032.

22. Barbillon, G. Plasmonic nanostructures prepared by soft UV nanoimprint lithography and their application in biological sensing. *Micromachines* **2012**, *3*, 21–27, doi:10.3390/mi3010021.

23. Ignatov, A.I.; Merzlikin, A.M.; Baryshev, A.V. Wood anomalies for s-polarized light incident on a one-dimensional metal gratingand their coupling with channel plasmons. *Phys. Rev. A* **2017**, *95*, 053843, doi:10.1103/PhysRevA.95.053843.

24. Maier, S.A.; Kik, P.G.; Atwater, H.A.; Meltzer, S.; Harel, E.; Koel, B.E.; Requicha, A.A.G. Local detection of electromagnetic energy transport below the diffraction limit in metal nanoparticle plasmon waveguides. *Nat. Mater.* **2003**, *2*, 229–232, doi:10.1038/nmat852.

25. Alu, A.; Engheta, N. Theory of linear chains of metamaterial/plasmonic particles as subdiffraction optical nanotransmission lines. *Phys. Rev. B* **2006**, *74*, 205436, doi:10.1103/PhysRevB.74.205436.

26. Markel, V.; Sarychev, A. Propagation of surface plasmons in ordered and disordered chains of metal nanospheres. *Phys. Rev. B* **2007**, *75*, 1–11, doi:10.1103/PhysRevB.75.085426.

27. Hadad, Y.; Steinberg, B.Z. Green's function theory for infinite and semi-infinite particle chains. *Phys. Rev. B* **2011**, *84*, 125402, doi:10.1103/PhysRevB.84.125402.

28. Auguie, B.; Barnes, W.L. Collective resonances in gold nanoparticle arrays. *Phys. Rev. Lett.* **2008**, *101*, 143902, doi:10.1103/PhysRevLett.101.143902.

29. Weber, D.; Albella, P.; Alonso-Gonzalez, P.; Neubrech, F.; Gui, H.; Nagao, T.; Hillenbrand, R.; Aizpurua, J.; Pucci, A. Longitudinal and transverse coupling in infrared gold nanoantenna arrays: Long range versus short range interaction regimes. *Opt. Express* **2011**, *19*, 15047–15061, doi:10.1364/OE.19.015047.

30. Shalaev, V.M.; Cai, W.; Chettiar, U.; Yuan, H.K.; Sarychev, A.K.; Drachev, V.P.; Kildishev, A.V. Negative index of refraction in optical metamaterials. *Opt. Lett.* **2005**, *30*, 3356–3358, doi:10.1364/OL.30.003356.

31. Kildishev, A.V.; Cai, W.; Chettiar, U.; Yuan, H.K.; Sarychev, A.K.; Drachev, V.P.; Shalaev, V.M. Negative refractive index in optics of metal-dielectric composites. *J. Opt. Soc. Am. B Opt. Phys.* **2006**, *23*, 423–433, doi:10.1364/JOSAB.23.000423.

32. Jen, Y.J.; Lakhtakia, A.; Yu, C.W.; Lin, C.T. Vapor-deposited thin films with negative real refractive index in the visible regime. *Opt. Express* **2009**, *17*, 7784–7789, doi:10.1364/OE.17.007784.

33. Ivanov, A.V.; Shalygin, A.N.; Galkin, V.Y.; Vedyayev, A.V.; Ivanov, V.A. Metamaterials from amorphous ferromagnetic microwires: interaction between microwires. *Solid State Phenom.* **2009**, *152–153*, 357–360, doi:10.4028/www.scientific.net/SSP.152-153.357.

34. Malyuzhinets, G.D. A note on the radiation principle. *Sov. Phys. Tech. Phys.* **1951**, *21*, 940–942.

35. Veselago, V. The electrodynamics of substances with simultaneously negative values of ϵ and μ. *Sov. Phys. Uspekhi* **1968**, *10*, 509–514, doi:10.1070/PU1968v010n04ABEH003699.

36. Bliokh, K.Y.; Bliokh, Y.P. Optical magnus effect as a consequence of berry phase anisotropy. *JETP Lett.* **2004**, *79*, 519–522.

37. Ivanov, A.V.; Shalygin, A.N.; Vedyaev, A.V.; Ivanov, V.A. Optical magnus effect in metamaterials fabricated from ferromagnetic microwires. *JETP Lett.* **2007**, *85*, 565–569, doi:10.1134/S0021364007110082.

38. Ono, A.; Kato, J.i.; Kawata, S. Subwavelength optical imaging through a metallic nanorod array. *Phys. Rev. Lett.* **2005**, *95*, 267407, doi:10.1103/PhysRevLett.95.267407.

39. Shvets, G.; Trendafilov, S.; Pendry, J.; Sarychev, A. Guiding, focusing, and sensing on the subwavelength scale using metallic wire arrays. *Phys. Rev. Lett.* **2007**, *99*, 053903, doi:10.1103/PhysRev. Lett. 99.053903.

40. Ikonen, P.; Simovski, C.; Tretyakov, S.; Belov, P.; Hao, Y. Magnification of subwavelength field distributions at microwavefrequencies using a wire medium slab operating in the canalization regime. *Appl. Phys. Lett.* **2007**, *91*, 104102, doi:10.1063/1.2767996.

41. Belov, P.A.; Zhao, Y.; Tse, S.; Ikonen, P.; Silveirinha, M.G.; Simovski, C.R.; Tretyakov, S.; Hao, Y.; Parini, C. Transmission of images with subwavelength resolution to distances of several wavelengthsin the microwave range. *Phys. Rev. B* **2008**, *77*, 193108, doi:10.1103/PhysRevB.77.193108.

42. Kawata, S.; Ono, A.; Verma, P. Subwavelength colour imaging with a metallic nanolens. *Nat. Photonics* **2008**, *2*, 438–442, doi:10.1038/nphoton.2008.103.

43. Wu, X.; Zhang, J.; Gong, Q. Metal-insulator-metal nanorod arrays for subwavelength imaging. *Opt. Express* **2009**, *17*, 2818–2825, doi:10.1364/OE.17.002818.

44. Silveirinha, M.G. Additional boundary conditions for nonconnected wire media. *New J. Phys.* **2009**, *11*, 113016. doi:10.1088/1367-2630/11/11/113016.

45. Genov, D.; Sarychev, A.K.; Shalaev, V.M.; Wei, A. Resonant field enhancements from metal nanoparticle arrays. *Nano Lett.* **2004**, *4*, 153–158, doi:10.1021/nl0343710.

46. Le, F.; Brandl, D.W.; Urzhumov, Y.A.; Wang, H.; Kundu, J.; Halas, N.J.; Aizpurua, J.; Nordlander, P. Metallic nanoparticle arrays: A common substrate for both surface-enhanced Raman scattering and surface-enhanced infrared absorption. *ACS Nano* **2008**, *2*, 707–718, doi:10.1021/nn800047e.

47. Aubry, A.; Lei, D.; Maier, S.; Pendry, J.P. Broadband plasmonic device concentrating the energy at the nanoscale: The crescent-shaped cylinder. *Phys. Rev. B* **2010**, *82*, 125430.

48. Vahala, K.J. Optical microcavities. *Nature* **2003**, *424*, 839–846.

49. Min, B.; Ostby, E.; Sorger, V.; Ulin-Avila, E.; Yang, L.; Zhang, X.; Vahala, K. High-Q surface plasmon polariton whispering gallery microcavity. *Nature* **2009**, *457*, 455–458, doi:10.1038/nature07627.

50. Johnson, P.B.; Christy, R.W. Optical constants of the noble metals. *Phys. Rev. B* **1972**, *6*, 4370–4379, doi:10.1103/PhysRevB.6.4370.

51. Tureci, H.; Schwefe, H.; Jacquod, P.; Stone, D. Modes of wave-chaotic dielectric resonators. *Prog. Opt.* **2005**, *47*, 75–108, doi:10.1016/S0079-6638(05)47002-X.

52. Armani, D.K.; Kippenberg, T.J.; Spillane, S.M.; Vahala, K.J. Ultra-high-Q toroid microcavityon a chip. *Nature* **2003**, *421*, 925–928.

53. Herr, T.; Brasch, V.; Jost, J.D.; Mirgorodskiy, I.; Lihachev, G.; Gorodetsky, M.L.; Kippenberg, T.J. Mode Spectrum and temporal soliton formation in optical microresonators. *Phys. Rev. Lett.* **2014**, *113*, 123901, doi:10.1103/PhysRevLett.113.123901.

54. Savchenkov, A.A.; Matsko, A.B.; Ilchenko, V.S.; Maleki, L. Optical resonators with ten million finesse. *Opt. Express* **2007**, *15*, 6768–6773, doi:10.1364/OE.15.006768.

55. Dumeige, Y.; Trebaol, S.; Ghisa, L.; Nguyen, T.K.; Tavernier, H.; Feron, P. Determination of coupling regime of high-Q resonators and optical gain of highly selective amplifiers. *J. Opt. Soc. Am. B Opt. Phys.* **2008**, *25*, 2073–2080, doi:10.1364/JOSAB.25.002073.

56. Gorodetsky, M.L.; Savchenkov, A.A.; Ilchenko, V.S. Ultimate Q of optical microsphere resonators. *Opt. Lett.* **1996**, *21*, 453–455, doi:10.1364/OL.21.000453.

57. Sumetsky, M. Mode localization and the Q-factor of a cylindrical microresonator. *Opt. Lett.* **2010**, *35*, 2385–2387, doi:10.1364/OL.35.002385.

58. Vollmer, F.; Arnold, S. Whispering-gallery-mode biosensing: Label-free detection down to single molecules. *Nat. Methods* **2008**, *5*, 591–596, doi:10.1038/NMETH.1221.

59. Dantham, V.R.; Holler, S.; Kolchenko, V.; Wan, Z.; Arnold, S. Taking whispering gallery-mode single virus detection and sizing to the limit. *Appl. Phys. Lett.* **2012**, *101*, 043704, doi:10.1063/1.4739473.

60. Kuznetsov, A.I.; Miroshnichenko, A.E.; Fu, Y.H.; Zhang, J.; Luk'yanchuk, B. Magnetic light. *Sci. Rep.* **2012**, *2*, 1–6, doi:10.1038/srep00492.

61. Cao, H.; Wiersig, J. Dielectric microcavities: Model systems for wave chaos and non-Hermitian physics. *Rev. Mod. Phys.* **2015**, *87*, 61–111, doi:10.1103/RevModPhys.87.61.

62. Albella, P.; Ameen Poyli, M.; Schmidt, M.K.; Maier, S.A.; Moreno, F.; Saenz, J.J.; Aizpurua, J. Low-loss electric and magnetic field-enhanced spectroscopy with subwavelength silicon dimers. *J. Phys. Chem. C* **2013**, *117*, 13573–13584, doi:10.1021/jp4027018.

63. Bakker, R.M.; Permyakov, D.; Yu, Y.F.; Markovich, D.; Paniagua-Dominguez, R.; Gonzaga, L.; Samusev, A.; Kivshar, Y.; Luk'yanchuk, B.; Kuznetsov, A.I. Magnetic and Electric Hotspots with Silicon Nanodimers. *Nano Lett.* **2015**, *15*, 2137–2142, doi:10.1021 /acs.nanolett.5b 00128.

64. Krasnok, A.E.; Miroshnichenko, A.E.; Belov, P.A.; Kivshar, Y.S. All-dielectric optical nanoantennas. *Opt. Express* **2012**, *20*, 20599–20604, doi:10.1364/OE.20.020599.

65. Hong, Y.; Pourmand, M.; Boriskina, S.V.; Reinhard, B.M. Enhanced light focusing in self-assembled optoplasmonic clusters with subwavelength dimensions. *Adv. Mater.* **2013**, *25*, 115–119, doi:10.1002/adma.201202830.

66. Hong, Y.; Qiu, Y.; Chen, T.; Reinhard, B.M. Rational assembly of optoplasmonic hetero-nanoparticle arrays with tunable photonic-plasmonic resonances. *Adv. Funct. Mater.* **2014**, *24*, 739–746, doi:10.1002/adfm.201301837.

67. Jahani, S.; Jacob, Z. All-dielectric metamaterials. *Nat. Nanotechnol.* **2016**, *11*, 23–36, doi:10.1038/nnano.2015.304.

68. Jahani, S.; Jacob, Z. Transparent subdiffraction optics: Nanoscale light confinement without metal. *Optica* **2014**, *1*, 96–100, doi:10.1364/OPTICA.1.000096.

69. Zhang, J.; Liu, W.; Zhu, Z.; Yuan, X.; Qin, S. Strong field enhancement and light-matter interactions with all-dielectric metamaterials based on split bar resonators. *Opt. Express* **2014**, *22*, 30889–30898, doi:10.1364/OE.22.030889.

70. Maslov, A.V.; Astratov, V.N. Imaging of sub-wavelength structures radiating coherently near microspheres. *Appl. Phys. Lett.* **2016**, *108*, 051104, doi:10.1063/1.4941030.

71. Shorokhov, A.S.; Melik-Gaykazyan, E.V.; Smirnova, D.A.; Hopkins, B.; Chong, K.E.; Choi, D.Y.; Shcherbakov, M.R.; Miroshnichenko, A.E.; Neshev, D.N.; Fedyanin, A.A.; et al. Multifold enhancement of third-harmonic generation in dielectric nanoparticles driven by magnetic Fano resonances. *Nano Lett.* **2016**, *16*, 4857–4861, doi:10.1021/acs.nanolett.6b01249.

72. Kim, S.J.; Fan, P.; Kang, J.H.; Brongersma, M.L. Creating semiconductor metafilms with designer absorption spectra. *Nat. Commun.* **2015**, *6*, 1–8, doi:10.1038/ncomms8591.

73. Kurochkin, I.N.; Sarychev, A.K.; Ryzhikov, I.A.; Budashov, I.A.; Maklakov, S.S.; Boyarintsev, S.O.; Lagarkov, A.N. Surface-enhanced Raman scattering-based biosensors. In *Portable Biosensing of Food Toxicants and Environmental Pollutants*; Series in Sensors; CRC Press-Taylor & Francis Group: Boca Raton, FL, USA, 2014; pp. 97–121.

74. Xu, H.; Aizpurua, J.; Kall, M.; Apell, P. Electromagnetic contributions to single-molecule sensitivity in surface-enhanced Raman scattering. *Phys. Rev. E* **2000**, *62*, 4318.

75. Kiran, M.S.; Itoh, T.; Yoshida, K.i.; Kawashima, N.; Biju, V.; Ishikawa, M. Selective Detection of HbA1c Using Surface Enhanced Resonance Raman Spectroscopy. *Anal. Chem.* **2010**, *82*, 1342–1348, doi:10.1021/ac902364h.

76. Sarychev, A.K.; Shalaev, V.M. *Electrodynamics of Metamaterials*; World Scientific Publishing Co. Pte. Ltd.: Singapore, 2007; pp. 1–247.

77. Brouers, F.; Blacher, S.; Lagarkov, A.N.; Sarychev, A.K.; Gadenne, P.; Shalaev, V.M. Theory of giant raman scattering from semicontinuous metal films. *Phys. Rev. B* **1997**, *55*, 13234–13245, doi:10.1103/PhysRevB.55.13234.

78. Genov, D.; Shalaev, V.; Sarychev, A. Surface plasmon excitation and correlation- induced localization-delocalization transition in semicontinuous metal films. *Phys. Rev. B* **2005**, *72*, 113102, doi:10.1103/PhysRevB.72.113102.

79. Seal, K.; Sarychev, A.K.; Noh, H.; Genov, D.A.; Yamilov, A.; Shalaev, V.M.; Ying, Z.C.; Cao, H. Near-field intensity correlations in semicontinuous metal-dielectric films. *Phys. Rev. Lett.* **2005**, *94*, 226101.

80. Seal, K.; Genov, D.A.; Sarychev, A.K.; Noh, H.; Shalaev, V.M.; Ying, Z.C.; Zhang, X.; Cao, H. Coexistence of localized and delocalized surface plasmon modes in percolating metal films. *Phys. Rev. Lett.* **2006**, *97*, 206103, doi:10.1103/PhysRevLett.97.206103.

81. Sarychev, A.K.; Boyarintsev, S.O.; Rakhmanov, A.L.; Kugel, K.I.; Sukhorukov, Y.P. Collective volume plasmons in manganites with nanoscale phase separation: Simulation of the measured infrared spectra of $La_{0.7}Ca_{0.3}MnO_3$. *Phys. Rev. Lett.* **2011**, *107*, 267401, doi:10.1103/PhysRevLett.107.267401.

82. Raman, C.V.; Krishnan, K.S. A New Type of Secondary Radiation. *Nature* **1928**, *121*, 501–502, doi:10.1038/121501c0.

83. Landsberg, G.; Mandelstam, L. Eine neue erscheinung bei der lichtzertreuung. *Die Naturwissenschaften* **1928**, *16*, 557–558.

84. Landsberg, G.S.; Mandelstam, L.I. Uber die lichtzerstrenung in kristallen. *Z. Fur. Phys.* **1928**, *50*, 769–780,

85. Balcytis, A.; Nishijima, Y.; Krishnamoorthy, S.; Kuchmizhak, A.; Stoddart, P.R.; Petruskevicius, R.; Juodkazis, S. From fundamental toward applied SERS: Shared principlesand divergent approaches. *Adv. Opt. Mater.* **2018**, *6*, 1800292, doi:10.1002/adom.201800292.

86. Fleischmann, M.; Hendra, P.; McQuillan, A. Raman spectra of pyridine adsorbed at a silver electrode. *Chem. Phys. Lett.* **1974**, *26*, 163–166, doi:10.1016/0009-2614(74)85388-1.

87. Zrimsek, A.B.; Wong, N.L.; Van Duyne, R.P. Single molecule surface-enhanced Raman spectroscopy: A critical analysis of the bianalyte vs. Isotopologue proof. *J. Phys. Chem. C* **2016**, *120*, 5133–5142, doi:10.1021/acs.jpcc.6b00606.

88. Henry, A.I.; Ueltschi, T.W.; McAnally, M.O.; Van Duyne, R.P. Spiers memorial lecture surface-enhanced Raman spectroscopy: From single particle/molecule spectroscopy to angstrom-scale spatial resolution and femtosecond time resolution. *Faraday Discuss.* **2017**, *205*, 9–30, doi:10.1039/c7fd00181a.

89. Panneerselvam, R.; Liu, G.K.; Wang, Y.H.; Liu, J.Y.; Ding, S.Y.; Li, J.F.; Wu, D.Y.; Tian, Z.Q. Surface-enhanced Raman spectroscopy: Bottlenecks and future directions. *Chem. Commun.* **2018**, *54*, 10–25, doi:10.1039/c7cc05979e.

90. Ding, S.Y.; You, E.M.; Tian, Z.Q.; Moskovits, M. Electromagnetic theories of surface-enhanced Raman spectroscopy. *Chem. Soc. Rev.* **2017**, *46*, 4042–4076, doi:10.1039/c7cs00238f.

91. Le Ru, E.C.; Blackie, E.; Meyer, M.; Etchegoin, P.G. Surface enhanced Raman scattering enhancement Factors: A comprehensive study. *J. Phys. Chem. C* **2007**, *111*, 13794–13803, doi:10.1021/jp0687908.

92. Freeman, L.M.; Pang, L.; Fainman, Y. Maximizing the electromagnetic and chemical resonances of surface-enhanced Raman scattering for nucleic acids. *ACS Nano* **2014**, *8*, 8383–8391, doi:10.1021/nn5028664.

93. McNay, G.; Eustace, D.; Smith, W.; Faulds, K.; Graham, D. Surface-enhanced Raman scattering (SERS) and surface-enhanced resonance Raman scattering (SERRS): A review of applications. *Appl. Spectrosc.* **2011**, *65*, 825–837, doi:10.1366/11-06365.

94. Morton, S.M.; Jensen, L. Understanding the molecule-surface chemical coupling in SERS. *J. Am. Chem. Soc.* **2009**, *131*, 4090–4098, doi:10.1021/ja809143c.

95. Laing, S.; Jamieson, L.E.; Faulds, K.; Graham, D. Surface-enhanced Raman spectroscopy for in vivo biosensing. *Nat. Rev. Chem.* **2017**, *1*, 1–19, doi:10.1038/s41570-017-0060.

96. Sharma, B.; Frontiera, R.R.; Henry, A.I.; Ringe, E.; Van Duyne, R.P. SERS: Materials, applications, and the future. *Mater. Today* **2012**, *15*, 16–25, doi:10.1016/S1369-7021(12)70017-2.

97. Alessandri, I.; Lombardi, J.R. Enhanced Raman scattering with dielectrics. *Chem. Rev.* **2016**, *116*, 14921–14981, doi:10.1021/acs.chemrev.6b00365.

98. McFarland, A.D.; Young, M.A.; Dieringer, J.A.; Van Duyne, R.P. Wavelength-scanned surface-enhanced Raman excitation spectroscopy. *J. Phys. Chem. B* **2005**, *109*, 11279–11285, doi:10.1021/jp050508u.

99. Perney, N.M.B.; Baumberg, J.J.; Zoorob, M.E.; Charlton, M.D.B.; Mahnkopf, S.; Netti, C.M. Tuning localized plasmons in nanostructured substrates for surface-enhanced Raman scattering. *Opt. Express* **2006**, *14*, 847–857, doi:10.1364/OPEX.14.000847.

100. Yan, B.; Thubagere, A.; Premasiri, W.R.; Ziegler, L.D.; Negro, L.D.; Reinhard, B.M. Engineered SERS Substrates with Multiscale Signal Enhancement: Nanoparticle Cluster Arrays. *ACS Nano* **2009**, *3*, 1190–1202, doi:10.1021/nn800836f.

101. Fan, M.; Andrade, G.F.S.; Brolo, A.G. A review on the fabrication of substrates for surface enhanced Raman spectroscopy and their applications in analytical chemistry. *Anal. Chim. Acta* **2011**, *693*, 7–25, doi:10.1016/j.aca.2011.03.002.

102. Banaee, M.G.; Crozier, K.B. Mixed dimer double resonance substrates for surface-enhanced Raman spectroscopy. *ACS Nano* **2011**, *5*, 307–314, doi:10.1021/nn102726j.

103. Li, W.D.; Ding, F.; Hu, J.; Chou, S.Y. Three-dimensional cavity nanoantenna coupled plasmonic nanodots for ultrahigh and uniform surface-enhanced Raman scattering over large area. *Opt. Express* **2011**, *19*, 3925–3936, doi:10.1364/OE.19.003925.

104. Stolcova, L.; Proska, J.; Prochazka, M. Hexagonally ordered gold semishells as tunable SERS substrates. In Proceedings of the NANOCON, Brno, Czech Republic, 23–25 October 2012; pp. 225–229.

105. Mattiucci, N.; D'Aguanno, G.; Everitt, H.O.; Foreman, J.V.; Callahan, J.M.; Buncick, M.C.; Bloemer, M.J. Ultraviolet surface-enhanced Raman scattering at the plasmonic band edge of a metallic grating. *Opt. Express* **2012**, *20*, 1868–1877, doi:10.1364/OE.20.001868.

106. Huang, J.A.; Zhao, Y.Q.; Zhang, X.J.; He, L.F.; Wong, T.L.; Chui, Y.S.; Zhang, W.J.; Lee, S.T. Ordered Ag/Si Nanowires Array: Wide-range surface-Enhanced Raman spectroscopy for reproducible biomolecule detection. *Nano Lett.* **2013**, *13*, 5039–5045, doi:10.1021/nl401920u.

107. Hu, F.; Lin, H.; Zhang, Z.; Liao, F.; Shao, M.; Lifshitz, Y.; Lee, S.T. Smart liquid SERS substrates based on Fe_3O_4/Au nanoparticles with reversibly tunable enhancement factor for practical quantitative detection. *Sci. Rep.* **2014**, *4*, 1–10, doi:10.1038/srep07204.

108. Lee, J.; Hua, B.; Park, S.; Ha, M.; Lee, Y.; Fan, Z.; Ko, H. Tailoring surface plasmons of high-density gold nanostar assemblies on metal films for surface-enhanced Raman spectroscopy. *Nanoscale* **2014**, *6*, 616–623, doi:10.1039/C3NR04752K.

109. Zhang, N.; Liu, K.; Liu, Z.; Song, H.; Zeng, X.; Ji, D.; Cheney, A.; Jiang, S.; Gan, Q. Ultrabroadband Metasurface for Efficient Light Trapping and Localization: A Universal Surface-Enhanced Raman Spectroscopy Substrate for "All" Excitation Wavelengths. *Adv. Mater. Interfaces* **2015**, *2*, 1–7. doi:10.1002/admi.201500142.

110. Jackson, J.B.; Westcott, S.L.; Hirsch, L.R.; West, J.L.; Halas, N.J. Controlling the surface enhanced Raman effect via the nanoshell geometry. *Appl. Phys. Lett.* **2003**, *82*, 257–259, doi:10.1063/1.1534916.

111. Campion, A.; Ivanecky, J.E.; Child, C.M.; Foster, M. On the mechanism of chemical enhancement in surface-enhanced Raman scattering. *J. Am. Chem. Soc.* **1995**, *117*, 11807–11808, doi:10.1021/ja00152a024.

112. Kambhampati, P.; Child, C.M.; Foster, M.C.; Campion, A. On the chemical mechanism of surface enhanced Raman scattering: Experiment and theory. *J. Chem. Phys.* **1998**, *108*, 5013–5026, doi:10.1063/1.475909.

113. Otto, A. The "chemical" (electronic) contribution to surface-enhanced Raman scattering. *J. Raman Spectrosc.* **2005**, *36*, 497–509, doi:10.1002/jrs.1355.

114. Jensen, L.; Aikens, C.; Schatz, G. Electronic structure methods for studying surface-enhanced Raman scattering. *Chem. Soc. Rev.* **2008**, *37*, 1061–1073, doi:10.1039/b706023h.

115. Jahn, M.; Patze, S.; Hidi, I.; Knipper, R.; Radu, A.; Muhlig, A.; Yuksel, S.; Peksa, V.; Weber, K.; Mayerhofer, T.; Cialla-May, D.; Popp, J. Plasmonic nanostructures for surface enhanced spectroscopic methods. *Analyst* **2016**, *141*, 756–793, doi:10.1039/C5AN02057C.

116. Wang, H.; Jiang, X.; Lee, S.; He, Y. Silicon nanohybrid-based surface-enhanced Raman scattering sensors. *Small* **2014**, *10*, 4455–4468, doi:10.1002/smll.201401563.

117. Zhang, M.; Fan, X.; Zhou, H.; Shao, M.; Antonio Zapien, J.; Wong, N.; Lee, S. A high-efficiency surface-enhanced Raman scattering substrate based on silicon nanowires array decorated with silver nanoparticles. *J. Phys. Chem. C* **2010**, *114*, 1969–1975, doi:10.1021/jp902775t.

118. Galopin, E.; Barbillat, J.; Coffinier, Y.; Szunerits, S.; Patriarche, G.; Boukherroub, R. Silicon nanowires coated with silver nanostructures as ultrasensitive interfaces for surface-enhanced Raman spectroscopy. *ACS Appl. Mater. Interfaces* **2009**, *1*, 1396–1403, doi:10.1021/am900087s.

119. Schmidt, M.; Hubner, J.; Boisen, A. Large Area Fabrication of Leaning Silicon Nanopillars for Surface Enhanced Raman Spectroscopy. *Adv. Mater.* **2012**, *24*, OP11–OP18, doi:10.1002/adma.201103496.

120. Bryche, J.F.; Belier, B.; Bartenlian, B.; Barbillon, G. Low-cost SERS substrates composed of hybrid nanoskittles for a highly sensitive sensing of chemical molecules. *Sens. Actuators B* **2017**, *239*, 795–799.

121. Magno, G.; Belier, B.; Barbillon, G. Gold thickness impact on the enhancement of SERS detection in low-cost Au/Si nanosensors. *J. Mater. Sci.* **2017**, *52*, 13650–13656, doi:10.1007/s10853-017-1445-3.

122. Magno, G.; Belier, B.; Barbillon, G. Al/Si nanopillars as very sensitive SERS substrates. *Materials* **2018**, *11*, 1534, doi:10.3390/ma11091534.

123. Sinha, G.; Depero, L.; Alessandri, I. Recyclable SERS substrates based on Au-coated ZnO nanorods. *ACS Appl. Mater. Interfaces* **2011**, *3*, 2557–2563, doi:10.1021/am200396n.

124. Khan, M.; Hogan, T.; Shanker, B. Gold-coated zinc oxide nanowire-based substrate for surface-enhanced Raman spectroscopy. *J. Raman Spectrosc.* **2009**, *40*, 1539–1545, doi:10.1002/jrs.2296.

125. Cheng, C.; Yan, B.; Wong, S.; Li, X.; Zhou, W.; Yu, T.; Shen, Z.; Yu, H.; Fan, H. Fabrication and SERS performance of silver-nanoparticle-decorated Si/ZnO nanotrees in ordered arrays. *ACS Appl. Mater. Interfaces* **2010**, *2*, 1824–1828, doi:10.1021/am100270b.

126. Cui, S.; Dai, Z.; Tian, Q.; Liu, J.; Xiao, X.; Jiang, C.; Wu, W.; Roy, V. Wetting properties and SERS applications of ZnO/Ag nanowire arrays patterned by a screen printing method. *J. Mater. Chem. C* **2016**, *4*, 6371–6379, doi:10.1039/C6TC00714G.

127. Shan, Y.; Yang, Y.; Cao, Y.; Fu, C.; Huang, Z. Synthesis of wheatear-like ZnO nano arrays decorated with Ag nanoparticles and its improved SERS performance through hydrogenation. *Nanotechnology* **2016**, *27*, 145502, doi:10.1088/0957-4484/27/14/145502.

128. Vo-Dinh, T.; Allain, L.R.; Stokes, D.L. Cancer gene detection using surface-enhanced Raman scattering (SERS). *J. Raman Spectrosc.* **2002**, *33*, 511–516, doi:10.1002/jrs.883.

129. Gniadecka, M.; Philipsen, P.A.; Sigurdsson, S.; Wessel, S.; Nielsen, O.F.; Christensen, D.H.; Hercogova, J.; Rossen, K.; Thomsen, H.K.; Gniadecki, R.; Hansen, L.K.; Wulf, H.C. Melanoma Diagnosis by Raman Spectroscopy and Neural Networks: Structure Alterations in Proteins and Lipids in Intact Cancer Tissue. *J. Investig. Dermatol.* **2004**, *122*, 443–449, doi:10.1046/j.0022-202X.2004.22208.x.

130. Kim, J.H.; Kim, J.S.; Choi, H.; Lee, S.M.; Jun, B.H.; Yu, K.N.; Kuk, E.; Kim, Y.K.; Jeong, D.H.; Cho, M.H.; Lee, Y.S. Nanoparticle probes with surface enhanced Raman spectroscopic tags for cellular cancer targeting. *Anal. Chem.* **2006**, *78*, 6967–6973, doi:10.1021/ac0607663.

131. Sha, M.; Xu, H.; Natan, M.J.; Cromer, R. Surface-enhanced Raman scattering tags for rapid and homogeneous detection of circulating tumor cells in the presence of human whole blood. *J. Am. Chem. Soc.* **2008**, *130*, 17214–17215, doi:10.1021/ja804494m.

132. Qian, X.; Peng, X.H.; Ansari, D.O.; Yin-Goen, Q.; Chen, G.Z.; Shin, D.M.; Yang, L.; Young, A.N.; Wang, M.D.; Nie, S. In vivo tumor targeting and spectroscopic detection with surface-enhanced Raman nanoparticle tags. *Nat. Biotechnol.* **2008**, *26*, 83–90, doi:10.1038/nbt1377.

133. Chon, H.; Lee, S.; Son, S.; Oh, C.H.; Choo, J. Highly sensitive immunoassay of lung cancer marker carcinoembryonic antigen using surface-enhanced Raman scattering of hollow gold nanospheres. *Anal. Chem.* **2009**, *81*, 3029–3034, doi:10.1021/ac802722c.

134. Lu, W.; Singh, A.K.; Khan, S.A.; Senapati, D.; Yu, H.; Ray, P.C. Gold Nano-popcorn-based targeted diagnosis, nanotherapy treatment, and in situ monitoring of photothermal therapy response of prostate cancer cells using surface-enhanced Raman spectroscopy. *J. Am. Chem. Soc.* **2010**, *132*, 18103–18114, doi:10.1021/ja104924b.

135. Wang, X.; Qian, X.; Beitler, J.J.; Chen, Z.G.; Khuri, F.R.; Lewis, M.M.; Shin, H.J.; Nie, S.; Shin, D.M. Detection of circulating tumor cells in human peripheral blood using surface-enhanced Raman scattering nanoparticles. *Cancer Res.* **2011**, *71*, 1526–1532, doi:10.1158/0008-5472.CAN-10-3069.

136. Samanta, A.; Maiti, K.K.; Soh, K.S.; Liao, X.; Vendrell, M.; Dinish, U.S.; Yun, S.W.; Bhuvaneswari, R.; Kim, H.; Rautela, S.; et al. Ultrasensitive near-infrared Raman reporters for SERS-based in vivo cancer detection. *Angew. Chem. Int. Ed.* **2011**, *50*, 6089–6092, doi:10.1002/anie.201007841.

137. Lin, D.; Feng, S.; Pan, J.; Chen, Y.; Lin, J.; Chen, G.; Xie, S.; Zeng, H.; Chen, R. Colorectal cancer detection by gold nanoparticle based surface-enhanced Raman spectroscopy of blood serum and statistical analysis. *Opt. Express* **2011**, *19*, 13565–13577, doi:10.1364/OE.19.013565.

138. Song, J.; Zhou, J.; Duan, H. Self-assembled plasmonic vesicles of SERS-encoded amphiphilic gold nanoparticles for cancer cell targeting and traceable intracellular drug delivery. *J. Am. Chem. Soc.* **2012**, *134*, 13458–13469, doi:10.1021/ja305154a.

139. Maiti, K.K.; Dinish, U.S.; Samanta, A.; Vendrell, M.; Soh, K.S.; Park, S.J.; Olivo, M.; Chang, Y.T. Multiplex targeted in vivo cancer detection using sensitive near-infrared SERS nanotags. *Nano Today* **2012**, *7*, 85–93, doi:10.1016/j.nantod.2012.02.008.

140. Jokerst, J.V.; Cole, A.J.; Van de Sompel, D.; Gambhir, S.S. Gold nanorods for ovarian cancer detection with photoacoustic imaging and resection guidance via Raman Imaging in living mice. *ACS Nano* **2012**, *6*, 10366–10377, doi:10.1021/nn304347g.

141. Feng, S.; Lin, D.; Lin, J.; Li, B.; Huang, Z.; Chen, G.; Zhang, W.; Wang, L.; Pan, J.; Chen, R.; Zeng, H. Blood plasma surface-enhanced Raman spectroscopy for non-invasive optical detection of cervical cancer. *Analyst* **2013**, *138*, 3967–3974, doi:10.1039/c3an36890d.

142. Vendrell, M.; Maiti, K.K.; Dhaliwal, K.; Chang, Y.T. Surface-enhanced Raman scattering in cancer detection and imaging. *Trends Biotechnol.* **2013**, *31*, 249–257, doi:10.1016/j.tibtech.2013.01.013.

143. Mert, S.; Ozbek, E.; Otunctemur, A.; Culha, M. Kidney tumor staging using surface-enhanced Raman scattering. *J. Biomed. Opt.* **2015**, *20*, 1–10, doi:10.1117/1.JBO.20.4.047002.

144. Velicka, M.; Urboniene, V.; Ceponkus, J.; Pucetaite, M.; Jankevicius, F.; Sablinskas, V. Detection of cancerous biological tissue areas by means of infrared absorption and SERS spectroscopy of intercellular fluid. *Proc. SPIE* **2015**, *9550*, 95500A, doi:10.1117/12.2186395.

145. Granger, J.H.; Schlotter, N.E.; Crawford, A.C.; Porter, M.D. Prospects for point-of-care pathogen diagnostics using surface-enhanced Raman scattering (SERS). *Chem. Soc. Rev.* **2015**, *45*, 3865–3882, doi:10.1039/c5cs00828j.

146. Rong, Z.; Wang, C.; Wang, J.; Wang, D.; Xiao, R.; Wang, S. Magnetic immunoassay for cancer biomarker detection based on surface-enhanced resonance Raman scattering from coupled plasmonic nanostructures. *Biosens. Bioelectron.* **2016**, *84*, 15–21, doi:10.1016/j.bios.2016.04.006.

147. Pazos, E.; Garcia-Algar, M.; Penas, C.; Nazarenus, M.; Torruella, A.; Pazos-Perez, N.; Guerrini, L.; Vazquez, M.; Garcia-Rico, E.; Mascarenas, J.; et al. Surface-Enhanced Raman Scattering Surface Selection Rules for the Proteomic Liquid Biopsy in Real Samples: Efficient Detection of the Oncoprotein c-MYC. *J. Am. Chem. Soc.* **2016**, *138*, 14206–14209, doi:10.1021/jacs.6b08957.

148. Oseledchyk, A.; Andreou, C.; Wall, M.A.; Kircher, M.F. Folate-targeted surface-enhanced resonance Raman scattering nanoprobe ratiometry for detection of microscopic ovarian cancer. *ACS Nano* **2016**, *11*, 1488–1497, doi:10.1021/acsnano.6b06796.

149. Litti, L.; Amendola, V.; Toffoli, G.; Meneghetti, M. Detection of low-quantity anticancer drugs by surface-enhanced Raman scattering. *Anal. Bioanal. Chem.* **2016**, *408*, 2123–2131. doi:10.1007/s00216-016-9315-4.

150. Schurmann, R.; Bald, I. Decomposition of DNA Nucleobases by laser irradiation of gold nanoparticles monitored by surface- enhanced Raman scattering. *J. Phys. Chem. C* **2016**, *120*, 3001–3009, doi:10.1021/acs.jpcc.5b10564.

151. Andreou, C.; Neuschmelting, V.; Tschaharganeh, D.F.; Huang, C.H.; Oseledchyk, A.; Iacono, P.; Karabeber, H.; Colen, R.R.; Mannelli, L.; Lowe, S.W.; et al. Imaging of liver tumors using surface-enhanced Raman scattering Nanoparticles. *ACS Nano* **2016**, *10*, 5015–5026, doi:10.1021/acsnano.5b07200.

152. Chen, Y.; Zhang, Y.; Pan, F.; Liu, J.; Wang, K.; Zhang, C.; Cheng, S.; Lu, L.; Zhang, W.; Zhang, Z.; et al. Breath analysis based on surface-enhanced Raman scattering sensors distinguishes early and advanced gastric cancer patients from healthy persons. *ACS Nano* **2016**, *10*, 8169–8179, doi:10.1021/acsnano.6b01441.

153. Chen, Y.; Ren, J.Q.; Zhang, X.G.; Wu, D.Y.; Shen, A.G.; Hu, J.M. Alkyne-modulated surface-enhanced Raman scattering-palette for optical interference-free and multiplex cellular imaging. *Anal. Chem.* **2016**, *88*, 6115–6119, doi:10.1021/acs.analchem.6b01374.

154. Cheng, Z.; Choi, N.; Wang, R.; Lee, S.; Moon, K.C.; Yoon, S.Y.; Chen, L.; Choo, J. Simultaneous detection of dual prostate specific antigens using surface-enhanced Raman scattering-based immunoassay for accurate diagnosis of prostate cancer. *ACS Nano* **2017**, *11*, 4926–4933, doi:10.1021/acsnano.7b01536.

155. Li, J.; Zhu, Z.; Zhu, B.; Ma, Y.; Lin, B.; Liu, R.; Song, Y.; Lin, H.; Tu, S.; Yang, C.J. Surface-enhanced Raman Scattering active plasmonic nanoparticles with ultrasmall interior nanogap for multiplex quantitative detection and cancer cell imaging. *Anal. Chem.* **2016**, *88*, 7828–7836, doi:10.1021/acs.analchem.6b01867.

156. Kneipp, J. Interrogating cells, tissues, and live animals with new generations of surface-enhanced raman scattering probes and labels. *ACS Nano* **2017**, *11*, 1136–1141, doi:10.1021/acsnano.7b00152.

157. Harmsen, S.; Wall, M.A.; Huang, R.; Kircher, M.F. Cancer imaging using surface-enhanced resonance Raman scattering nanoparticles. *Nat. Protoc.* **2017**, *12*, 1400–1414, doi:10.1038/nprot.2017.031.

158. Xu, H.; Bjerneld, E.J.; Kall, M.; Borjesson, L. Spectroscopy of single hemoglobin molecules by surface enhanced Raman Scattering. *Phys. Rev. Lett.* **1999**, *83*, 4357–4360, doi:10.1103/ PhysRevLett.83.4357.

159. Dingari, N.C.; Horowitz, G.L.; Kang, J.W.; Dasari, R.R.; Barman, I. Raman spectroscopy provides a powerful diagnostic tool for accurate determination of albumin glycation. *PLoS ONE* **2012**, *7*, e32406, doi:10.1371/journal.pone.0032406.

160. Barman, I.; Dingari, N.C.; Kang, J.W.; Horowitz, G.L.; Dasari, R.; Feld, M.S. Raman spectroscopy- based sensitive and specific detection of glycated hemoglobin. *Anal. Chem.* **2012**, *84*, 2474–2482, doi:10.1021/ac203266a.

161. Lin, J.; Lin, J.; Huang, Z.; Lu, P.; Wang, J.; Wang, X.; Chen, R. Raman Spectroscopy of human hemoglobin for diabetes detection. *J. Innov. Opt. Health Sci.* **2014**, *7*, 1350051, doi:10.1142/S179354581350051X.

162. Das, G.; Mecarini, F.; Angelis, F.D.; Prasciolu, M.; Liberale, C.; Patrini, M.; Fabrizio, E.D. Attomole (amol) myoglobin Raman detection from plasmonic nanostructures. *Microelectron. Eng.* **2008**, *85*, 1282–1285, doi:10.1016/j.mee.2007.12.082.

163. Benford, M.E.; Wang, M.; Kameoka, J.; Cote, G.L. Detection of Cardiac Biomarkers Exploiting Surface Enhanced Raman Scattering (SERS) using a Nanofluidic Channel Based Biosensor towards Coronary Point-of-Care Diagnostics. In *Plasmonics in Biology and Medicine VI*; SPIE: Bellingham, WA, USA, 2009; p. 719203, doi:10.1117/12.809661.

164. Chon, H.; Lee, S.; Yoon, S.Y.; Lee, E.K.; Chang, S.I.; Choo, J. SERS-based competitive immunoassay of troponin I and CK-MB markers for early diagnosis of acute myocardial infarction. *Chem. Commun.* **2014**, *50*, 1058–1060, doi:10.1039/c3cc47850e.

165. Wang, R.; Chon, H.; Lee, S.; Cheng, Z.; Hong, S.H.; Yoon, Y.H.; Choo, J. Highly sensitive detection of hormone estradiol E2 using SERS-based immunoassays for the clinical diagnosis of precocious puberty. *ACS Appl. Mater. Interfaces* **2016**, *8*, 10665–10672, doi:10.1021/acsami.5b10996.

166. Bodelon, G.; Montes-Garcia, V.; Lopez-Puente, V.; Hill, E.H.; Hamon, C.; Sanz-Ortiz, M.N.; Rodal-Cedeira, S.; Costas, C.; Celiksoy, S.; Perez-Juste, I.; et al. Detection and imaging of quorum sensing in Pseudomonas aeruginosa biofilm communities by surface-enhanced resonance Raman scattering. *Nat. Mater.* **2016**, *15*, 1203–1211, doi:10.1038/NMAT4720.

167. Duan, N.; Chang, B.; Zhang, H.; Wang, Z.; Wu, S. Salmonella typhimurium detection using a surface-enhanced Raman scattering-based aptasensor. *Int. J. Food Microbiol.* **2016**, *218*, 38–43, doi:10.1016/j.ijfoodmicro.2015.11.006.

168. Duan, N.; Yan, Y.; Wu, S.; Wang, Z. Vibrio parahaemolyticus detection aptasensor using surface-enhanced Raman scattering. *Food Control* **2016**, *63*, 122–127, doi:10.1016/j.foodcont.2015.11.031.

169. Yang, T.; Zhang, Z.; Zhao, B.; Hou, R.Y.; Kinchla, A.; Clark, J.M.; He, L. Real-time and in situ monitoring of pesticide penetration in edible leaves by surface-enhanced Raman scattering mapping. *Anal. Chem.* **2016**, *88*, 5243–5250, doi:10.1021/acs.analchem.6b00320.

170. Chen, J.; Huang, Y.; Kannan, P.; Zhang, L.; Lin, Z.; Zhang, J.; Chen, T.; Guo, L. Flexible and adhesive SERS Active tape for rapid detection of pesticide residues in fruits and vegetables. *Anal. Chem.* **2016**, *88*, 2149–2155, doi:10.1021/acs.analchem.5b03735.

171. Zhou, Q.; Meng, G.; Wu, N.; Zhou, N.; Chen, B.; Li, F.; Huang, Q. Dipping into a drink: Basil-seed supported silver nanoparticles as surface-enhanced Raman scattering substrates for toxic molecule detection. *Sens. Actuators B* **2016**, *223*, 447–452, doi:10.1016/j.snb.2015.09.115.

172. Tian, L.; Jiang, Q.; Liu, K.K.; Luan, J.; Naik, R.R.; Singamaneni, S. Bacterial nanocellulose- based flexible surface enhanced Raman scattering substrate. *Adv. Mater. Interfaces* **2016**, *3*, 1–8, doi:10.1002/admi.201600214.

173. Liu, Y.; Zhou, H.; Hu, Z.; Yu, G.; Yang, D.; Zhao, J. Label and label-free based surface-enhanced Raman scattering for pathogen bacteria detection: A review. *Biosens. Bioelectron.* **2017**, *94*, 131–140, doi:10.1016/j.bios.2017.02.032.

174. Chen, N.; Ding, P.; Shi, Y.; Jin, T.; Su, Y.; Wang, H.; He, Y. Portable and reliable Surface-Enhanced Raman scattering silicon chip for signal-on detection of trace trinitrotoluene explosive in real systems. *Anal. Chem.* **2017**, *89*, 5072–5078, doi:10.1021/acs.analchem.7b00521.

175. Hakonen, A.; Rindcevicius, T.; Schmidt, M.S.; Andresson, P.O.; Juhlin, L.; Svedendahl, M.; Boisen, A.; Kall, M. Detection of nerve gases using surface-enhanced Raman scattering substrates with high droplet adhesion. *Nanoscale* **2016**, *8*, 1305–1308, doi:10.1039/c5nr06524k.

176. Boyarintsev, S.O.; Sarychev, A.K. Computer simulation of surface-enhanced Raman scattering in nanostructured metamaterials. *J. Exp. Theor. Phys.* **2011**, *113*, 963–971, doi:10.1134/S1063776111140123.

177. Ma, R.M.; Ota, S.; Li, Y.; Yang, S.; Zhang, X. Explosives detection in a lasing plasmon nanocavity. *Nat. Nanotechnol.* **2014**, *9*, 600–604, doi:10.1038/nnano.2014.135.

178. Soteropulos, C.E.; Hunt, H.K.; Armani, A.M. Determination of binding kinetics using whispering gallery mode microcavities. *Appl. Phys. Lett.* **2011**, *99*, 103703, doi:10.1063/1.3634023.

179. Shopova, S.I.; Rajmangal, R.; Holler, S.; Arnold, S. Plasmonic enhancement of a whispering-gallery-mode biosensor for single nanoparticle detection. *Appl. Phys. Lett.* **2011**, *98*, 243104, doi:10.1063/1.3599584.

180. Ozer, N. Optical properties and electrochromic characterization of sol-gel deposited ceria films. *Sol. Energy Mater. Sol. Cells* **2001**, *68*, 391–400, doi:10.1016/S0927-0248(00)00371-8.

181. Patsalas, P.; Logothetidis, S.; Metaxa, C. Optical performance of nanocrystalline transparent ceria films. *Appl. Phys. Lett.* **2002**, *81*, 466–468, doi:10.1063/1.1494458.

182. Krogman, K.; Druffel, T.; Sunkara, M. Anti-reflective optical coatings incorporating nanoparticles. *Nanotechnology* **2005**, *16*, S338–S343, doi:10.1088/0957-4484/16/7/005.

183. Verma, A.; Karar, N.; Bakhshi, A.K.; Chander, H.; Shivaprasad, S.M.; Agnihotry, S.A. Structural, morphological and photoluminescence characteristics of sol-gel derived nano phase CeO$_2$ films deposited using citric acid. *J. Nanopart. Res.* **2007**, *9*, 317–322, doi:10.1007/s11051-006-9085-6.

184. Mansilla, C. Structure, microstructure and optical properties of cerium oxide thin films prepared by electron beam evaporation assisted with ion beams. *Solid State Sci.* **2009**, *11*, 1456–1464, doi:10.1016/j.solidstatesciences.2009.05.001.

185. Balakrishnan, G.; Sundari, S.T.; Kuppusami, P.; Mohan, P.C.; Srinivasan, M.; Mohandas, E.; Ganesan, V.; Sastikumar, D. A study of microstructural and optical properties of nanocrystalline ceria thin films prepared by pulsed laser deposition. *Thin Solid Films* **2011**, *519*, 2520–2526, doi:10.1016/j.tsf.2010.12.013.

186. Oh, T.S.; Tokpanov, Y.S.; Hao, Y.; Jung, W.; Haile, S.M. Determination of optical and microstructural parameters of ceria films. *J. Appl. Phys.* **2012**, *112*, 103535, doi:10.1063/1.4766928.

187. Murugan, R.; Vijayaprasath, G.; Mahalingam, T.; Hayakawa, Y.; Ravi, G. Effect of RF power on the properties of magnetron sputtered CeO$_2$ thin films. *J. Mater. Sci. Mater. Electron.* **2015**, *26*, 2800–2809, doi:10.1007/s10854-015-2761-5.

188. Tribelsky, M.I.; Luk'yanchuk, B.S. Anomalous light scattering by small particles. *Phys. Rev. Lett.* **2006**, *97*, 263902–263906, doi:10.1103/PhysRevLett.97.263902.

189. Vergeles, S.S.; Sarychev, A.K.; Tartakovsky, G.T. All-dielectric light concentrator to subwavelength volume. *Phys. Rev. B* **2017**, *95*, 085401, doi:10.1103/PhysRevB.95.085401.

190. Ru, E.C.L.; Etchegoin, P.G. *Principles of Surface-Enhanced Raman Spectroscopy and Related Plasmonic Effects*; Elsevier: Amsterdam, The Netherlands, 2009.

191. Rahmani, M.; Lukyanchuk, B.; Ng, B.; Tavakkoli, A.K.G.; Liew, Y.F.; Hong, M. Generation of pronounced Fano resonances and tuning of subwavelength spatial light distribution in plasmonic pentamers. *Opt. Express* **2011**, *19*, 4952–4956, doi:10.1364/OE.19.004949.

192. Li, K.; Stockman, M.I.; Bergman, D.J. Self-Similar Chain of Metal Nanospheres as an Efficient Nanolens. *Phys. Rev. Lett.* **2003**, *91*, 227402, doi:10.1103/PhysRevLett.91.227402.

193. Schultz, D.A. Plasmon Resonant particles for biological detection. *Curr. Opin. Biotech.* **2003**, *14*, 13–22, doi:10.1016/S0958-1669(02)00015-0.

194. Prodan, E.; Radloff, C.; Halas, N.J.; Nordlander, P. A hybridization model for the plasmon response of complex nanostructures. *Science* **2003**, *302*, 419–422, doi:10.1126/science.1089171.

195. Kristensen, A.; Yang, J.K.W.; Bozhevolnyi, S.I.; Link, S.; Nordlander, P.; Halas, N.J.; Mortensen, N.A. Plasmonic colour generation. *Nat. Rev. Mater.* **2016**, *2*, 16088, doi:10.1038/natrevmats.2016.88.

196. Hong, Y.; Reinhard, B.M. Collective photonic-plasmonic resonances in noble metal-dielectric nanoparticle hybrid arrays. *Opt. Mat. Express* **2014**, *4*, 2409–2422, doi:10.1364/OME.4.002409.

197. Pi, S.; Zeng, X.; Zhang, N.; Ji, D.; Chen, B.; Song, H.; Cheney, A.; Xu, Y.; Jiang, S.; Sun, D.; Song, Y.; Gan, Q. Dielectric-grating-coupled surface plasmon resonance from the back side of the metal film for ultrasensitive sensing. *IEEE Photonics J.* **2016**, *8*, 1–7, doi:10.1109/JPHOT.2015.2509870.

198. Santiago-Cordoba, M.A.; Boriskina, S.V.; Vollmer, F.; Demirel, M.C. Nanoparticle-based protein detection by optical shift of a resonant microcavity. *Appl. Phys. Lett.* **2011**, *99*, 073701, doi:10.1063/1.3599706.

199. Gromova, M.S.; Sigolaeva, L.V.; Fastovets, M.A.; Evtushenko, E.G.; Babin, I.A.; Pergushov, D.V.; Amitonov, S.V.; Eremenko, A.V.; Kurochkin, I.N. Improved adsorption of choline oxidase on a polyelectrolyte LBL film in the presence of iodide anions. *Soft Matter.* **2011**, *7*, 7404–7409, doi:10.1039/c1sm05655g.

200. Tamer, U.; Boyaci, I.; Temur, E.; Zengin, A.; Dincer, I.; Elerman, Y. Fabrication of magnetic gold nanorod particles for immunomagnetic separation and SERS application. *J. Nanopart. Res.* **2011**, *13*, 3167–3176, doi:10.1007/s11051-010-0213-y.

201. Bandarenka, H.V.; Girel, K.V.; Bondarenko, V.P.; Khodasevich, I.A.; Panarin, A.Y.; Terekhov, S.N. Formation regularities of plasmonic silver nanostructures on porous silicon for effective surface-enhanced Raman scattering. *Nanoscale Res. Lett.* **2016**, *11*, 262, doi:10.1186/s11671-016-1473-y.

202. Cottat, M.; Lidgi-Guigui, N.; Tijunelyte, I.; Barbillon, G.; Hamouda, F.; Gogol, P.; Aassime, A.; Lourtioz, J.M.; Bartenlian, B.; Lamy de la Chapelle, M. Soft UV nanoimprint lithography-designed highly sensitive substrates for SERS detection. *Nanoscale Res. Lett.* **2014**, *9*, 623, doi:10.1186/1556-276X-9-623.

203. Hentschel, M.; Saliba, M.; Vogelgesang, R.; Giessen, H.; Alivisatos, A.P.; Liu, N. Transition from isolated to collective modes in plasmonic oligomers. *Nano Lett.* **2010**, *10*, 2721–2726, doi:10.1021/nl101938p.

204. Atay, T.; Song, J.H.; Nurmikko, A. Strongly interacting plasmon nanoparticles: From dipole interaction to conductively coupled regime. *Nano Lett.* **2004**, *4*, 1627–1631, doi:10.1364/IQEC.2004.IPDB7.

205. Naik, G.V.; Shalaev, V.M.; Boltasseva, A. Alternative plasmonic materials: Beyond gold and silver. *Adv. Mat.* **2013**, *25*, 1–31, doi:10.1002/adma.201205076.

206. Swiontek, S.; Faryad, M.; Lakhtakia, A. Surface plasmonic polaritonic sensor using a dielectric columnar thin film. *J. Nanophotonics* **2014**, *8*, 083986, doi:10.1117/1.JNP.8.083986.

207. Lakhtakia, A.; Faryad, M. Theory of optical sensing with Dyakonov-Tamm waves. *J. Nanophotonics* **2014**, *8*, 083072, doi:10.1117/1.JNP.8.083072.

208. Sumetsky, M. Whispering-gallery-bottle microcavities: The three-dimensional etalon. *Opt. Lett.* **2004**, *29*, 8–10, doi:10.1364/OL.29.000008.

209. Sumetsky, M. Localization of light on a cone: Theoretical evidence and experimental demonstration for an optical fiber. *Opt. Lett.* **2011**, *36*, 145–147, doi:10.1364/OL.36.000145.

210. Barannik, A.A.; Bunyaev, S.A.; Cherpak, N.T. Conical quasioptical dielectric resonator. *Tech. Phys. Lett.* **2005**, *31*, 811–812.

211. Kishk, A.A.; Yin, Y.; Glisson, A.W. Conical dielectric resonator antennas for wide-band applications. *IEEE Trans. Antennas Propag.* **2002**, *50*, 469–474, doi:10.1109/TAP.2002.1003382.

212. Kishk, A.A.; Zhang, X.; Glisson, A.W.; Kajfez, D. Numerical analysis of stacked dielectric resonator antennas excited by a coaxial probe for wideband applications. *IEEE Trans. Antennas Propag.* **2003**, *51*, 1996–2006.

213. Chern, G.D.; Tureci, H.E.; Stone, A.D.; Chang, R.K.; Kneissl, M.; Johnson, N.M. Unidirectional lasing from InGaN multiple-quantum-well spiral-shaped micropillars. *Appl. Phys. Lett.* **2003**, *83*, 1710–1712, doi:10.1063/1.1605792.

214. Gmachl, C.; Capasso, F.; Narimanov, E.E.; Nockel, J.U.; Stone, A.D.; Faist, J.; Sivco, D.L.; Cho, A.Y. High-Power Directional Emission from Microlasers with Chaotic Resonators. *Science* **1998**, *280*, 1556–1564, doi:10.1126/science.280.5369.1556.

215. Gmachl, C.; Narimanov, E.E.; Capasso, F.; Baillargeon, J.N.; Cho, A.Y. Kolmogorov-Arnold-Moser transition and laser action on scar modes in semiconductor diode lasers with deformed resonators. *Opt. Lett.* **2002**, *27*, 824–826, doi:10.1364/OL.27.000824.

216. Fang, W.; Cao, H.; Podolskiy, V.A.; Narimanov, E.E. Dynamical localization in microdisk lasers. *Opt. Express* **2005**, *13*, 5641–5652, doi:10.1364/OPEX.13.005641.

217. Bykov, V.; Gologanov, A.; Shevyakov, V. Test structure for SPM tip shape deconvolution. *Appl. Phys. A* **1998**, *66*, 499–502, doi:10.1142/9789814280365_0127.

218. Green, M.A.; Keevers, M.J. Optical Properties of Intrinsic Silicon at 300 K. *Prog. Photovolt.* **1995**, *3*, 189–192, doi:10.1002/pip.4670030303.

219. Hussain, S.; Bhatia, C.S.; Yang, H.; Danner, A.J. Characterization of C-apertures in a successful demonstration of heat-assisted magnetic recording. *Opt. Lett.* **2015**, *40*, 3444–3447, doi:10.1364/OL.40.003444.

220. Schuller, J.; Barnard, E.; Cai, W.; Jun, Y.; White, J.; Brongersma, M. Plasmonics for extreme light concentration and manipulation. *Nat. Mater.* **2010**, *9*, 193–204, doi:10.1038/nmat2630.

221. Novotny, L.; Hulst, N. Antennas for light. *Nat. Photonics* **2011**, *5*, 83–90, doi:10.1038/nphoton.2010.237.

222. Fedorov, I.A.; Parfenyev, V.M.; Vergeles, S.S.; Tartakovsky, G.T.; Sarychev, A.K. Allowable number of plasmons in nanoparticle. *JETP Lett.* **2014**, *100*, 530–534, doi:10.1134/S0021364014200053.

223. Vergeles, S.S.; Sarychev, A.K. Silicon plasmonics and optical field concentration at nanometer scale. In *Metamaterials, Metadevices, and Metasystems 2015*; SPIE: Bellingham, WA, USA, 2015; p. 954415.

224. Yuan, G.; Rogers, E.; Roy, T.; Shen, Z.; Zheludev, N. Flat super-oscillatory lens for heat-assisted magnetic recording with sub-50 nm resolution. *Opt. Express* **2014**, *22*, 6428–6437, doi:10.1364/OE.22.006428.

225. Bohn, J.L.; Nesbitt, D.J.; Gallagher, A. Field enhancement in apertureless near-field scanning optical microscopy. *JOSA A* **2001**, *18*, 2998–3006, doi:10.1364/JOSAA.18.002998.

226. Bouhelier, A.; Beversluis, M.; Hartschuh, A.; Novotny, L. Near-Field Second-Harmonic Generation Induced by Local Field Enhancement. *Phys. Rev. Lett.* **2003**, *90*, 013903, doi:10.1103/PhysRevLett.90.013903.

227. Cancado, L.; Hartschuh, A.; Novotny, L. Tip-enhanced Raman spectroscopy of carbon nanotubes. *J. Raman Spectrosc.* **2009**, *40*, 1420–1426, doi:10.1002/jrs.2448.

228. Gerton, J.M.; Wade, L.A.; Lessard, G.A.; Ma, Z.; Quake, S.R. Tip-enhanced fluorescence microscopy at 10 nanometer resolution. *Phys. Rev. Lett.* **2004**, *93*, 180801, doi:10.1103/PhysRevLett.93.180801.

229. Kato, S.; Chonan, S.; Aoki, T. High-numerical-aperture microlensed tip on an air-clad optical fiber. *Opt. Lett.* **2014**, *39*, 773–776, doi:10.1364/OL.39.000773.

230. Wang, Y.; Du, Z.; Park, Y.; Chen, C.; Zhang, X.; Pan, L. Quasi-3D plasmonic coupling scheme for near-field optical lithography and imaging. *Opt. Lett.* **2015**, *40*, 3918–3921, doi:10.1364/OL.40.003918.

231. Song, Q.; Ge, L.; Redding, B.; Cao, H. Channeling Chaotic Rays into Waveguides for Efficient Collection of Microcavity Emission. *Phys. Rev. Lett.* **2012**, *108*, 243902, doi:10.1103/PhysRevLett.108.243902.

232. Sato, K.; Mizusawa, A.; Ishida, K.; Seki, T.; Shima, T.; Takanashi, K. Magneto-Optical Spectra of Ordered and Disordered FePt Films Prepared at Reduced Temperatures. *Trans. Magn. Soc. Jpn.* **2004**, *4*, 297–300, doi:doi.org/10.3379/tmjpn2001.4.297.

233. Bergmann, J.; Heusinger, M.; Andra, G.; Falk, F. Temperature dependent optical properties of amorphous silicon for diode laser crystallization. *Opt. Express* **2012**, *20*, A856–A863, doi:10.1364/OE.20.00A856.

234. Singh, S.; Pandey, M.; Chand, N.; Biswas, A.; Bhattacharya, D.; Dash, S.; Tyagi, A.; Dey, R.; Kulkarni, S.; Patil, D. Optical and mechanical properties of diamond like carbon films deposited by microwave ECR plasma CVD. *Bull. Mater. Sci.* **2008**, *31*, 813–818,

235. Mednikarov, B.; Spasov, G.; Babeva, T.; Pirov, J.; Sahatchieva, M.; Popova, C.; Kulischa, W. Optical properties of diamond-like carbon and nanocrystalline diamond films. *J. Optoelectron. Adv. Mater.* **2005**, *7*, 1407–1413.

236. Almeida, V.R.; Xu, Q.; Barrios, C.A.; Lipson, M. Guiding and confining light in void nanostructure. *Opt. Lett.* **2004**, *29*, 1209–1211, doi:10.1364/OL.29.001209.

237. Koos, C.; Vorreau, P.; Vallaitis, T.; Dumon, P.; Bogaerts, W.; Baets, R.; Esembeson, B.; Biaggio, I.; Michinobu, T.; Diederich, F.; et al. All-optical high-speed signal processing with silicon-organic hybrid slot waveguides. *Nat. Photonics* **2009**, *3*, 216–219, doi:10.1038/nphoton.2009.25.

238. Liu, Q.; Tu, X.; Kim, K.W.; Kee, J.S.; Shin, Y.; Han, K.; Yoon, Y.J.; Lo, G.Q.; Park, M.K. Highly sensitive Mach-Zehnder interferometer biosensor based on silicon nitride slot waveguide. *Sens. Actuators B* **2013**, *188*, 681–688, doi:10.1016/j.snb.2013.07.053.

239. Mongia, R.K.; Bhartia, P. Dielectric resonator antennas-A review and general design relations for resonant frequency and bandwidth. *Int. J. Microw. Millim.-Wave Comput.-Aided Eng.* **1994**, *4*, 230–247, doi:10.1002/mmce.4570040304.

240. Permyakov, D.; Sinev, I.; Markovich, D.; Ginzburg, P.; Samusev, A.; Belov, P.; Valuckas, V.; Kuznetsov, A.I.; Lukanchuk, B.S.; Miroshnichenko, A.E.; et al. Probing magnetic and electric optical responses of silicon nanoparticles. *Appl. Phys. Lett.* **2015**, *106*, 171110, doi:10.1063/1.4919536.

241. Weller, D.; Mosendz, O.; Parker, G.; Pisana, S.; Santos, T.S. L1 0 FePtX-Y media for heat-assisted magnetic recording. *Phys. Status Solidi (a)* **2013**, *210*, 1245–1260, doi:10.1002/pssa.201329106.

242. Hu, J.; Cher, K.M.; Varghese, B.; Xu, B.; Lim, C.; Shi, J.; Chen, Y.; Ye, K.; Zhang, J.; An, C.; et al. FePt-based hamr media with a function layer for better thermal control. *IEEE Trans. Magn.* **2016**, *52*, 1–6, doi:10.1109/TMAG.2015.2478901.

![materials logo] *materials*

MDPI

Article

AFM-Nano Manipulation of Plasmonic Molecules Used as "Nano-Lens" to Enhance Raman of Individual Nano-Objects

Angélina D'Orlando [1] , **Maxime Bayle** [2] , **Guy Louarn** [2] **and Bernard Humbert** [2,*]

[1] INRA BIA UR 1268, 44300 Nantes, France; angelina.dorlando@inra.fr
[2] IMN J. Rouxel, UMR 6502 CNRS-Univ Nantes, BP 32229 44322 Nantes, France;
   maxime.bayle@cnrs-imn.fr (M.B.); guy.louarn@cnrs-imn.fr (G.L.)
[*] Correspondence: bernard.humbert@cnrs-imn.fr; Tel.: +33-(0)2-40-37-39-89 (B.H. & A.D.)

Received: 30 March 2019; Accepted: 25 April 2019; Published: 27 April 2019

Abstract: This paper explores the enhancement of Raman signals using individual nano-plasmonic structures and demonstrates the possibility to obtain controlled gold plasmonic nanostructures by atomic force microscopy (AFM) manipulation under a confocal Raman device. By manipulating the gold nanoparticles (Nps) while monitoring them using a confocal microscope, it is possible to generate individual nano- structures, plasmonic molecules not accessible currently by lithography at these nanometer scales. This flexible approach allows us to tune plasmonic resonance of the nanostructures, to generate localized hot spots and to circumvent the effects of strong electric near field gradients intrinsic to Tip Enhanced Raman Spectroscopy (TERS) or Surface Enhanced Raman Spectroscopy (SERS) experiments. The inter Np distances and symmetry of the plasmonic molecules in interaction with other individual nano-objects control the resonance conditions of the assemblies and the enhancement of their Raman responses. This paper shows also how some plasmonic structures generate localized nanometric areas with high electric field magnitude without strong gradient. These last plasmonic molecules may be used as "nano-lenses" tunable in wavelength and able to enhance Raman signals of neighbored nano-object. The positioning of one individual probed nano-object in the spatial area defined by the nano-lens becomes then very non-restrictive, contrary to TERS experiments where the spacing distance between tip and sample is crucial. The experimental flexibility obtained in these approaches is illustrated here by the enhanced Raman scatterings of carbon nanotube.

Keywords: enhanced Raman spectroscopy; plasmonic nanoparticles; AFM-nanomanipulations; optical near-field; plasmonic molecules

---

1. Introduction

The strong demand to miniaturize to improve the spatial resolution and to decrease the detection limits in optical spectroscopic devices has initiated a significant amount of research in the field of the plasmonics [1–5]. The recent propositions in the literature are based on the nano-structuration of nanoparticles (Nps) of metal (Reference [5] and references therein). Indeed, electrons on a metal surface, excited by electromagnetic fields (for instance by focusing light beams on Nps), are able to enhance the electric field locally up to several orders of magnitude. The spatial area where the field is "concentrated" is as small as just a few nanometer squares which provides the opportunity to design the electric field of light with a spatial resolution of 2–3 orders of magnitude below the diffraction limit. This effect has been at the origin of the Surface Enhancement Raman Spectroscopy SERS spectroscopy and of Tip Enhanced Raman spectroscopy (TERS) that presents an exponential decrease. The enhancement was predicted to be increased on to originate in the near-field generated in inter-particle gaps, for instance within the colloidal aggregates [1–5]. However, because the experiments involved wide distributions of aggregates of different sizes and shapes or of many different nano-structured surfaces, the correlations between their near-field properties and SERS spectra were highly challenging. Then many experiments have tried to rationalize these nano-structuration by lithography, colloidal depositions, self-similar assembling [5,6], often in a "top-down" way. In the field of molecular fluorescence detections, the combination of the process assembly simplicity with the fluorescence enhancement made the self-assembled colloidal nanoparticle gap antennas optimal to extend a wide variety of single-molecule applications toward the micromolar concentrations [7]. This self-assembly may be helped by a pre-step of lithography [8]. Thus, for instance, spatially programmable Au nanorod dimers have been obtained with enhanced capacities for fluorescence applications, highlighting the opportunities for precise tunability of the plasmonic modes in larger assemblies. F. Koendernicks recently published a review on the different possibilities of this kind of approach to obtain single photon nano-antennas on large spatial areas [9]. These kinds of approaches have also been used to build substrates for surface enhanced Raman spectroscopy uses [10,11]. For instance, large-area multifunctional wedge and pyramid arrays directly onto planar substrates via template stripping [10] allowing in the same time to magnetically trap the species and to characterize them by SERS in the optical near-field. By lithography, substrates constituted of gold nano-stripes have been proposed and deeply studied in SERS applications [11]. The SERS efficiency of probe molecules was investigated on non-annealed and annealed lithographic gold stripes. The SERS intensity is not linked to the far-field response of the non-annealed stripes. This mismatch between the far-field and near-field response comes from a significant contribution of the surface roughness of the stripes. The annealing of lithographed substrates by decreasing roughness resulted in a strong weakening of the Raman signals. Their results further suggested that the SERS effect was more pronounced in the red part of the visible range, far from the plasmon resonance of the structures and was essentially generated by the sub-20nm structures coming from the roughness than more by the organized lithographed over 20 nm. Therefore, nano-shapes, sub-20 nm, for SERS applications remains an important key to investigate. However, another bottom-up approach was proposed, in particular by Chuntonovs' team [12,13]. The idea was to describe small aggregates of some metallic Nps as plasmonic molecules [5,12,13]. Indeed, well-defined assemblies of Nps that sustain surface-plasmon resonances have been labelled "plasmonic molecules" [12–14], where the individual "atoms" of these plasmonic molecules are then the individual metallic Nps, called "plasmonic atoms". In this paper, the individual gold Nps will correspond to "plasmonic atoms". In the dipole approximation, the "plasmonic atoms" may be described as individual Nps dipoles and the inter-strength depends then on the single-particle polarizability, which scales as the cube of the diameter of the individual Nps. Moreover, the energy of the interaction of the two dipoles induced on the surfaces of the individual Nps upon excitation scales as the inverse of the cube of the gap-distance between the particle centers. Nordlander and co-workers developed a plasmon hybridization description [12–16] which predicts the modes of interacting Nps (plasmonic atoms) and calculates their corresponding energies, i.e., the resonance spectra.

In this application field of individual plasmonic molecules, we have developed since 2013 [17] an experimental approach where molecules are built atom per atom with the help of an Atomic Force Microscope (AFM) coupled directly with an optical spectrometer in order to monitor the optical changes. Tong et al. initiated this approaching 2008 to understand the enhancement of Raman signals of nanotubes of carbon (Single wall nanotubes of carbon SWCNT) by approaching an individual Au Np to one isolated target SWCNT via AFM [18]. The advantage of this method is to use the same Au Nps in all the experiments, thus all the physical changes are caused by changes of the shape of plasmonic molecule or to the distances between plasmonic atoms but never due to intrinsic Np changes (the individual diameters, individual shapes, surface chemistry are fixed). Thus the discussion of the experimental results are facilitated. The remaining experimental problem in this approach is to have the perfect control of distances between the plasmonic molecule and the interacting nano-objects. Indeed, due to the strong decrease of the optical near fields (as previously evoked in inverse of the cube of the gap distances between Nps) the signals strongly depend on the different relative spacings. Consequently, if the goal is to use one plasmonic molecule as a nano-sensor, its response varies with only spatial parameters. This paper proposes a shaping of plasmonic molecule to solve this problem and to avoid the strong gradient effect of the near field around metallic Nps. This original approach will be applied to follow the Raman spectra of individual nanotubes of carbon. This kind of object is now a "quasi" academic object widely described and used in the same kind of experimental conditions [19,20]. For instance, Reichs' team realized the coupling of carbon nanotubes as a one-dimensional model system to lithographed gold nanodimers acting as near-field cavities. Their plasmonic cavities enhanced the Raman signal of a small nanotube bundle by three order of magnitude. Our paper will show that this enhancement may also be reached by nano-manipulating one single particle from the assembly of Au Nps constituting the cavity.

2. Materials and Methods

2.1. Materials

We synthesized AuNPs with very precise protocols ensuring that the only parameters to be modified were the reactant molar ratio (citrate to gold ratio) and the sequence of reagent addition [21]. We followed the 'Turkevich' protocol introduced in 1951 [22] rationalized recently by Li Shi et al. [21]. Our target for this work has been to obtain a diameter around 20 nm for the manipulations. The gold (III) mother solution was diluted with water to obtain 50 mL of "yellow orange" solutions at [AuIII] = 0.25 mM in 100 mL double-necked round flasks. The flasks were immersed in an oil bath to maintain the temperature around 95 °C. Then 2.5 mL of a citrate solution (molar ratio fixed 2) preheated during 10 min at about the reaction temperature were all added at once. After 15 to 20 minutes of reaction, the solutions were then cooled slowly to the room temperature. These colloidal suspensions were characterized by dynamic light scattering (DLS), ultraviolet visible (UV-vis) absorption and electronic microscopy. Respectively, DLS experiments were carried out using on a NanoZS apparatus (Malvern Panalytical, Malvern, UK) operating at k = 632.8 nm; UV–vis spectroscopy was performed on Cary 5000 Scan UV-Visible Spectrometer (Agilent, Santa Clara, CA, USA) and the primary size distributions and the shape of particles were determined by transmission electron microscopy (Hitachi H-9000 NAR, Tokyo, Japan, operating at 300 kV with a Scherzer resolution of 0.18 nm). Our dispersion were then characterized by a single modal distribution around 22 nm ± 5 nm. Figure 1 displays an example of a series of characterizations carried out on one of our synthesizes.

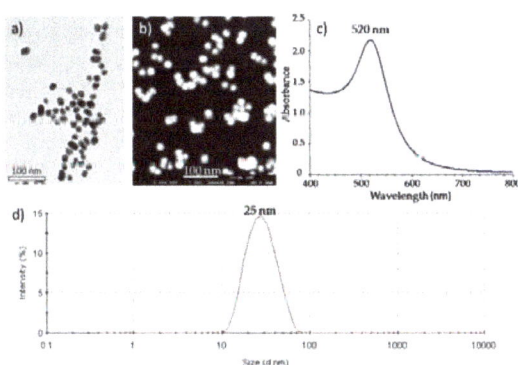

Figure 1. Example of characterizations carried out on Au Np colloidal suspensions synthesized during this work: (**a**) TEM images (scale bar of 100 nm) of one deposited drop on carbon sheet, (**b**) scan electronic image of deposited drops from a spray on SEM substrates (**c**) UV-vis spectrum recorded for this colloidal solution and (**d**) DLS characterization.

Purified arc-discharge (>90%) SWCNTs retreated through three temperature cycles under Argon atmosphere, allowing removal of residual amorphous carbon and repair of some defects, were used in our experiments. Ultra-sonic baths were used to increase the probabilities to obtain individual SWCNTs, before being drop casted on a clean mica substrate. No surfactants were used. To avoid the spread of water on the substrate, which might be an inconvenience during nano-manipulation, we had to control the relative humidity (RH), keeping it around 35% RH and at 22 °C. Finally, the Raman spectra of SWCNT sample, were recorded and compared to the Resonant Raman fingerprints of the free standing well index-identified SWCNTs.

2.2. Experimental Methods

Intermittent contact mode imaging and nanotubes manipulations are carried out with a NanoWizard®AFM (JPK Instruments, Berlin, Germany). For these experiments, a standard rectangular cantilever (PPP-NCL-W probes, Nanosensors, Neuchâtel, Switzerland is used with a free resonance frequency of 250 kHz and a typical spring constant of about 40 N/m.

Raman spectra were recorded with a triple subtractive monochromator T64000 Horiba-JobinYvon (Kyoto, Japan) spectrometer equipped with an Olympus confocal microscope with a motorized 80 nm step XY stage. The spectrometer was also equipped with a notch filter to eliminate the Rayleigh scattering with a 100 cm^{-1} cut-off. The detector was a CCD cooled by liquid nitrogen. The samples were excited with either an argon laser line at 514.53 nm or with a 561 nm solid laser or yet with a Helium-Neon laser at 633 nm. The details of the experimental setup are given elsewhere [23]. The laser beam with a controlled power was focused on samples with a spot size diameter of about 0.8 µm. The Raman backscattering was collected through the objective of the microscope (×100, numerical aperture of 0.95 or ×20 numerical aperture of 0.35) and dispersed by a 1800 grooves·mm^{-1} gratings to obtain 2.7 cm^{-1} spectral resolutions for the 514.53 nm excitation beam and around 2.5 cm^{-1} with 561 nm excitation. The wavenumber in vacuum accuracy was better than 0.8 cm^{-1}. The polarization response of the optical device was checked by measuring the depolarization ratios for the perfectly known bands of reference liquid pure products. For instance, the experimental depolarization ratio for the 459 cm^{-1} symmetric component of the CCl_4 spectrum is 0.02 to 0.005 for the different wavenumber positions of the centered CCD camera or 0.03 to 0.01 for the C–H stretching mode of CH_2Cl_2. During all experiments, the signal of a single crystal of silicon has been systematically checked. The results displayed here were obtained in such a way that the setup worked in the confocal microscope mode that was aligned with the automated XYZ table. To improve spatial and spectral resolutions, appropriate gratings according to the excitation wavelength and confocal mode were used. The focused power of the laser beam was checked for each wavelength and for each sample to avoid any transformation or heating of the samples. Here, only the spectra obtained at the limit of detection (defined here as the ratio signal on noise equal or superior to 3) will be displayed in order to discuss and compare the order of magnitude of enhancement of our results. The enhancement effects in near fields were too high to use the same focused laser power to estimate these comparisons; the usual power used to record conventional Raman spectra applied for all nanostructures damaged indeed either SWCNT or plasmonic molecules. Typically, if a power at about 1 mW was used for conventional Raman, only few micro-watt powers had to be used in the resonant nanostructures.

2.3. Numerical Modeling

For each experimentally studied structure, 3D finite element (FE) calculations were performed using COMSOL software (Multiphysics version 4.5), with Radio Frequency (RF) and Wave Optics modules adapted to the electromagnetic waves affecting objects at the nanoscale. Our calculations were beforehand confirmed by comparing numerical results with analytical results obtained in the Mie's theory field. Dimensions of each modeled Au Np were based on the corresponding AFM topography images. The environment surrounding nanoparticles has been assimilated to the air (with an optical index of 1.00015).

The absorption and scattering cross sections are defined by the rate of electromagnetic energy W that is absorbed or scattered across the surface of an imaginary sphere (centered on the particle) divided by the incident irradiance P_{inc} (W/m^2):

$$\sigma_{abs} = \frac{W_{abs}}{P_{inc}} \text{ and } \sigma_{diff} = \frac{W_{diff}}{P_{inc}} \tag{1}$$

With W_{abs} total absorbed power rate (W), obtained by the integration of the energy loss Q_{loss} (W/m^3) in the particle volume V_p:

$$W_{abs} = \iiint_{V_p} Q_{loss} \, dV \tag{2}$$

and W_{diff} the total scattered power rate [W], obtained by the integration of the Poynting vector of the scattered field P_{dif} (W/m^2) on the surface of a virtual sphere around the particle, between air domain and a Perfectly Matched Layer (PML):

$$W_{diff} = \oiint_S P_{dif} \cdot n \, dS \tag{3}$$

where n is the unit normal vector to the virtual sphere. The Q_{loss} parameter is calculated by COMSOL. Finally, the far-field variable, calculated on the internal boundary of the PML, is defined by:

$$E_{far} = \lim_{r \to \infty} \left(r \, E_{diff} \right) \tag{4}$$

The far-field variable E_{far} represents the scattering amplitude rather than the physical electric field, and is measured in units $[E_{far}] = (m \cdot V/m) = (V)$ [24].

Having only one medium and relatively simple structures, the direct resolution generates the total field, whose incident field can be subtracted when processing the results. In almost all of the calculations in this work, the amplitude of the incident field is taken at 1 V/m. The dielectric properties of gold are derived from those proposed by Etchegoin, Le Ru and Meyer [25].

2.4. Combining Raman and AFM-Manipulation

The objects studied in this work were obtained by deposing a 20 µL drop of CNTs suspension followed by the drop of a second 20 µL volume near to the first area of deposition. Thus small "reservoirs" of Au Nps were constituted not far from the area where some CNTs have been deposited. In a first step AFM images were collected to find isolated individual CNTs or small bundles of CNTs. In a second step, some nano-manipulations have been undertaken on the CNTs themselves, for example to separate two individual CNT objects. The last steps consisted in Au Nps nanomanipulations: from the Au Np reservoirs, one by one Au Np was driven in the near field of the chosen CNT in order to design the shapes or the relative spacing inter the plasmonic molecule in interaction with the selected CNT (Figure 2).

Figure 3 displays examples of the kind of structures that we built in less than one hour for each one of them, on areas of around 5 × 5 µm^2, i.e., larger than the focused laser beam used in optical measurement. Thus for this paper, via the AFM manipulation of gold nanoparticles, the plasmonic nanostructures of interest were constituted such as the visible cross or line in Figure 3, with always the same Au Np individuals. We were thus able to adapt the plasmon modes according to our wishes via the hybridization of the plasmonic modes, as mentioned in the previous part, avoiding the much used lithography process which is less precise at the nanometer scale. Numerous two-dimensional geometries are thus conceivable.

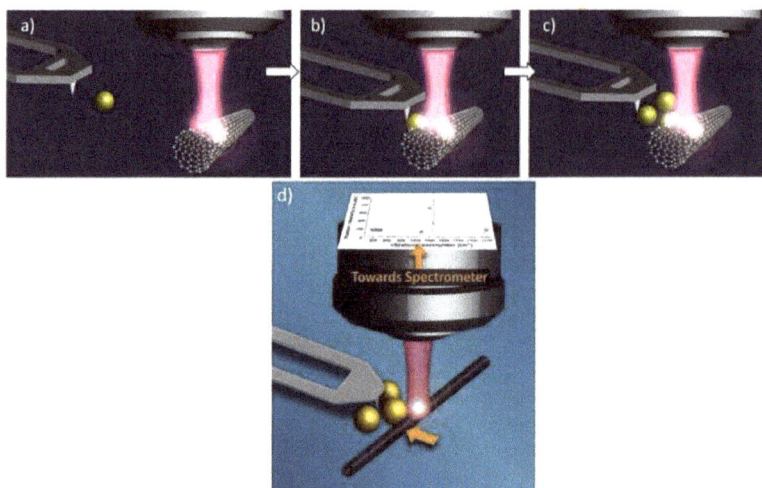

Figure 2. Scheme of our experiments of Raman-AFM Manipulation combining. (**a**) Manipulation of the first individual Au Np, coming from the reservoir; (**b**) Approach of the SWCNT by the Au Np pushed by the AFM tip; (**c**) Designing of the nanostructure (plasmonic molecule) with three plasmonic atoms and (**d**) Monitoring by Raman spectroscopy at several wavelengths of excitation.

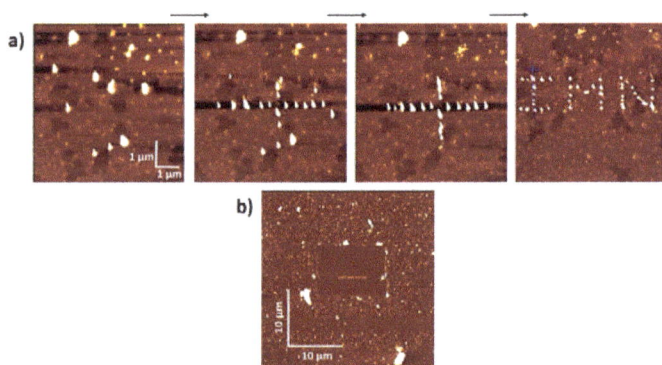

Figure 3. Examples of gold structures created by AFM manipulation; (**a**) different steps leading to a cross, then to the IMN logo; (**b**) long gold NP line of about 6 µm, the surrounding area being cleaned by the AFM-Tip in contact mode. (bar scales at top are 1 µm and down 10 µm).

Before Au Np nano-manipulations, the Raman signals at different excitation wavelengths were recorded to characterize the optical response of the selected CNT. If the wavelength corresponded to a possible electronic transition of this CNT, a spectrum was measurable (Raman Resonance), but when the wavelength did not match with a characteristic transition of the CNT individual (the general case) no spectrum was detectable. After Au Np AFM manipulations but before the optical studies of the nano-structurations, the investigated area was cleaned of other gold particles and of any dust particle surrounding the tubes by the AFM-tip scanning in contact mode. Thus the cleaned area was consistent with the spatial area probed by optical microscopy (Figure 3b). Since the same AFM-tip was used to manipulate and to image at each step the area of interest, the tip was progressively damaged resulting in sometimes AFM image artifacts. For example, Figure 3 presents the deformation of the spherical shapes of Au Nps in prismatic shapes after 500 nano-manipulations for writing "IMN". Then in this case the spatial resolution has been degraded from better than 1 nm to higher than 3 nm.

3. Results

First, we have localized by AFM some isolated CNT on a Mica substrate (Figure 3), called CNT 1 and CNT 2. According to the AFM measurements, these CNTs whose heights are less than 3 nm are almost individual CNTs (at most, small sets of two to four individual CNTs). Then gold structures such as the diamond in Figure 4 have been created around these CNTs. Raman spectroscopy analysis have then been done before and after this gold manipulation step with the same wavelengths. Raman spectroscopy of the CNT 2 and CNT 1 are presented respectively in Figures 5 and 6.

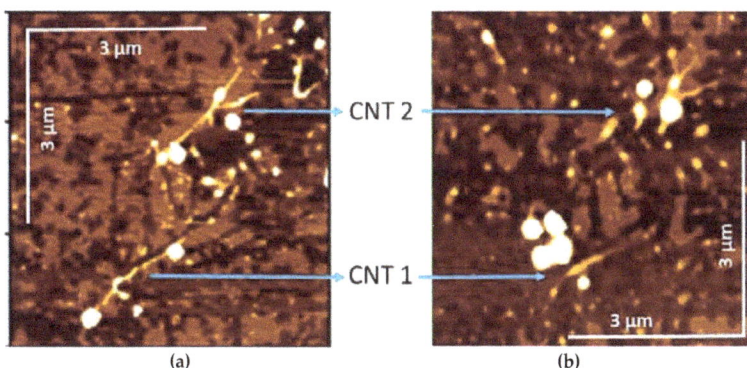

Figure 4. (**a**) Example of an AFM image of two isolated CNTs (or small bundles of SWCNTs) in presence of dispersed Au Nps in the surrounding. This area was after cleaned of dust and gold particles surrounding the SWCNTs to obtain the configuration displayed at right. (**b**) The same SWCNTs embellished with some Au Nps plasmonic molecules.

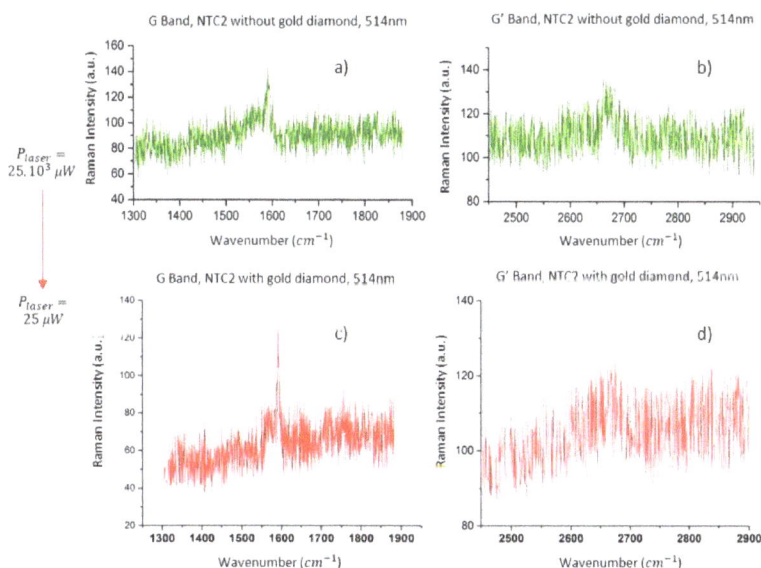

Figure 5. Spectra obtained (6 accumulations of 10 s) at the limit of detections, for the CNT 2 tube of the Figure 3; (**a**,**b**) G and G′ bands of CNT 2 without gold excited at 514.53 nm, for a significant focused power of 25 mW, on around 1 μm^2; (**c**,**d**) the Raman signals at the detection threshold of CNT 2 dressed with two gold particles on each side, recorded with an incident laser power of 25 μW, i.e., 1000 times lower than in the previous conditions.

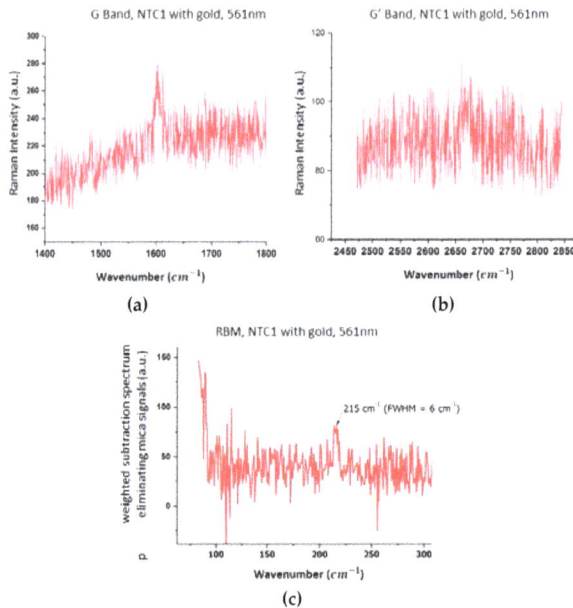

Figure 6. G (**a**) and G′ (**b**) bands of the CNT 1, excited at 561 nm for a low power of 3.3 µW; (**c**) RBM of the CNT 1 excited at the same power and wavelength, obtained after removing the mica substrate bands.

3.1. Raman Spectroscopy Results for CNT 2

Isolated and without gold, this set (CNT 2) does not resound at 633 nm and very weakly at 561 nm. With an excitation at 514.53 nm, a Raman signal is only detected, with a "high" incident power, from 25 to 50 mW: the bands G (~1591 cm^{-1}) and G′ (~2670 cm^{-1}) (green curves on the top graphs of Figure 5) were displayed (with six accumulations of 10 s). Now, in the presence of two gold particles (on each side of the CNT 2—Figure 4), the signals for G band at 1591 cm^{-1}, G′ band at 2670 cm^{-1} are detected with a power of only 25 µW, 1000 times lower than the previous one. The G band seems to be slightly more amplified and especially thinner than the G′ band (verified for higher incident powers at 50 µW). The measurement of the mid-height band indicates a width of about 6 to 7 cm^{-1}. This last result, combined with the fact the resonant Raman spectra were obtained only for one wavelength, confirmed the hypothesis of probing one individual tube that could be stated on the basis of the AFM height measurements.

3.2. Raman Spectroscopy Results for CNT 1

Red and yellow (561 nm) excitations of isolated and gold-free CNT 1 does not generate any Raman scattering. Only an excitation at 594 nm with 5 mW (six accumulations of 5 s) focused around the CNT 1, gives a Raman spectrum. By contrast, when CNT 1 is embellished with a diamond-shaped gold plasmonic molecule and a single nano-particle on its other side (Figure 3b), a Raman fingerprint spectrum appeared at 561nm and under a remarkable weak focused power of 3.3 µW (Figure 6) (6 accumulations of 10 s). In this case, all the usual modes are observed: the radial breathing mode, RBM, at 215 cm^{-1}, the fine G band at 1600 cm^{-1} with a full width at half maximum (FWHM) of only 6 cm^{-1}, and the G′ band at 2668 cm^{-1}. Even if the RBM intensity is weak here, almost hidden in noise and in many bands of mica in this spectral region, the corresponding spectrum of Figure 6a, where the RBM appeared, was obtained by weighted subtraction of the substrate signals. Thus, due to the presence of the plasmonic molecule, the enhanced Raman spectrum excited at this wavelength can become possible with an enhancement factor higher than 10^3. This estimation based only on the ratio of detection thresholds could have appeared unsatisfactory and we have tried to compare Raman intensities collected with the same focused powers recorded in conventional configuration and with plasmonic structures. In these cases, the power (some mW) used to excite the conventional Raman spectrum, damaged the plasmonic molecules, either by moving Au Nps or by changing relative spacing or yet by trapping the Au Nps on the CNTs. Therefore in this work, we will use only the thresholds to give estimations of the near field effects obtained with the plasmonic molecules. We are developing studies on these different phenomena involving the interactions CNT-Au Nps.

4. Discussion

The enhancement factor of the Raman signals excited at 514.53 nm, of CNT 2 by the Au Np dimer (Figure 5) from either side of CNT 2 was estimated between 500 and 1000 from the Raman experiments. This estimation is based on the value of the ratio of the detection thresholds, collected on the same area probed on the same optical objective, with the same spectroscopic configuration. This enhancement is then produced only by the presence of the plasmonic structure. Some experiments with an excitation wavelength outside of the resonances (of CNTs and AuNps) at 785 nm have been tried to estimate the backscattering effect of the AuNp structures presence on our Raman signals. We had not been able to record any signals without damaging our structures (the Finite Element Method (FEM) simulations gave only a backscattering power outside of the resonance multiplied by around 2). Thus, we assumed that the essential effect in these experiments comes from resonance plasmonic effects. In the CNT 2 case, the hybridization of plasmonic levels of the two nanoparticles is obtained with a gap spacing of around 10 nm, between the two particles of 22 nm, without chemical contact between CNT and Au Nps. In this configuration, the numerical Comsol FEM modeling gives an enhancement of the near electric field of about 4, at 525 nm. If we assumed that the SERS signal is as the power 4 of the electric field amplitude, then the SERS will be multiplied by 256. If we compare the dimer to the monomer, the maximum of the plasmonic resonance is slightly shifted from only 5 nm and broadened by a factor 2 as described in the literature on dimer plasmonic molecules [5]. This result is comparable in particular with those obtained by Zhu et al in 2014 [26] that measured SERS enhancement factor for gap distances of 10 nm between two 90nm disks between 2 and 3 orders of magnitude. Then in the case dimer-CNT 2 assembly, the Raman measurement may be explained only by the enhancement of local electric field created by the dimer, without other important effect involving shift of resonance or coupling between structures or yet electronic transfer. However, this enhancement is strongly dependent on the gap spacing and on the position of the Raman scatterer, because the gradient of the electric field magnitude in the gap is strong. Therefore, we had proposed other plasmonic molecules to diminish this gradient effect and tested them with CNT 1.

The CNT 1 (Figure 6) trapped in the gap between the diamond shape molecule and an isolated plasmonic atom gives us other pieces of information than the CNT 2. The enhanced Raman spectrum of CNT 1 allowed us to identify its chirality nature in coherence with the no direct observed resonance effect at 561 nm of the CNT 1 without Au Nps. Indeed, few CNTs on Kataura diagrams [27–29] correspond to the RBM wavenumber observed for the CNT 1. The "closest CNT" on the Kataura plot of our observations would be the metallic (8,8) SWCNT characterized by a RBM wavenumber at 216 cm^{-1}. It is characterized by an electronic transition of around 18,300 cm^{-1} (i.e., 545 nm), not too far of 561 nm excitation. The second kind of possible SWCNT would be the metallic (12,3) tube. This tube is characterized by its electronic transitions at around 16,500 cm^{-1} (i.e., 605 nm) and 19,900 cm^{-1} (502 nm) and whose RBM wavenumber is expected at 217 cm^{-1}. These second possibility of SWCNTs is then not characterized by an electronic transition excitable at 561 nm.

The second experimental fact concerning CNT 1, is that the G band intensity, usually not very relatively strong for metallic CNTs by comparison with RBM signal, is more enhanced, with the diamond shaped plasmonic molecule than the RBM and G' signals. The usual Raman profile is then deformed by the plasmonic resonant molecules.

The enhancement observed around 560 nm is clearly explained by the plasmonic molecule electronic structure. As displayed in the Figure 7, the computed scattering cross section of the plasmonic molecule studied here, shows a resonance at 555 nm with the incident field parallel to the gap direction. The local electric field in the gap at the resonance wavelength is multiplied by 10, i.e., the Raman scattering could be expected to be multiply by 10^4.

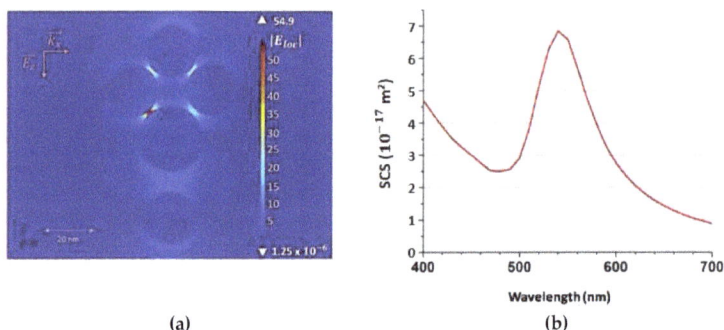

Figure 7. (**a**) Near-field mapping around the gold structure. (**b**) The plasmonic structure Scattering Cross Section (SCS).

Now to understand the change of profile of the Raman spectrum, the coupling between the CNT and this resonant plasmonic molecule must be investigated. Indeed, the CNT Raman resonances (see for example Reference [30]) may be obtained either at the wavelength corresponding at the electronic transition (Raman resonance at the excitation) or at the inelastic scattered wavelength (Raman resonance at the scattering). Raman excitation profiles display these two resonances [30] versus laser wavelength, resonances with the incident or with scattered photons.

Taking in account both these resonance effects of CNTs gives the principle scheme shown in Figure 8 applied to our case. In this scheme, the Raman spectra may be excited at (i) the wavelength corresponding to the energy allowing the electronic transition (blue lines in Figure 8) in order to induce the enhancement of the whole inelastic spectrum without strong change in the relative profile or (ii) also to a wavelength that could favor, not the excitation, but the inelastic scattering processes (orange and green lines in Figure 8). The first (i) condition could be favored in the (8,8) CNT and then the Raman signal would be dominated by the RBM signal with a weaker intensity for G, as usual for metallic CNT. However, that does not correspond to our observation. Now let us consider the second (ii) condition for (12,3) chirality. For example, to favor the G band, the wavenumber of the incident laser beam must correspond to the electronic transition wavenumber plus the vibration wavenumber characteristic of the G mode (here around 1600 cm^{-1}). Then if CNT 1 is assigned to a SWCNT (12,3), this last inelastic amplified process will be obtained at its maximum for 16,500 cm^{-1} (electronic transition) + 1600 cm^{-1} (G mode), i.e., 18,100 cm^{-1} (corresponding to 552 nm). In our experimental case, we have obtained this kind of enhancement with an incident wavelength at 561nm. If we add that the Raman fingerprint is deformed with a G band more enhanced, we assign the observed enhancement to a Raman resonance effect obtained with inelastic scattered photons induced by the plasmonic enhancement at this resonant wavelength due to the peculiar shape of the diamond plasmonic molecules. If the experiment was conducted to look into the resonance at the inelastic scattered G', then the plasmonic molecule would have to be deformed to shift its plasmonic resonance to 510 nm. The experimental Raman observation of Figure 5 is then only interpreted by the enhancement effect produced by the diamond plasmonic molecule coupled with the single plasmonic atom dressing the (12,3) CNT, where the signal G is favored by a resonance at the inelastic scattering energy.

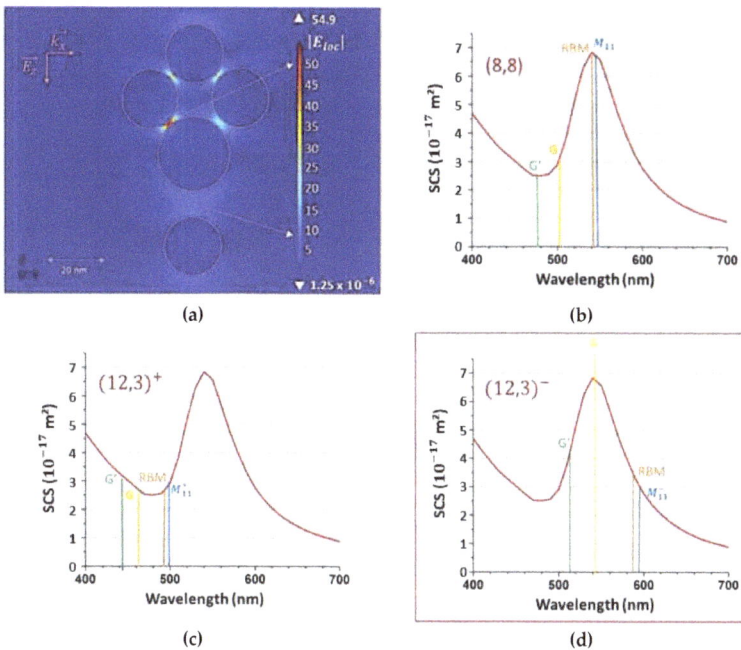

Figure 8. FEM computations coupled with metallic CNT behaviors. (**a**) the FEM simulations of the electric field show (indicated with the white arrows) two strong enhanced field area. The CNT 1 is in the region where the field is enhanced by 10. The different possibilities to generate Raman resonance effects (at the excitation, transition M 11 in blue or at the scattered resonances for the three principal Raman CNT signals). (**b**) If CNT 1 was a (8,8) SWCNT the resonance at 561 nm would enhance the excitation Raman process, in this case, all the spectrum would be enhanced. (**c**) For the first electronic transition of a (12,3) SWCNT, neither the transition at the excitation and no inelastic scatterings could be enhanced by the resonance of the plasmonic molecule. (**d**) For the second transition of a (12,3) SWCNT, the Raman resonance at the excitation is weakly enhanced, while the Raman resonance at the scattering would favor the G-signal. This latter case allows interpreting the signals observed in Figure 5.

A second interesting point with this plasmonic molecule with a spatial gap with plasmonic structures from either side of the CNT 1, comes from the quasi absence of gradient of local electric field in this gap. Figure 8 shows that the plasmonic molecule presents several hot points (where the electric field multiplied by 50) localized inside the structure and not in the gap between the molecule and the atom. In this spatial spacing, the local electric field at the resonance is only multiplied by 10 but at all positions in this gap. This kind of behavior is encountered when two plasmonic molecules are separated by a small distance. The Figure 9 gives another example of one possible assembly and compares it with a gold tip usable in a TERS experiment. The exponential decrease of the amplitude of the local electric field observed for TERS configuration is then replaced by a quasi-constant electric field amplitude for a gap of 5 nm. The value of the electric amplitude obtained in this gap is equal to this one obtained at 2nm of the apex of the tip in TERS mode. Now, when the distance is increased to 10 nm the amplitude decreases and mostly a gradient of the amplitudes reappears. Of course, when there is not important gradient effect, the position of the CNT trapped in the gap does not have consequences on the measurements. This point could be important in the conception of nano-Raman sensors based on plasmonic molecules.

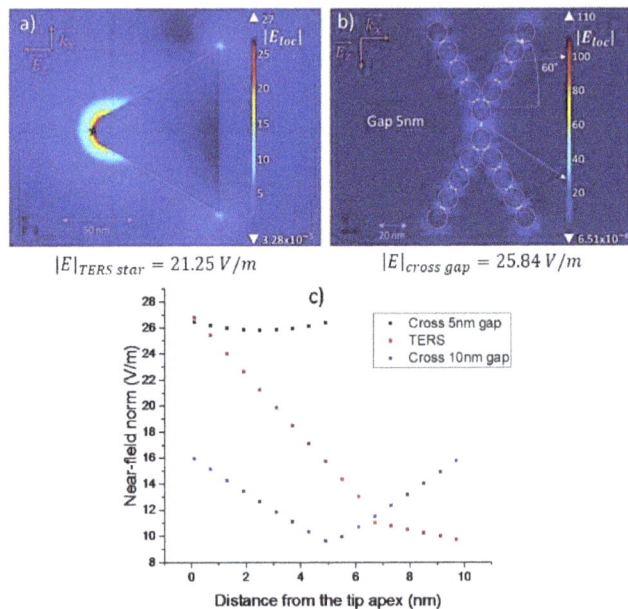

Figure 9. Computed near-field mappings (**a**) of a TERS tip; (**b**) a tip consisting of 9 NPs of 20 nm in diameter. The angles with respect to the normal are identical for the two structures, i.e., 60°. The respective values of the near-field norm were taken at the star point for the TERS tip, and at the white mark for the gold cross gap. (**c**) Amplitudes of the local electric field vs either the distance at the apex of the tip (TERS in red points) or the positions in the gap when the gap is only 5 nm (black) or 10 nm (blue points).

In our experiments, we have nano-manipulated 20nm-diameter plasmonic atoms, consequently the range of wavelength accessible by tuning of the shape and inter-distances of our plasmonic nano-structures is limited between 510 and 570 nm. To increase this range (to 700 nm), we suggest manipulating 40 or 50 nm diameter gold particles, always synthesized by a Turkevich's way by changing synthesis conditions [21,22]. We are developing different nano-manipulations coupled with pre-micro manipulations to generate different natures of plasmonic molecules connected more or less with each other's, in order to use different reserves of Au Nps around the area of interest. Thus, we hope to expand the possibilities of designing plasmonic molecules and without chemical direct bound between molecule and Au surfaces to be generate Raman enhancement. Finally, the possibility to design shapes and symmetry of plasmonic molecules opens the possibilities to control the physical coupling between them and other nano-objects, such as here CNTs to generate new nano-emitters based on the Raman effect.

5. Conclusions

This paper demonstrates the feasibility to build and to control nanostructures of Au Nps. These structures can be regarded as plasmonic molecules whose optical resonances are tuned by modifying the shape, the symmetry and the inter-particle (inter plasmonic atom) distances with the help of an AFM device coupled with an optical spectrometer. Originally, we have worked with a Raman spectrometer equipped with many excitation wavelengths in order to monitor enhanced Raman fingerprint signals of single nano-objects. For each plasmonic molecules, our approach allowed us to prospect actively their responses by changing interparticle distances and then to test several configurations with the same Au particles as well as to interpret without ambiguity the recorded changes. Indeed, when you compare plasmonic molecules constituted by different plasmonic atoms, a part of the responses can come from changes of the atoms themselves (sphericity, more or less oblong, not constant diameter, angular shape, ...) whereas this work avoids this difficulty. Let us note that our approach is applicable for nanorods too (gold or silver). For gold, this work shows that, instead of using diameters around 20 nm, an average diameter around 40 nm of plasmonic atoms will allow to tune the resonances by deforming shape of plasmonic molecule on a larger spectral range. Furthermore, for Raman applications (nano-sensor) the existence of large spacing between two plasmonic molecules allows to have a such enhanced signal as in a Tip Enhanced experiment but with a weak gradient. In this latter case, the distance between the object to be probed and the plasmonic molecules is less crucial and easier to manage than SERS and TERS experiments.

Author Contributions: Conceptualization, B.H. and A.D.; Methodology, B.H.; Validation, A.D., M.B., G.L. and B.H.; Formal Analysis, G.L., A.D.; Investigation, A.D. and B.H.; Data Curation, A.D.; Writing-Original Draft Preparation, A.D. and B.H.; Writing-Review & Editing, A.D., M.B., G.L. and B.H.; Supervision, G.L. and B.H.; Project Administration, B.H. **Angélina D'Orlando** was the PhD student to whom the project was to be assigned. She headed the experimental works and modeling computations. She co-wrote in depth the paper after having prepared the original draft.**Maxime Bayle** expert in SERS has contributed substantially to this project and co-wrote this paper.**Guy Louarn** is our expert in plasmonic physics and is an expert in its modeling. He supervised the different steps of modeling.**Bernard Humbert** is at the origin of this study, in particular to use the nano-manipulation coupled with the optical spectroscopy. He supervised the all of the work. He is an expert in the near field and developed one of the first setups of Raman Near Field device (JRS 99). He co-wrote in depth this paper.

Funding: This research received no external funding.

Acknowledgments: The authors of this paper acknowledge Jean Yves Mevellec, physicist and engineer for his experimental help in Raman spectroscopy and for discussion about Raman assignments of spectra of CNTs.

Conflicts of Interest: The authors declare no conflict of interest.

References

1. Willets, K.A.; Van Duyne, R.P. Localized Surface Plasmon Resonance Spectroscoy and Sensing. *Annu. Rev. Phys. Chem.* **2007**, *58*, 267–297. [CrossRef]

2. Jain, P.K.; Huang, X.; El-Sayed, I.H.; El-Sayed, M.A. Review of Some Interesting Surface Plasmon Resonance-Enhanced Properties of Noble Metal Nanoparticles and Their Applications to Biosystems. *Plasmonics* **2007**, *2*, 107–118. [CrossRef]

3. Schuller, J.A.; Barnard, E.S.; Cai, W.; Jun, Y.C.; White, J.S.; Brongersma, M.L. Plasmonics for Extreme Light Concentration andManipulation. *Nat. Mater.* **2010**, *9*, 193–204. [CrossRef]

4. Halas, N.J.; Lal, S.; Chang, W.-S.; Link, S.; Nordlander, P. Plasmons in Strongly Coupled Metallic Nanostructures. *Chem. Rev.* **2011**, *111*, 3913–3961. [CrossRef]

5. Haran, G.; Chuntonov, L. Artificial Plasmonic Molecules and Their Interaction with Real Molecules. *Chem. Rev.* **2018**, *118*, 5539–5580. [CrossRef] [PubMed]

6. Bouvrée, A.; D'Orlando, A.; Makiabadi, T.; Martin, S.; Louarn, G.; Mevellec, J.Y.; Humbert, B. Nanostructured and nanopatterned gold surfaces: Application to the surface-enhanced Raman spectroscopy. *Gold Bull.* **2013**, *46*, 283–290. [CrossRef]

7. Punj, D.; Regmi, R.; Devilez, A.; Plauchu, R.; Moparthi, S.; Stout, B.; Bonod, N.; Rigneault, H.; Wenger, J. Self-Assembled Nanoparticle Dimer Antennas for Plasmonic-Enhanced Single-Molecule Fluorescence Detection at Micromolar Concentrations. *ACS Photonics* **2015**, *2*, 1099–1107. [CrossRef]

8. Flauraud, V.; Mastrangeli, M.; Bernasconi, G.D.; Butet, J.; Alexander, D.T.L.; Shahrabi, E.; Martin, O.J.F.; Brugger, J. Nanoscale topographical control of capillary assembly of nanoparticles. *Nat. Nanotechnol.* **2017**, *12*, 73–80. [CrossRef] [PubMed]

9. Koenderink, A.F. Single Photon Antennas. *ACS Photonics* **2017**, *4*, 710–722. [CrossRef]

10. Kumar, S.; Johnson, W.; Wood, C.K.; Qu, T.; Wittenberg, N.J.; Otto, L.M.; Shaver, J.; Long, N.J.; Victora, R.H.; Edel, J.B.; Oh, S.H. Template-Stripped Multifunctional Wedge and Pyramid Arrays for Magnetic Nanofocusing and Optical Sensing. *ACS Appl. Mater. Interfaces* **2016**, *8*, 9319–9326. [CrossRef]

11. Sow, I.; Grand, J.; Lévi, G.; Aubard, J.; Félidj, N.; Tinguely, J.C.; Hohenau, A.; Krenn, J.R. Revisiting Surface-Enhanced Raman Scattering on Realistic Lithographic Gold Nanostripes. *J. Phys. Chem. C* **2013**, *117*, 25650–25658. [CrossRef]

12. Chuntonov, L.; Haran, G. Trimeric Plasmonic Molecules: The Role of Symmetry. *Nano Lett.* **2011**, *11*, 2440–2445. [CrossRef] [PubMed]

13. Chuntonov, L.; Haran, G. Effect of Symmetry Breaking on the Mode Structure of Trimeric Plasmonic Molecules. *J. Phys. Chem. C.* **2011**, *115*, 19488–19495. [CrossRef]

14. Prodan, E.; Radloff, C.; Halas, N.J.; Nordlander, P. A Hybridization Model for the Plasmon Response of Complex Nanostructures. *Science* **2003**, *302*, 419–422. [CrossRef] [PubMed]

15. Nordlander, P.; Oubre, C.; Prodan, E.; Li, K.; Stockman, M.I. Plasmon Hybridization in Nanoparticle Dimers. *Nano Lett.* **2004**, *4*, 899–903. [CrossRef]

16. Prodan, E.; Nordlander, P. Plasmon Hybridization in Spherical Nanoparticles. *J. Chem. Phys.* **2004**, *120*, 5444–5454. [CrossRef]

17. D'Orlando, A. Nano-structuration de nanoparticules métalliques pour exaltation de champs électromagnétiques locaux en spectroscopie Raman. Ph.D. Thesis, University of Nantes, Nantes, France, 11 December 2015.

18. Tong, L.; Li, Z.; Zhu, T.; Xu, H.; Liu, Z. Single Gold-Nanoparticle-Enhanced Raman Scattering of Individual Single-Walled Carbon Nanotubes via Atomic Force Microscope Manipulation. *J. Phys. Chem. C* **2008**, *112*, 7119–7123. [CrossRef]

19. Heeg, S.; Oikonomou, A.; Fernandez-Garcia, R.; Lehmann, C.; Maier, S.A.; Vijayaraghavan, A.; Reich, S. Plasmon-Enhanced Raman Scattering by Carbon Nanotubes Optically Coupled with Near-Field Cavities. *Nano Lett.* **2014**, *14*, 1762–1768. [CrossRef]

20. Mueller, N.S.; Heeg, S.; Kusch, P.; Gaufres, E.; Tang, N.Y.-W.; Hubner, U.; Martel, R.; Vijayaraghavan, A.; Reich, S. Plasmonic enhancement of SERS measured on molecules in carbon nanotubes. *Faraday Discuss.* **2017**, *205*, 85–103. [CrossRef] [PubMed]

21. Shi, L.; Buhler, E.; Boué, F.; Carn, F. How does the size of gold nanoparticles depend on citrate to gold ratio in Turkevich synthesis? Final answer to a debated question. *JCIS* **2017**, *492*, 191–198. [CrossRef]

22. Turkevich, J.; Stevenson, P.C.; Hillier, J. A study of the nucleation and growth processes in the synthesis of colloidal gold. *Discuss. Faraday Soc.* **1951**, *11*, 55–75. [CrossRef]

23. Grausem, J.; Humbert, B.; Spajer, M.; Courjon, D.; Burneau, A.; Oswalt, J. Near field Raman. *J. Raman. Spectrosc.* **1999**, *30*, 833–840. [CrossRef]

24. Yushanov, S.; Crompton, J.S.; Koppenhoefer, K.C. MieSolution-Paper-Crompton_paper.pdf. Available online: https://cn.comsol.com/paper/download/181101/crompton_paper.pdf (accessed on 1 November 2018).

25. Etchegoin, P.G.; Le Ru, E.C.; Meyer, M. An analytic model for the optical properties of gold. *J. Chem. Phys.* **2006**, *125*, 164705. [CrossRef] [PubMed]

26. Zhu, W.; Crozier, K.B. Quantum Mechanical Limit to Plasmonic Enhancement as Observed by Surface-Enhanced Raman Scattering. *Nat. Commun.* **2014**, *5*, 5228. [CrossRef]

27. Kataura, H.; Kumazawa, Y.; Maniwa, Y.; Umezub, I.; Susuki, S.; Ohtsuka, Y.; Achiba, Y. Optical Properties of Single-Wall Carbon Nanotubes. *Synth. Met.* **1999**, *103*, 2555–2558. [CrossRef]

28. Liu, K.; Deslippe, J.; Xiao, F.; Capaz, R.B.; Hong, X.; Aloni, S.; Zettl, A.; Wang, W.; Bai, X.; Louie, S.G.; Wang, E.; Wang, F. An atlas of carbon nanotube optical transitions. *Nat. Nanotechnol.* **2012**, *7*, 325–329. [CrossRef]

29. Zhang, D.; Yang, J.; Yang, F.; Li, R.; Li, M.; Ji, D.; Li, Y. (n,m) Assignments and quantification for single walled carbon nanotubes on SiO_2/Si substrates by resonant Raman spectroscopy. *Nanoscale* **2015**, *7*, 10719–10727. [CrossRef] [PubMed]

30. Moura, L.G.; Moutinho, M.V.O.; Venezuela, P.; Fantini, C.; Righi, A.; Strano, M.S.; Pimenta, M.C. Raman excitation profile of the G band in single-chirality carbon nanotubes. *Phys. Rev. B* **2014**, *89*, 035402. [CrossRef]

materials

MDPI

Review

Plasmonics for Biosensing

Xue Han * , **Kun Liu and Changsen Sun**

School of Optoelectronic Engineering and Instrumentation Science, Dalian University of Technology,
Dalian 116024, China; liukun@dlut.edu.cn (K.L.); suncs@dlut.edu.cn (C.S.)
* Correspondence: xue_han@dlut.edu.cn; Tel.: +86-0411-84706240

Received: 30 March 2019; Accepted: 24 April 2019; Published: 30 April 2019

Abstract: Techniques based on plasmonic resonance can provide label-free, signal enhanced, and real-time sensing means for bioparticles and bioprocesses at the molecular level. With the development in nanofabrication and material science, plasmonics based on synthesized nanoparticles and manufactured nano-patterns in thin films have been prosperously explored. In this short review, resonance modes, materials, and hybrid functions by simultaneously using electrical conductivity for plasmonic biosensing techniques are exclusively reviewed for designs containing nanovoids in thin films. This type of plasmonic biosensors provide prominent potential to achieve integrated lab-on-a-chip which is capable of transporting and detecting minute of multiple bio-analytes with extremely high sensitivity, selectivity, multi-channel and dynamic monitoring for the next generation of point-of-care devices.

Keywords: plasmonics; resonance modes; biosensing; plasmonic materials; hybrid function; multi-channel sensing; spectroelectrochemistry

1. Introduction

Along with the fast development in nanofabrication techniques, nano-optics has been boosted and applied in various research and industry areas. Plasmonics is one major branch of nano-optics. Due to its ability to generate nanoscale hot spots which are close to the size of bioparticles, it has been applied in biosensing broadly with enhanced sensitivity for refractive index (RI) changes and enhanced light/matter interactions [1–5]. For example, prostate-specific antigen has been detected with a visual limit of detection (LOD) as low as 0.0093 ng/mL [6]; with the mechanism that glucose oxidase can control the growth of nanoparticles, an inverse sensitivity has been achieved as low as 4×10^{-20} M [7]; nanohole array has been used to detect exosomes with the LOD as approximately 670×10^{-18} M and potential cancer diagnose without biopsy is possible based on this technique [8], and other applications in bio-interfacial research [9], heavy ions in water [10], foodborne pathogen [11], and drug delivery [12] have been carried out. Some of the plasmonic devices have already been developed into portable manner toward point-of-care (PoC) applications [13–15].

Here in this short review, plasmonic biosensing techniques based on nanovoid-type designs are discussed. In Section 2, propagating plasmonic resonance mode based on planar design is mentioned briefly for the introduction on sensing methodologies based on RI changes. For various nanovoid (array) designs, extraordinary transmission effect (EOT), Fabry-Perot (FP) like resonance, and Fano resonance can be excited and used for specific biosensing purposes. In Section 3, plasmonic materials used for biosensing techniques are discussed according to their working frequency range. In Section 4, hybrid sensing techniques by applying the electric conductive function to plasmonic devices are discussed. Plasmonics based on nanoparticles is as important as nanovoid-type plasmonic sensors. These nanoparticles have been used to enhance fluorescence [16–18] and Raman spectroscopy [19–32], and to generate heat to introduce convection flows for particle transportation [33]. Currently, hot electron is another hot topic focused on these nanoparticles for plasmonic photocatalytic studies [34,35]. There are comprehensive reviews of plasmonic nanoparticles can be used as reference resources [36–45].

2. Surface Plasmon Resonance

Surface plasmon resonance (SPR) phenomenon happens at the interface of a conductive material and a dielectric medium. Under the resonance condition, the incident light is used to generate a collective charge density wave propagating at this interface and this wave is called surface plasmon polariton (SPP). This phenomenon was first observed by Wood in 1902 [46], and in 1941 Fano proposed a theoretical explanation for Wood's anomaly by considering a SP as a superficial wave [47]. In 1968 Kretschmann [48] and Otto [49] proposed using a prism coupler to generate SPR respectively, and the first SPR sensor was invented by Liedberg, et al. for gas sensing and antibody-antigen binding events until 1983 [50]. Initially, SPR sensors were mainly based on metal thin films. With the development in nanofabrication techniques, nanostructures have been fabricated in metal thin films and geometric factors have been introduced to control generation conditions of SPPs. In addition, different plasmonic resonance modes and modes coupling have been proposed theoretically and explored experimentally to enhance the nearfield to reduce LOD [51,52]. In this section, film-type sensor is discussed on the purpose to introduce the generation of SPR, plasmonic biosensing mechanisms and interrogation methodologies. Localized SPR (LSPR), EOT effect, FP-like and Fano resonance modes are discussed for nanovoid-type plasmonic sensors.

2.1. Film-Type Plasmonic Sensors

2.1.1. SPP Generation

Optical thin noble metal films are most used in film-type plasmonic sensors. When Equation (1) is satisfied, a SPP can be generated at the metal/dielectric interface and this type of SPP is also called propagating SPP [50], which is demonstrated as Figure 1.

$$k_{spp,d}(\omega) = k_d \left(\frac{1}{\varepsilon_m(\omega)} + \frac{1}{\varepsilon_d(\omega)} \right)^{-1/2} \tag{1}$$

where k_d is the wave vector of the incident light in a dielectric medium, ε_m and ε_d is the dielectric constant of the metal material and the dielectric medium respectively, and $k_{spp,d}$ is the wave vector of the SPP which propagates at the interface of this metal and this dielectric medium.

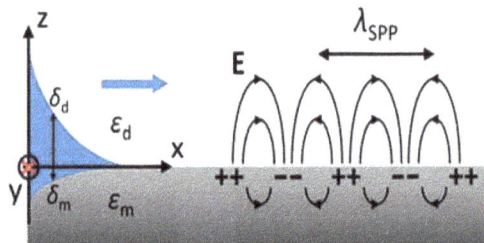

Figure 1. Demonstration of the layer geometry for propagating SPP generation. Adapted from [52], with permission from 2015 © The Royal Society of Chemistry.

The demonstrative dispersion curve of SPP ($k_{spp} = k_{spp,0}$) for a metal/free space (dielectric medium, k_0) interface is plotted as Figure 2a. At the same frequency, k_{spp} is always larger than k_0. From the free space, simply shining a laser beam at this interface cannot excite a SPP. However, a prism made of glass ($k_p > k_0$) can be used to match k_{spp} with proper coupling methodologies. In Kretschmann (Figure 2b) and Otto (Figure 2c) coupling methodologies, an evanescent wave from the total internal reflection (TIR) at the metal/prism interface is used to generate SPP at the metal/free space interface. In addition, the evanescent wave of the SPP penetrates into both the metal layer and the dielectric medium with a defined depth δ_m and δ_d, as shown in Figure 1. Take a gold/air interface, penetration depths are 328 nm and 28 nm into air and gold separately at 633 nm incident light [53]. These numbers demonstrate that SPP is highly confined at the interface. Gratings also can be used to match the wave vector of a SPP, as shown in Figure 2d. Later coupling techniques based on fibers/waveguides have also been developed [54].

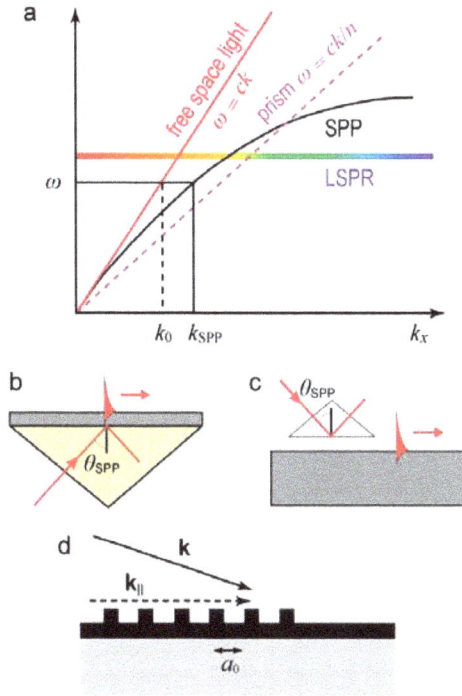

Figure 2. (**a**) Dispersion curves for SPP along metal/free space interface (black curve), light in free space (red curve) and prism material (pink dashed curve); (**b**) Kretschmann; (**c**) Otto and (**d**) grating coupling method. Adapted from [55], with permission © 2017 American Chemical Society.

As shown in Equation (1), the dispersion of SPP depends on $\varepsilon_d(\omega)$. When a RI change happens to the dielectric medium that close to the metal surface, which could be a bulk RI change or thin layers added on the surface of the metal, $k_{spp,d}(\omega)$ is modified to $k_{spp,d'}(\omega)$ and the resonance condition/coupling condition is altered. Varied interrogation methodologies can be used to monitor the resonance condition and extract the RI changes of $\varepsilon_d(\omega)$.

2.1.2. Angular Interrogation

For an angular interrogation based on a prism coupler, the incident angle should satisfy Equation (2) to generate a SPP and a minimal reflection can be observed at this coupling angle θ_r which is larger than the critical angle of TIR. When there is a RI change happens next to the metal thin film, the coupling angle θ_r is modified to θ'_r. As shown in Figure 3a, an angular interrogation method is used for bulk RI change from 1.32 to 1.37 at 850 nm and 630 nm incident light.

$$k_p \sin(\theta_r) = k_{spp,d}(\omega) = k_d \left(\frac{1}{\varepsilon_m(\omega)} + \frac{1}{\varepsilon_d(\omega)} \right)^{-1/2} \tag{2}$$

In an angular interrogation setup, there are two strategies. One is that the incident laser beam is collimated, and a rotation stage is involved to change the incident angle. The incident beam needs to be guaranteed shining on the same spot during a rotation. In this scenario the mechanical vibration is the major factor for the limited sensitivity. The other strategy is that the incident light has a range of angles, and the portion corresponding to the coupling angle is used to generate a SPP and a dark bar is resulted in the reflected angular spectrum. For this scenario, a camera is needed to align incident angles with pixels and the mechanical vibration is deleted from the background noise.

2.1.3. Wavelength Interrogation

In wavelength interrogation, a range of incident wavelengths is used and the incident angle is fixed. At the coupling wavelength, a minimal reflection is observed. As shown in Figure 3b, interrogations using the incident angle at 51.65 degree and 56.12 degree are demonstrated separately to show the shift of the coupling wavelength when the bulk RI changed from 1.320 to 1.325. Compared to an angular interrogation, a wavelength interrogation is more complex in data analysis since a range of refractive indices of the control system (the coupling component, metal thin film and dielectric medium without analyte) is needed. Since no continuously stage rotation is needed in the progress of the measurement, a wavelength interrogation is mechanical stable for sensing applications. Although the mentioned difficulty, spectral SPR sensors has been used to determine the RI dispersion of a variety of biomaterials [56].

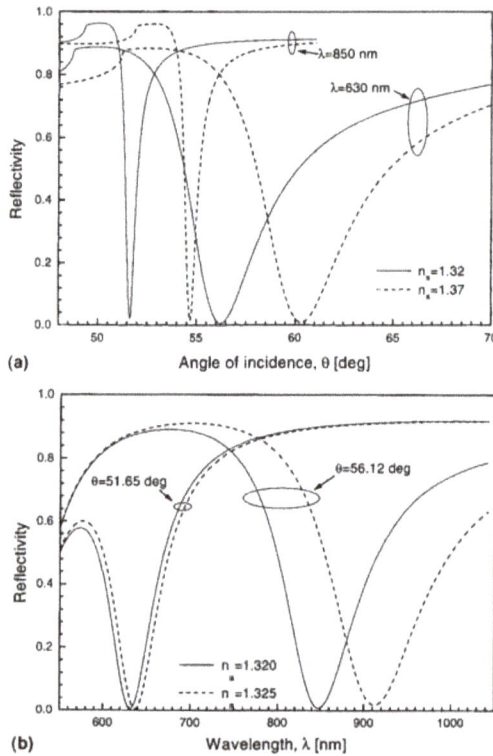

Figure 3. Reflectivity for a system using a prism coupler. (**a**) Angular interrogation. (**b**) Wavelength interrogation. Adapted from [57], with permission © 1999 Elsevier Science S.A.

The main problem of angular and wavelength interrogation methods is associated with the natural low detection limit of amplitude sensing schemes. This limit is conditioned by the level of noises in measurements and LOD normally is estimated as 10^{-5} to 10^{-6} RI units (RIU) [58].

2.1.4. Phase Interrogation

To increase the detection sensitivity for biomolecules and their interactions at one spot, a phase interrogation was also proposed at the very early stage. In 1976, F. Abeles clearly proposed to use phase interrogation for SPR sensing [59]. To excite SPPs, only p-polarized (TM) incident light can be used for film-type plasmonics, and a dramatic phase shift occurs at the resonance. Interferometry can be used to generate interference pattern if a proper reference beam is selected. However, this methodology only works for film-type SPR when biomolecules are uniformly immobilized on the device surface in a large area which is not practical for biosensing. Another scheme to use this phase-shift character is polarimetry based on an ellipsometer. An incident beam containing both s- (TE) and p-polarized components is used, and the ratio between the reflectance of p- and s-polarization is shown as in Equation (3),

$$\frac{r_p}{r_s} = \tan \Psi e^{i\Delta} = \frac{|r_p|}{|r_s|}e^{i\Delta}, \; \Delta = \delta_p - \delta_s \tag{3}$$

where r_p and r_s are the complex reflectance for p- and s-polarized components, and Δ is the phase difference (shift) between p- and s-polarized components. When measure $\tan \Psi = \frac{|r_p|}{|r_s|}$ against incident angle or wavelength, a minimal value can be obtained at the resonance condition which is the same as an angular or a wavelength interrogation mentioned previously. For phase interrogation, $e^{i\Delta}$ is retracted from the ellipsometry measurement. Compared to angular or wavelength interrogation, signal/noise ratio can be efficiently increased and improved phase treatments can be applied to obtain the phase shift information. One drawback would be a rotation analyzer or other expensive phase modulation component is needed for the measurement.

In Figure 4a, the reflected electric vector $\overline{E}_r$ at the initial resonance condition and the one $\overline{E}_r'$ after a Δn happened next to the metal thin film for an altered resonance condition are presented. Figure 4b compares the reflection intensity and the phase shift against incident angle. A much steeper change around the resonance angle can be observed for the phase curve. This work demonstrates that the LOT of a phase interrogation on the RI change could reach 10^{-8} and even 10^{-9} RIU [60]. Later, novel phase interrogation methodologies have been proposed to further increase the detection sensitivity [61,62] and have been applied in prominent biosensing devices [63,64].

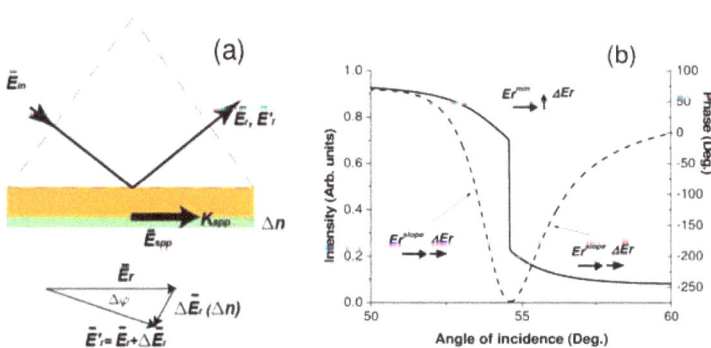

Figure 4. (a) SPP generation presented in electric vectors and (b) Reflection spectrum of intensity/angle (dashed line) and phase/angle (solid line). Adapted from [60], with permission from © 2009 Optical Society of America.

To increase the sensitivity for a RI change introduced by tiny amounts of biomolecules attached on the surface of a SPR sensor, multi-layer designs have also been proposed for propagation SPP to make the resonance curve narrower and deeper [65]. A major design is to add another dielectric waveguide layer on top of the metal thin film and plasmon waveguide resonance (PWR) can be generated. This approach was first proposed by Salamon et al. in 1997 [66]. With both p- and s-polarized incident light, properties of attached biomolecules can be characterized [67,68]. The exceptional narrow line widths of PWR spectra yield enhanced sensitivities which is approximately 20 times better when compared to the conventional SPR sensors on bulk RI changes [69]. Besides this major improvement in sensitivity, the dielectric waveguide coating also provides protection for the metal layer and brings an extra surface immobilization function for various molecules [70]. Interrogation methods mentioned above is based on the RI change introduced by the surface immobilized molecules, hence they are not selective. The selectivity can be introduced by coating a thin layer of specific binding molecules. Along with the interest to improve the sensitivity, introducing selectivity by surface modifications for specific biomolecular interactions and binding events is also under focus for film-type SPR sensors [71].

Film-type SPR sensors have disadvantages in using complex and bulky coupling systems and limited sensitivity due to the poor mechanical and thermal stability. Another major limitation is the narrow range of tunability in the resonance frequency. Determined mainly by the dispersion characteristics of the metal layer and the dielectric bulk medium, there exists a thickness of the metal layer that provides the highest sensitivity. To guarantee a high sensitivity, the resonance frequency only can be tuned in a really narrow range. Practically, a metal thin film with an exact designed thickness cannot be fabricated is the reason that the experimentally determined sensitivity of a film-type SPR sensor is not as good as the theoretical expectation. Based on this extremely narrow tunable range of resonance frequencies, light-matter interactions cannot be used to include intrinsic selectivity naturally into the sensing mechanisms.

2.2. Nanovoid-Type Plasmonics

By fabricating nanostructures in conductive thin films, nanoscale geometric factors and/or lattice factors have been introduced to SPPs generation conditions [4,72]. The major advantages of these nanostructures are: (1) they can generate locally enhanced hot spots confined in 3D with a similar size as biological particles which is possible for single biomolecule studies; (2) they has the capability to tune the resonance frequency for a specific light-matter interaction, e.g., Raman spectroscopy, infrared vibrational spectroscopy, or fluorescence; (3) besides various types of resonance modes can be used, modes coupling also can be manipulated to generate much narrower peaks/dips in spectra which can be applied in the detection of minute RI change for a further increased sensitivity [73]; (4) an incident laser beam can be used directly to excite plasmonic resonance without using an external coupling element which reduces the complexity of the entire system. Among versatile designs of nanovoid structures, nanohole [74,75], double nanohole [76] and bowtie aperture structures [77,78] are a few of the most theoretically and experimentally examined ones.

2.2.1. Localized Surface Plasmon Resonance

LSPR mode is mainly used for plasmonics based on nanoparticles. Compared to protruding isolated nanoparticles which generate heat accumulation, nanovoid designs can avoid this issue since the heat dissipates through the rest large area of conductive films. This is beneficial for some biomolecule studies which need to avoid heating effects. For LSPR, the analytical result of the electrical dipole moment for nanoparticles can be used for the nanovoid by exchanging positions of the dielectric constant of the metal nanoparticle and the one of the surrounding medium [53].

Various geometries have been applied to generate LSPR. Single circular nanoholes in optical thin gold films exhibit a distinct tunability in LSPR frequency as the size of the hole changes [79] or as the thickness of the metal film changes [80]. LSPR based on single elongated nanoholes have also been examined and compared to nanodisks [81].

To enhance the light-matter interaction, a double nanohole plasmonic design has been used to enhance fluorescence intensity by Regmi et al. [82]. As shown in Figure 5a, LSPR generates nanoscale hot spots at the tips of the double nanohole. With an incident laser beam polarized parallel to the apex region, the locally enhanced nearfield can enhance the fluorescence. Dye molecules near these tips have enhanced fluorescence, and molecules in the double nanohole void are also excited by the incident light without any enhancement, but molecules on top of the gold surface cannot be excited. As shown in Figure 5b, the fluorescence correlation spectroscopy measurements for polarization parallel (red color) and perpendicular to the apex region (blue color), and the fluorescence measurement using a confocal microscope (green color) are plotted to demonstrate the enhancement of fluorescence by using a double nanohole at LSPR. In addition, in Figure 5c, the brightness of fluorescence for these three cases are compared at different incident powers. For clearness, the count was doubled for the perpendicular polarization case and was multiplied by 10 for the confocal measurement. With a double nanohole, the enhancement of fluorescence can be as high as 100-fold.

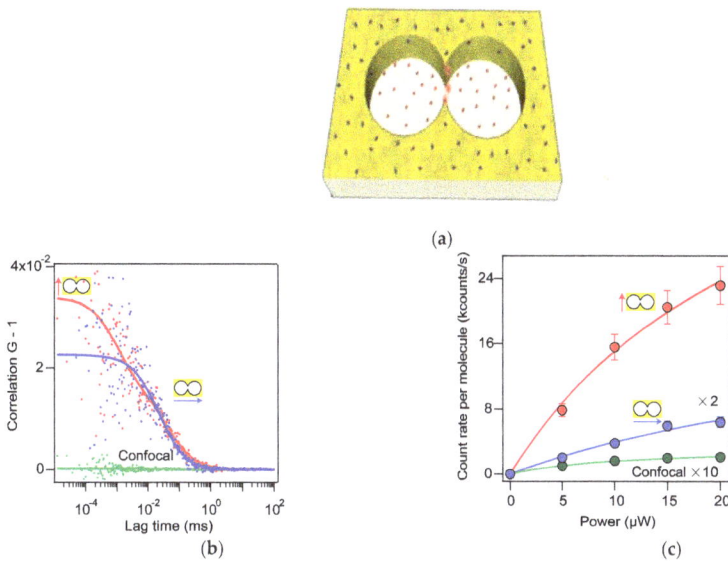

Figure 5. (a) Sketch of a double nanohole under an incident light polarized along with the two tips. (b) Fluorescence correlation spectroscopies for with double nanohole using incident light polarized parallel (red color) and perpendicular (blue color) to the apex region, and with a confocal microscope (green color). Dotted data is the measurement result and the solid curve is the corresponding fitting result for each case. (c) Comparison in brightness of fluorescence among these three cases at different incident power. Adapted from [82], with permission © 2015 Springer Nature Publishing A.G.

2.2.2. Magnetic Dipole

As shown in the previous section, an electric dipole can be generated in nanovoids which is a symmetric plasmon mode. When two nanoparticles or nanovoids brought close enough, plasmons hybridization occurs to generate two resonances: one is symmetric that the electric dipolar moments are parallel, and the other one is antisymmetric that the electric dipolar moments are antiparallel. In the second mode, a loop-like current is generated and it can be treated as a magnetic dipolar moment. Currently, most of the research on magnetic dipole is mainly on standing nanostructures [83–85], and for modes coupling to generate Fano resonance to have narrow spectra features for RI sensing [86].

2.2.3. Toroidal Dipole

Most biomolecules are chiral, and chirality can be crucially important, e.g., for a drug molecule. Plasmonic planar chiral metamaterials has been researched to generate chiral electromagnetic (EM) fields to probe chiral molecular structure [87]. As shown in Figure 6a, the reflection spectra from the back-face illumination, the front-face illumination, and its optical rotatory dispersion (ORD) are plotted for the left-handed and right-handed gammadions respectively. Demonstration of both illumination schemes and the sketch of the plasmonic design are also shown. In Figure 6b,c, the measured reflectivity using left-handed and right-handed structures are plotted respectively for 3 types of proteins. These proteins do not show clearly difference from buffer solution when using left-handed structure, while clearly shifts are observed with right-handed structure for natural BSA. Comparing between the natural and denatured BSA protein, it is clearly that this design is sensitive to the second structural information of proteins. The measured differences in the effective refractive indices of chiral samples exposed to left- and right-handed chiral fields generated by these plasmonic nanostructures are found to be up to 10^6 times greater than those observed in optical polarimetry measurements.

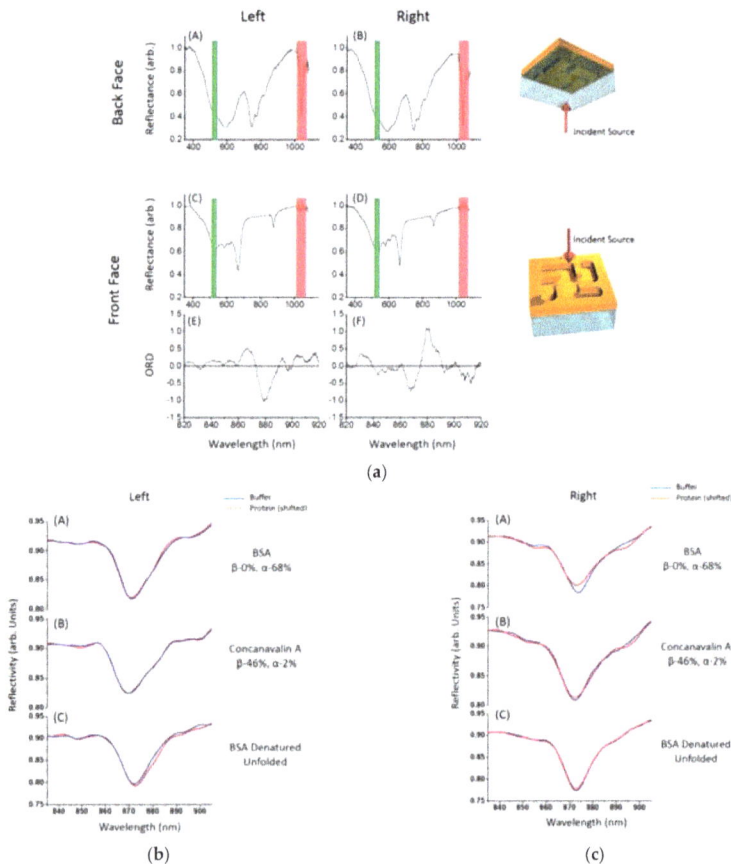

Figure 6. (a) Reflectance of left-handed and right-handed structure from back face and front face illuminations, and the ORD is plotted for front face illumination. Red bar represents the laser source, and green bar represents the second harmonic generation (SHG). Reflectivity of 3 types of proteins are plotted for (b) left-handed structure and (c) right-handed structure. Adapted from [87], with permission © 2016 American Chemical Society.

2.2.4. Extraordinary Optical Transmission

When single nanoholes fabricated with an array pattern, EOT phenomenon can be observed easily. The first observation of EOT effect was in 1998 by Ebbesen et al. [88]. They found that sub-micron cylindrical cavities in metallic films displayed highly unusual zero-order transmission spectra at wavelength larger than the array period. Since then theoretical and experimental research have been continuously done on understanding this phenomenon [89] from aspects, e.g., the shape of the nanohole [90] and the width of the nanohole [91].

For an array of nanoholes, EOT frequency has also been tuned to a fluorescence emission wavelength of the measured Cy-5 molecules to enhance the sensitivity recently by Baburin et al., as shown in Figure 7. Nanoholes with 175 nm diameter are fabricated in 100 nm Ag film with varied periods. In addition, the period which has an EOT peak corresponding to the emission frequency of Cy-5 is selected. Hence the nanohole array is used as an optical filter and the LOD is improved. This application proves the benefit of tuning resonance frequency for a specific light-matter interaction process [92]. For a single nanohole, the EOT effect has also been theoretically described and experimentally achieved [93,94]. With increased mechanical and thermal stabilities, EOT effect from a single nanohole has the potential to be applied in single biomolecule studies.

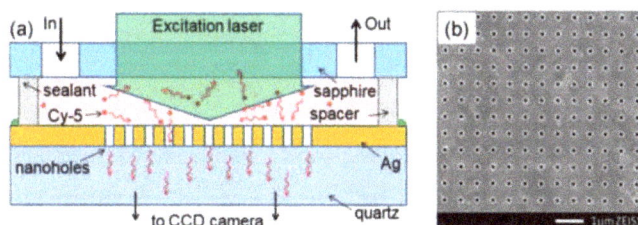

Figure 7. (**a**) Schematic diagram of the sensor, (**b**) SEM image of the nanohole array based on a 100 nm Ag thin film. Adapted from [92], with permission © 2018 Optical Society of America.

2.2.5. Fabry-Perot Modes

FP like resonance as a guided mode can be generated in a metallic cavity and has a behavior which is well described by the FP formalism. The double nanohole [95], bowtie aperture [96], nanogroove [97] and coaxial nanoring aperture [98] have been explored for FP like resonance. The zeroth-order of FP like resonance has been applied for surface enhanced infrared absorption spectroscopy [99]. As shown in Figure 8, a nanogap with 10 nm (a) and 7 nm (b) width are compared. These nanogaps are originally filled with Al_2O_3 as no etching samples. The nanogaps are then etched with H_3PO_4, and the transmission spectra are measured after 4 minutes. Since the local RI is reduced by removing the high RI Al_2O_3 away, the resonance is observed shifted toward larger wavenumber and the transmission is higher for both nanogaps. Then 5 nm of silk is spin coated on. Dips in transmission spectra are observed due to absorption of silk proteins. A fitting is used to obtain the transmission profile which represents the case without absorption and the fitted resonance frequency is shifted towards the original location. Simulation results are provided for 10 nm wide nanogap for no etching, after etching and silk coating situations, as shown in Figure 8c. The absorption is calculated based on data from (a) and (b) for two nanogap designs, and absorption peaks matching the two infrared vibrational absorption peaks, amide I and II, of silk protein are observed for both designs. The schematic nanogap design is shown in Figure 8e. A 10^4 to 10^5 times enhancement on the absorption measurement is shown in Figure 8f.

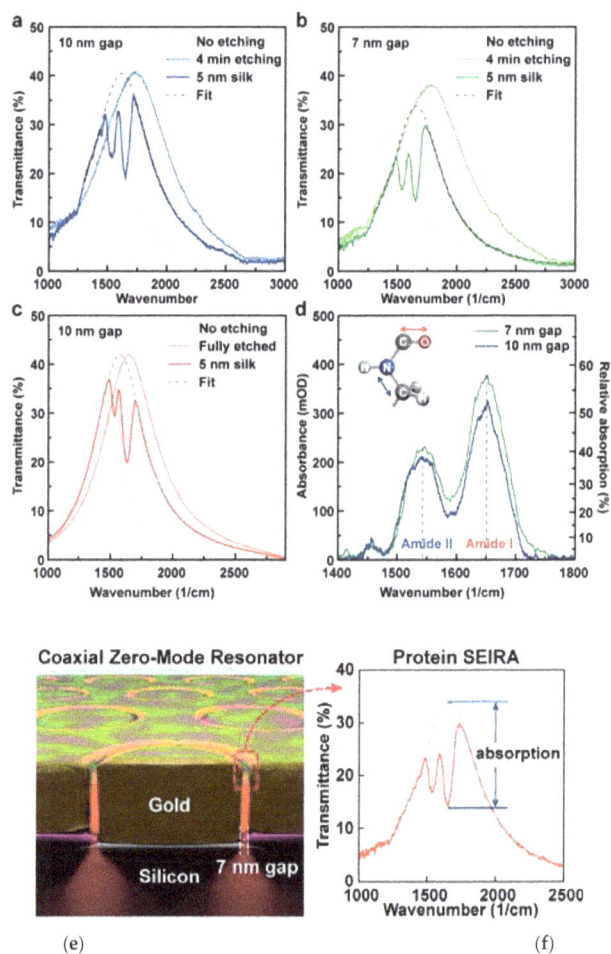

Figure 8. Transmission spectra for no etching, after etching and silk coated conditions for (**a**) 10 nm and (**b**) 7 nm wide nanogap respectively. (**c**) Simulated transmission spectra using 10 nm wide nanogaps under three conditions are plotted for comparison. (**d**) Absorption of 5 nm silk by using 10 nm and 7 nm wide nanogap. (**e**) Schematic presentation of nanogap-based coaxial zero-mode resonator. (**f**) Fitting curve is used in transmission spectra to demonstrate the absorption intensity. Adapted from [99], with permission © 2018 American Chemical Society.

2.2.6. Fano Resonance

Fano resonance is not the same as Lorentz type of resonance with a symmetric profile. It has an asymmetric profile with a narrow bandwidth. For plasmonics to achieve a Fano resonance, an effective approach is to employ the hybridization of different plasmonic modes [100,101]. Fano resonance has been studied based on multi-bowtie apertures [102] and nanocavity combined with waveguide [103].

Fano resonance based on an array of nanoholes has been applied in biosensing with phase interrogation methodology [104]. The Fano resonance is generated by coupling between the surface plasmon mode generated from the grating effect of the nanohole array and the LSPR mode of the single holes. The Fano resonance profile is plotted for both intensity (yellow color) and phase (green color) against incident wavelength, as shown in Figure 9a. In Figure 9b, the optical path difference (OPD) is compared by using a transparent substrate (grey color) and a nanohole array fabricated in Au thin film (orange color) for RI change detection. This change is introduced by changing the thickness of SiO$_2$ layer. A thin SiO$_2$ layer could represent the situation of a molecular thin film deposited on the surface of the sensor. OPD contrast curves for plasmonic device (left top plot) and a transparent substrate (left bottom plot) are plotted separately. With this plasmonic sensor, the sensitivity for RI detection is as high as 9000 nm/RIU.

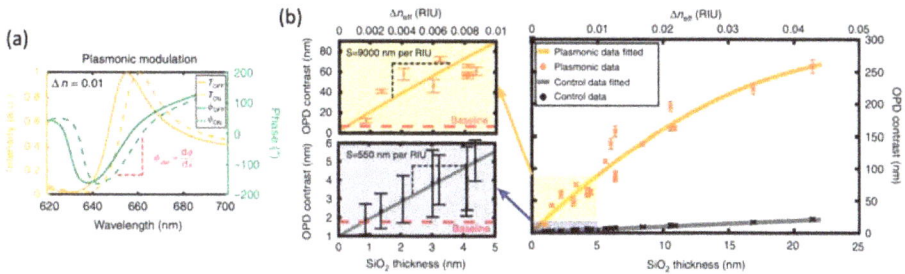

Figure 9. (**a**) Intensity (yellow color) and phase (green color) profiles for 0.01 RIU change. (**b**) OPD contrast comparison between this plasmonic sensor (orange color) and a transparent substrate (grey color) for different thickness of SiO$_2$ cover layer. Zoomed in range for thin SiO$_2$ is plotted separately for this plasmonic sensor (left top plot) and a transparent substrate (left bottom plot). Adapted from [104], with permission © 2018 Springer Nature Publishing AG.

The interrogation methodologies for nanovoid type of plasmonic devices are mainly on wavelength interrogation, while other interrogation methods mentioned previously are all can be used to suit a sensing condition.

3. Materials for Plasmonics

As demonstrated in Section 2, the dispersion of the conductive material is the essence to generate plasmonic resonance, and it is also the limitation of the resonance frequency tuning range that a plasmonic device can provide. Natural conductive materials, e.g., metal and graphene, and synthesized materials, e.g., semiconductors and metamaterials, can be used to generate resonance located at different frequency regimes. In this section, materials used for plasmonics are discussed to demonstrate the necessity to select proper materials when a specific range of frequencies is required for biosensing.

3.1. Metal

3.1.1. Near Infrared to Long-Wavelength Portion of Visible

Metals are the most used materials for plasmonics [105–107]. At the early stage of the development, research was focused on Ag and Au because of the favorable bulk dielectric properties of these metals.

When the frequency is lower than near infrared (NIR) regime, noble metal can be treated as ideal conductor. Light can be perfectly reflected, and EM wave cannot propagate. In the NIR and visible regime, EM wave penetrates into the metal and loss is increased. SPPs can be generated in this frequency regime [108]. In the ultra violet (UV) regime, metals are transparent, but high energy photons can generate photoelectrons which causes loss [109]. Au and Ag have strong absorption due to the electron bandgap transition. Although Ag has a better dispersion curve in the visible range for plasmonics, it needs protection layers to prevent chemical and mechanical damages which limits its application in biosensing. Due to the chemical and mechanical stability, Au has been more used for plasmonic sensors. In this frequency range, surface plasmonic enhanced infrared vibrational spectroscopy and Raman scattering can be used for biosensing [110,111].

3.1.2. UV to Short-Wavelength Portion of Visible

Motivated by intriguing prospects of combining plasmonic activity with interesting intrinsic targeted materials properties, the plasmonics by using novel metals have received more attention. Currently, research on UV range plasmonics is under focus [112].

Al represents an interesting material both from a fundamental and an applications point of view. It is an abundant and cheap material compared to noble metals. Due to its low price, Al is studied relatively more than the other possible UV plasmonic metals. A reasonably strong interband transition in Al is localized in a narrow energy range around 1.5 eV. Below and above this energy, Al is very much Drude-like. Currently, more research on Al-based plasmonics is focused on optical devices operated in the UV regime, e.g., for optical and plasmonic integrated circuits [113] and color filters [114,115]. In this range of frequencies, surface plasmon enhanced Raman scattering, absorption, fluorescence can be used for biosensing. Al material is not only used for UV plasmonics, but also used for absorbance enhancement in long-wavelength portion of visible regime plasmonics [116,117]. The application of plasmonics based on Al material in biosensing can be expected to increase steeply since related design and fabrication technologies are getting mature [118].

Currently plasmonics based on these UV plasmonic metal materials is in the developing stage. For example, Mg [119–121], Ga [122,123], Rh or a few metals combination designs [124] are mainly based on nanoparticle form which are made from chemical synthesis methods. Synthesized nanoparticles have disadvantages in low uniformity, repeatability, and difficulty in manipulation of single nanoparticles for sensing. However, the study of plasmonics using these nanoparticles provides insights for the future nanovoid-type plasmonic devices based on these materials in aspects of geometry design, modes coupling, etc.

3.2. Other Materials

Plasmonic biosensors working in infrared regime based on graphene is one the hottest topic recently [125–127]. It has been applied as a gas sensor [128] and used for vibrational spectroscopy [129]. Graphene has also been used to combine with other metal materials [130,131]. By adding graphene on top of a thin layer of Au, the sensitivity was improved to 10^{-18} M for single-stranded DNA using phase interrogation [132]. Graphene has also been deposited on top of a nanohole array to enhance the Raman signal 2×10^5 times [133].

Semiconductor materials and graphene combined with semiconductor are other research directions to control the plasmonic resonance in the mid-infrared regime [134–136].

Plasmonics based on magnetic materials is another topic could be interesting for biosensing applications. Nanohole array fabricated in magnetic thin film has been explored for magneto-optical activity introduced by a SPP [137]. Theoretical work on nanohole arrays in Au-Co-Au multi-layer design has been examined for improved sensitivity for RI change [138].

Currently, both novel natural materials and metamaterials are under research for plasmonics [139]. In Table 1, plasmonic sensors based on different materials are listed. Tunable range of the resonance frequency and the sensing mechanism for various analytes are shown to demonstrate the capabilities of current nanovoid-type plasmonic biosensors. Based on specific light-matter interaction mechanisms, materials for the plasmonic biosensor should be selected properly to match the working frequency range to the interaction frequency.

Table 1. Plasmonic biosensors based on different materials.

Materials		Design	Bio-Analyte	Resonance Mode	Experimental RI Sensing	Enhanced Light-Matter Interactions
Metal	Au	Thin-film: 50 nm (2007) [13]	Rocin	Propagating SPP: 635 nm	Angular Interrogation; LOD: $\sim$3.3 $\times$ 10^{-6} RIU	N/A
		Nanohole array: hole diameter $\sim$ 200 nm; array period $\sim$ 450 nm; thickness of Au $\sim$ 200 nm (2014) [8]	Exosome (50–100 nm); Selectivity: polyethylene glycol (PEG) +monoclonal antibodies	EOT effect: long-wavelength portion in visible regime	Wavelength Interrogation; LOD: $\sim$670 aM	N/A
		Coaxial nanoring aperture array: depth $\sim$ 50 nm; inner ring diameter $\sim$ 170 nm; outer ring diameter $\sim$ 290 nm; array period $\sim$ 720 nm; thickness of Au $\sim$ 250 nm (2018) [72]	poly(allylamine)hydrochloride (PAH, 65 kDa); poly(styrenesulfonate) (PSS, 75 kDa)	Modes coupling of dipolar moments from nanohole and nanodisk; long-wavelength portion in visible regime	Wavelength Interrogation; LOD: $\sim$1.4 $\times$ 10^{-4} RIU (estimated)	N/A
		Nanohole array: hole diameter $\sim$ 200 nm; array period $\sim$ 600 nm; thickness of Au $\sim$ 120 nm (2018) [104]	A/G $\sim$ 50 kDa; IgG $\sim$ 150 kDa	EOT effect: long-wavelength portion in visible regime	Phase interrogation; LOD: $\sim$4.5 $\times$ 10^{-6} RIU (estimated)	N/A
		Coaxial nanoring aperture array: ring aperture $\sim$ 7 nm; inner diameter $\sim$ 710 nm; array period $\sim$ 720 nm; thickness of Au $\sim$ 80 nm (2018) [99]	Silk protein: absorbance peaks $\sim$ 1650 and 1546 cm^{-1}	Zeroth-order FP resonance: near infrared regime	N/A	IR absorption enhancement: 10^4 $\sim$ 10^5
	Ag	Nanohole array: hole diameter $\sim$ 175 nm; array period $\sim$ 450 nm; thickness of Ag $\sim$ 100 nm (2019) [92]	Cy-5 dye molecules; Excitation wavelength: 628 nm	EOT effect: long-wavelength portion in visible regime	N/A	LOD: Attogram $\sim$ single molecule counting sensors
	Al	Nano bowtie aperture: outline $\sim$ 45) nm; gap $\sim$ 30 nm; thickness of Al $\sim$ 17) nm (2012) [78]	Alexa Fluor 647 molecules; Excitation wavelength: 632.8 nm	Modes coupling of dipolar moments from two arms of the bowtie aperture; visible to near infrared regimes	N/A	Fluorescence enhancement: $\sim$ 12 fold
Graphene		Graphene on 50 nm Au (2015) [132]	24-mer ssDNA (7.3 kDa)	Propagating SPP: 785 nm	Phase interrogation; LOD: $\sim$10^{-9} RIU	N/A
		Ribbon array on gold substrate: 80 nm width (2016) [129]	polyethylene oxide (PED)	Graphene plasmonic resonance: mid-infrared	N/A	Infrared absorption enhancement: $\sim$ 20 fold

4. Hybridization with Electric Conductivity

Due to natural properties of materials used for plasmonic sensors, i.e., electrical conductivity, nanovoid-type plasmonic sensors can hybridize this function for active plasmonics, dielectrophoresis, multi-channel sensing, and electrochemistry which can be beneficial to achieve lab-on-a-chip.

4.1. Active Plasmonics

Nanovoid-type plasmonic sensors can have the resonance frequency modified by tuning parameters of the designed geometry before the nanofabrication procedure and this tuning is not dynamic after the sensor is manufactured. An active plasmonic sensor can be beneficial for a biosensing technique to tune the resonance frequency closer to the specific light-matter interaction frequency to enhance detection signal [55].

Electrical tuning is one of the major methods to have an active plasmonic sensor. With a fixed design, the resonance condition(s) of the sensor can be fine-tuned under an external electrical modulation. There are a few ways to achieve this type of active plasmonic sensors. For example, a layer of liquid crystals (LC) is added on top of a plasmonic nanohole array and by applying an electric field on these crystals the transmission is modified [140]. As shown in Figure 10a, when an electrical modulation is applied between this nanohole array and the upper electrode, the alignment of LC is modified. In Figure 10b, simulated enhanced E-fields are demonstrated for two cases. An image of a nanosquare array fabricated in Al thin film is shown in Figure 10c. When $n_o = n_e$, the orthogonal modes excited by two linear polarized incident light in a single nanosquare have the same intensity. The light transmits through the nanoholes under these two modes with the same velocity. When $n_o \neq n_e$, with the same incident wavelength the vertical polarization still excites the vertical resonance mode, while the horizontal polarization cannot excite the horizontal resonance due to the modified RI. When an electrical modulation is applied, the alignment of the LC can be controlled. As a result, the resonance frequencies for these two modes and the appeared color of the whole array are altered, as shown in Figure 10d. This methodology has been proposed for color filter purpose, but the potential to apply this technique in biosensing is obvious.

Research has also been done on electrically tunable imbedded indium tin oxide (ITO) layer to modulate the phase and amplitude of the reflected light respectively which depends on the angle of incidence at the targeted wavelength [141]. This technique can be used for infrared absorption spectroscopy for biosensing. Another possible method is to use graphene as the plasmonic material and apply an electric voltage to alter the carrier density for resonance frequency tuning [129]. Currently, most of active plasmonic devices are targeting optoelectronic applications. Active plasmonic techniques have a huge potential to be used for biosensing to dynamically tune the sensitivity and target different bioparticles or bioprocesses detections.

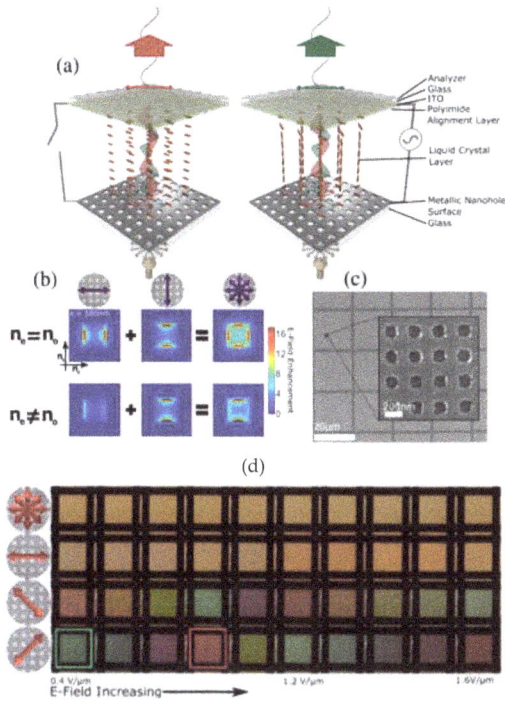

Figure 10. (**a**) Schematic of the LC plasmonic device without and with applied voltage. (**b**) FDTD simulation results of two orthogonal resonance modes. (**c**) SEM image of nanohole array fabricated in Al thin film. (**d**) Appeared color of nanohole arrays under different applied voltages. Adapted from [140], with permission © 2017 Optical Society of America.

4.2. Particles Transportation

A nanohole array has been explored to combine with a microfluidic system to transport biomolecules [142]. However, for a nanosize hot spot, bioparticles from a low concentration solution still need long waiting times to get close to the nanoscale sensing area to be detected. To solve this issue, dielectrophoresis (DEP) technique has already been combined with plasmonic devices. A sensor consists of a Au thin film as the top electrode and a Au thin film fabricated with nanohole arrays as the bottom electrode, and it uses DEP to transport biomolecules [143]. In Figure 11a, a nanohole (200 nm in diameter) array in 100 nm thick Au film is used as the lower electrode and the upper Au electrode has a thickness of 20 nm. For plasmonic sensing, EOT effect is applied. In Figure 11b, DEP is demonstrated for the cases without (top figure) and with an electrical potential (bottom figure). The DEP effect on transporting BSA proteins is demonstrated in Figure 11c and different concentrations are tested. In the first 500 s, no electrical potential is applied and no protein is detected. Then DEP effect is activated, and proteins in different concentrations are all detected in a really short time. This result demonstrates an efficient protein transportation using DEP with this design. These two Au layers are also applied as reflection surfaces to form a FP resonance cavity. As shown in Figure 11d, the transmission profile is shown with a dash-dot line with two EOT peaks, and FP resonances can be observed as the blue curve. A zoomed in range of the transmission peak on the long wavelength side of the transmission spectrum is shown in Figure 11e. With the combination of using a micro FP cavity and the EOT effect of a nanohole array, LOD of this design is approximately 0.14 pM by assuming the noise level is 0.05 nm for the wavelength spectrum.

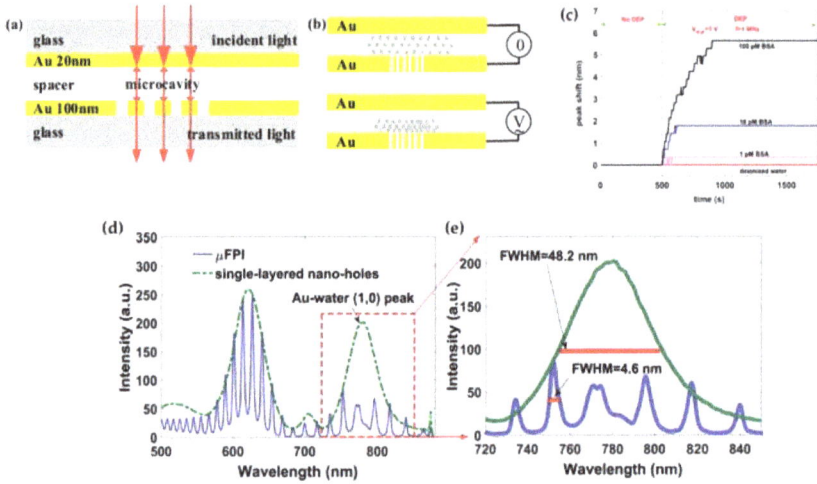

Figure 11. (**a**) Schematic representation of the plasmonic sensing device. (**b**) Conceptual representation of DEP on particle transportation. (**c**) DEP effect on transporting BSA proteins in different concentration solutions. (**d**) Transmission spectrum for the nanohole array with two peaks from the EOT effect (green dash-dot curve) and FP resonances within the transmission profile (blue solid curve). (**e**) Zoomed in curve for the transmission peak in the long wavelength range. Adapted from [143], with permission © 2018 Elsevier B.V.

4.3. Multi-Channel Sensing

If a metal thin film fabricated with nanohole arrays is lifted as a membrane suspended in a liquid medium, nanoholes can be used as tunnels for bioparticles to go through. When a particle goes through a tunnel, both the optical and electrical responses of the sensor are modified by this bioparticle occupying the nanovoid. A single bowtie aperture has been explored to transport DNA through, and both optical and electrical signals were examined for its transportation through the hole [144]. As shown in Figure 12a, when a DNA molecule goes through the feed gap of this bowtie aperture, both the optical transmission and electrical current signals has a dip. For the optical signal, the presence of the DNA molecule modified the local RI and hence the resonance frequency is shifted away. In addition, for the electrical signal, the DNA blocks the ion current which causes a drop in the current signal. Figure 12b shows a TEM image of the bowtie aperture fabricated in a 100 nm Au thin film. The across width of the bowtie is 160 nm and the side length is 100 nm. The feed gap is 20 nm. In Figure 12c, the simulated nearfield enhancement is demonstrated over the bowtie aperture. Optical and electrical signals can be used simultaneously to make a sensing technique more comprehensive.

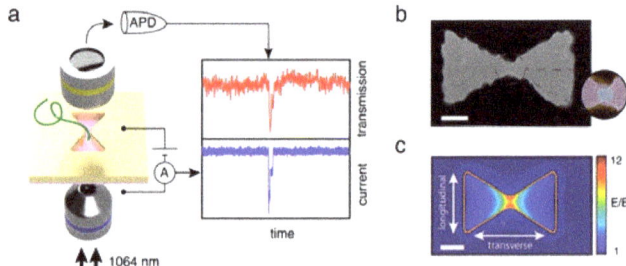

Figure 12. (**a**) Schematic representation of the plasmonic and electrical sensing mechanisms. (**b**) TEM image of the bowtie aperture. (**c**) Simulation result on E-field enhancement. Adapted from [144], with permission © 2018 American Chemical Society.

4.4. Spectroelectrochemistry

As early as the film-type plasmonics, electrochemistry technique was combined with SPR sensing technologies for redox protein studies to avoid the electrical background and further enhance the signal/noise ratio for surface enhanced Raman spectroscopy [145,146]. With the development of plasmonics based on nanostructures, electrochemistry has been coupled to LSPR from nanohole arrays to detect neurotransmitters [147] and DNA-based structure-switching [148]. In this work, nanohole arrays in 100 nm thick Au films with 150 nm diameter and different periods were tested. Three-electrode system was used for the electrochemical experiment. As shown in Figure 13, a redox tag represented in blue color can adsorbed on the sensor surface (A) or attached to a DNA molecule (B). In the first case, the redox tag can have electrochemical reaction. As shown in the bottom Figure 13C, the transmission for this case can be monitored to retract the information of the redox state of the tag. The reduced and oxidized states of the tag have different refractive indices, so the resonance frequency swings back and forth during the redox reaction which appears as an oscillation in the transmission when a fix incident wavelength is used. For the second case, the redox tag is attached to DNA molecule which causes the distance between the redox tag and the sensor surface increased, so the electrochemical reaction is hindered.

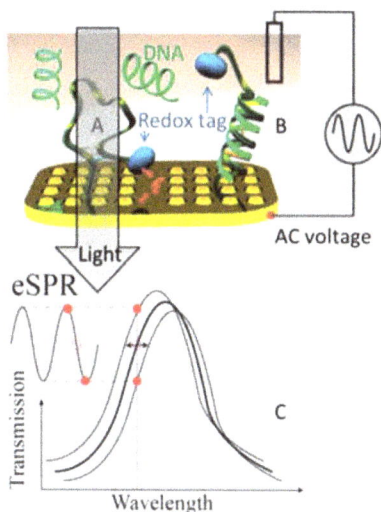

Figure 13. Schematic presentation for electrochemical SPR sensor. (**A**): redox tag attached directly on the surface of the sensor, (**B**): redox tag attached on a DNA, (**C**): transmission for A under a redox reaction. Adapted from [148], with permission © 2015 by WILEY-VCH Verlag GmbH & Co.

As mentioned previously in Section 2, plasmonic sensors based on nanovoid geometries can generate various resonance modes, e.g., toroidal dipole that can be used to detect structural information for biomolecules. These sensors have huge potential to be combined with electrochemistry for redox protein studies to reveal their structural properties dynamically in the process of the redox action.

5. Conclusions

Plasmonics, based on materials science and nanostructures, have provided a new playground for researchers to generate desired resonance modes and modes coupling with enhanced nearfield at target frequency which can be used to enhance specific light-matter interactions to further understand the properties of biomolecules and bioprocesses. With its intrinsic electrical property, hybrid functions of active plasmonics, multi-channel sensing, particle transportation and electrochemical study can be added to its nano-optical features for nanovoid-type plasmonic biosensors to be integrated on one chip with high sensitivity, selectivity, simultaneous multiplicity and real time monitoring. In addition, these labs-on-a-chip have huge potential to be used as commercial health instruments for PoC purposes in the near future [149,150].

Author Contributions: Each author contributed equally.

Funding: Supported by the Fundamental Research Funds for the Central Universities (DUT18RC(3)065).

Acknowledgments: School of Optoelectronic Engineering and Instrumentation Science at Dalian University of Technology.

Conflicts of Interest: The authors declare no conflict of interest.

References

1. Agarwal, K.; Hwang, S.; Bartnik, A.; Buchele, N.; Mishra, A.; Cho, J.-H. Small-Scale Biological and Artificial Multidimensional Sensors for 3D Sensing. *Small* **2018**, *14*, 1801145. [CrossRef] [PubMed]

2. Yang, X.; Sun, Z.; Low, T.; Hu, H.; Guo, X.; Garcia de Abajo, F.J.; Avouris, P.; Dai, Q. Nanomaterial-Based Plasmon-Enhanced Infrared Spectroscopy. *Adv. Mater.* **2018**, *30*, e1704896. [CrossRef]

3. Xavier, J.; Vincent, S.; Meder, F.; Vollmer, F. Advances in optoplasmonic sensors—Combining optical nano/microcavities and photonic crystals with plasmonic nanostructures and nanoparticles. *Nanophotonics* **2018**, *7*, 1–38. [CrossRef]

4. Mejia-Salazar, J.R.; Oliveira, O.N., Jr. Plasmonic Biosensing. *Chem. Rev.* **2018**, *118*, 10617–10625. [CrossRef] [PubMed]

5. Taylor, A.B.; Zijlstra, P. Single-Molecule Plasmon Sensing: Current Status and Future Prospects. *ACS Sens.* **2017**, *2*, 1103–1122. [CrossRef]

6. Guo, L.; Xu, S.; Ma, X.; Qiu, B.; Lin, Z.; Chen, G. Dual-color plasmonic enzyme-linked immunosorbent assay based on enzyme-mediated etching of Au nanoparticles. *Sci. Rep.* **2016**, *6*, 32755. [CrossRef] [PubMed]

7. Rodriguez-Lorenzo, L.; de la Rica, R.; Alvarez-Puebla, R.A.; Liz-Marzan, L.M.; Stevens, M.M. Plasmonic nanosensors with inverse sensitivity by means of enzyme-guided crystal growth. *Nat. Mater.* **2012**, *11*, 604–607. [CrossRef]

8. Im, H.; Shao, H.; Park, Y.I.; Peterson, V.M.; Castro, C.M.; Weissleder, R.; Lee, H. Label-free detection and molecular profiling of exosomes with a nano-plasmonic sensor. *Nat. Biotechnol.* **2014**, *32*, 490–495. [CrossRef] [PubMed]

9. Jackman, J.A.; Ferhan, A.R.; Cho, N.J. Nanoplasmonic sensors for biointerfacial science. *Chem. Soc. Rev.* **2017**, *46*, 3615–3660. [CrossRef] [PubMed]

10. Wang, S.; Forzani, E.S.; Tao, N. Detection of heavy metal ions in water by high-resolution surface plasmon resonance spectroscopy combined with anodic stripping voltammetry. *Anal. Chem.* **2007**, *79*, 4427–4432. [CrossRef]

11. Koubova, V.; Brynda, E.; Karasova, L.; Skvor, J.; Homola, J.; Dostalek, J.; Tobiska, P.; Rosicky, J. Detection of foodborne pathogens using surface plasmon resonance biosensors. *Sens. Actuators B Chem.* **2001**, *74*, 100–105. [CrossRef]

12. Kurzatkowska, K.; Santiago, T.; Hepel, M. Plasmonic nanocarrier grid-enhanced Raman sensor for studies of anticancer drug delivery. *Biosens. Bioelectron.* **2017**, *91*, 780–787. [CrossRef]

13. Feltis, B.N.; Sexton, B.A.; Glenn, F.L.; Best, M.J.; Wilkins, M.; Davis, T.J. A hand-held surface plasmon resonance biosensor for the detection of ricin and other biological agents. *Biosens. Bioelectron.* **2008**, *23*, 1131–1136. [CrossRef]

14. Skottrup, P.D.; Nicolaisen, M.; Justesen, A.F. Towards on-site pathogen detection using antibody-based sensors. *Biosens. Bioelectron.* **2008**, *24*, 339–348. [CrossRef]

15. Ahmadivand, A.; Gerislioglu, B.; Manickam, P.; Kaushik, A.; Bhansali, S.; Nair, M.; Pala, N. Rapid Detection of Infectious Envelope Proteins by Magnetoplasmonic Toroidal Metasensors. *ACS Sens.* **2017**, *2*, 1359–1368. [CrossRef]

16. Bauch, M.; Dostalek, J. Collective localized surface plasmons for high performance fluorescence biosensing. *Opt. Express* **2013**, *21*, 20470–20483. [CrossRef]

17. Luan, J.Y.; Morrisse, J.J.; Wang, Z.Y.; Derami, H.G.; Liu, K.K.; Cao, S.S.; Jiang, Q.S.; Wang, C.Z.; Kharasch, E.D.; Naik, R.R.; et al. Add-on plasmonic patch as a universal fluorescence enhancer. *Light-Sci. Appl.* **2018**, *7*, 7129. [CrossRef]

18. Fothergill, S.M.; Joyce, C.; Xie, F. Metal enhanced fluorescence biosensing: From ultra-violet towards second near-infrared window. *Nanoscale* **2018**, *10*, 20914–20929. [CrossRef]

19. Luo, S.-C.; Sivashanmugan, K.; Liao, J.-D.; Yao, C.-K.; Peng, H.-C. Nanofabricated SERS-active substrates for single-molecule to virus detection in vitro: A review. *Biosens. Bioelectron.* **2014**, *61*, 232–240. [CrossRef]

20. Feng, S.; Wang, W.; Tai, I.T.; Chen, G.; Chen, R.; Zeng, H. Label-free surface-enhanced Raman spectroscopy for detection of colorectal cancer and precursor lesions using blood plasma. *Biomed. Opt. Express* **2015**, *6*, 3494–3502. [CrossRef]

21. Vo-Dinh, T.; Wang, H.N.; Scaffidi, J. Plasmonic nanoprobes for SERS biosensing and bioimaging. *J. Biophotonics* **2010**, *3*, 89–102. [CrossRef]

22. Huang, J.A.; Zhao, Y.Q.; Zhang, X.J.; He, L.F.; Wong, T.L.; Chui, Y.S.; Zhang, W.J.; Lee, S.T. Ordered Ag/Si nanowires array: Wide-range surface-enhanced Raman spectroscopy for reproducible biomolecule detection. *Nano Lett.* **2013**, *13*, 5039–5045. [CrossRef]

23. Sivapalan, S.T.; DeVetter, B.M.; Yang, T.K.; van Dijk, T.; Schulmerich, M.V.; Carney, P.S.; Bhargava, R.; Murphy, C.J. Off-Resonance Surface-Enhanced Raman Spectroscopy from Gold Nanorod Suspensions as a Function of Aspect Ratio: Not What We Thought. *ACS Nano* **2013**, *7*, 2099–2105. [CrossRef]

24. Xu, L.-J.; Zong, C.; Zheng, X.-S.; Hu, P.; Feng, J.-M.; Ren, B. Label-Free Detection of Native Proteins by Surface-Enhanced Raman Spectroscopy Using Iodide-Modified Nanoparticles. *Anal. Chem.* **2014**, *86*, 2238–2245. [CrossRef]

25. Wang, F.; Joshi, B.P.; Chakrabarty, A.; Zhang, H.; Wei, Q.-H. Plasmonic Patch Nanoantennas for Reproducible and High-Sensitivity Chemical Detection with Surface-Enhanced Raman Spectroscopy. *ECS Trans.* **2018**, *85*, 77–85. [CrossRef]

26. Gao, J.; Zhang, N.; Ji, D.; Song, H.; Liu, Y.; Zhou, L.; Sun, Z.; Jornet, J.M.; Thompson, A.C.; Collins, R.L.; et al. Superabsorbing Metasurfaces with Hybrid Ag-Au Nanostructures for Surface-Enhanced Raman Spectroscopy Sensing of Drugs and Chemicals. *Small Methods* **2018**, *2*, 1800045. [CrossRef]

27. Sivashanmugan, K.; Liao, J.D.; Liu, B.H.; Yao, C.K.; Luo, S.C. Ag nanoclusters on ZnO nanodome array as hybrid SERS-active substrate for trace detection of malachite green. *Sens. Actuators B Chem.* **2015**, *207*, 430–436. [CrossRef]

28. Zrimsek, A.B.; Wong, N.L.; Van Duyne, R.P. Single Molecule Surface-Enhanced Raman Spectroscopy: A Critical Analysis of the Bianalyte versus Isotopologue Proof. *J. Phys. Chem. C* **2016**, *120*, 5133–5142. [CrossRef]

29. Zhang, Y.; Shen, J.; Xie, Z.; Dou, X.; Min, C.; Lei, T.; Liu, J.; Zhu, S.; Yuan, X. Dynamic plasmonic nano-traps for single molecule surface-enhanced Raman scattering. *Nanoscale* **2017**, *9*, 10694–10700. [CrossRef]

30. Darby, B.L.; Etchegoin, P.G.; Le Ru, E.C. Single-molecule surface-enhanced Raman spectroscopy with nanowatt excitation. *Phys. Chem. Chem. Phys.* **2014**, *16*, 23895–23899. [CrossRef]

31. Le Ru, E.C.; Grand, J.; Sow, I.; Somerville, W.R.; Etchegoin, P.G.; Treguer-Delapierre, M.; Charron, G.; Felidj, N.; Levi, G.; Aubard, J. A scheme for detecting every single target molecule with surface-enhanced Raman spectroscopy. *Nano Lett.* **2011**, *11*, 5013–5019. [CrossRef]

32. Garcia-Rico, E.; Alvarez-Puebla, R.A.; Guerrini, L. Direct surface-enhanced Raman scattering (SERS) spectroscopy of nucleic acids: From fundamental studies to real-life applications. *Chem. Soc. Rev.* **2018**, *47*, 4909–4923. [CrossRef]

33. Yanik, A.A.; Huang, M.; Kamohara, O.; Artar, A.; Geisbert, T.W.; Connor, J.H.; Altug, H. An optofluidic nanoplasmonic biosensor for direct detection of live viruses from biological media. *Nano Lett.* **2010**, *10*, 4962–4969. [CrossRef]

34. Wu, N. Plasmonic metal-semiconductor photocatalysts and photoelectrochemical cells: A review. *Nanoscale* **2018**, *10*, 2679–2696. [CrossRef]

35. Cushing, S.K.; Chen, C.J.; Dong, C.L.; Kong, X.T.; Govorov, A.O.; Liu, R.S.; Wu, N. Tunable Non-Thermal Distribution of Hot Electrons in a Semiconductor Injected from a Plasmonic Gold Nanostructure. *ACS Nano* **2018**, *12*, 7117–7126. [CrossRef]

36. Sagle, L.B.; Ruvuna, L.K.; Ruemmele, J.A.; Van Duyne, R.P. Advances in localized surface plasmon resonance spectroscopy biosensing. *Nanomedicine* **2011**, *6*, 1447–1462. [CrossRef]

37. Petryayeva, E.; Krull, U.J. Localized surface plasmon resonance: Nanostructures, bioassays and biosensing—A review. *Anal. Chim. Acta* **2011**, *706*, 8–24. [CrossRef]
38. Liu, J.; He, H.; Xiao, D.; Yin, S.; Ji, W.; Jiang, S.; Luo, D.; Wang, B.; Liu, Y. Recent Advances of Plasmonic Nanoparticles and their Applications. *Materials* **2018**, *11*, 1833. [CrossRef]
39. Hanske, C.; Sanz-Ortiz, M.N.; Liz-Marzan, L.M. Silica-Coated Plasmonic Metal Nanoparticles in Action. *Adv. Mater.* **2018**, *30*, 1707003. [CrossRef]
40. Fan, W.; Leung, M.K. Recent Development of Plasmonic Resonance-Based Photocatalysis and Photovoltaics for Solar Utilization. *Molecules* **2016**, *21*, 180. [CrossRef]
41. Okamoto, H.; Narushima, T.; Nishiyama, Y.; Imura, K. Local optical responses of plasmon resonances visualised by near-field optical imaging. *Phys. Chem. Chem. Phys.* **2015**, *17*, 6192–6206. [CrossRef]
42. Austin, L.A.; Kang, B.; El-Sayed, M.A. Probing molecular cell event dynamics at the single-cell level with targeted plasmonic gold nanoparticles: A review. *Nano Today* **2015**, *10*, 542–558. [CrossRef]
43. Zhang, S.; Geryak, R.; Geldmeier, J.; Kim, S.; Tsukruk, V.V. Synthesis, Assembly, and Applications of Hybrid Nanostructures for Biosensing. *Chem. Rev.* **2017**, *117*, 12942–13038. [CrossRef]
44. Carregal-Romero, S.; Caballero-Diaz, E.; Beqa, L.; Abdelmonem, A.M.; Ochs, M.; Huhn, D.; Suau, B.S.; Valcarcel, M.; Parak, W.J. Multiplexed sensing and imaging with colloidal nano- and microparticles. *Annu. Rev. Anal. Chem.* **2013**, *6*, 53–81. [CrossRef]
45. Amendola, V.; Pilot, R.; Frasconi, M.; Marago, O.M.; Iati, M.A. Surface plasmon resonance in gold nanoparticles: A review. *J. Phys. Condens. Matter* **2017**, *29*, 203002. [CrossRef]
46. Wood, R.W. On a Remarkable Case of Uneven Distribution of Light in a Diffraction Grating Spectrum. *Proc. Phys. Soc. Lond.* **1902**, *18*, 269–275. [CrossRef]
47. Fano, U. The Theory of Anomalous Diffraction Gratings and of Quasi-Stationary Waves on Metallic Surfaces (Sommerfeld's Waves). *J. Opt. Soc. Am.* **1941**, *31*, 213–222. [CrossRef]
48. Kretschmann, E.; Raether, H. Radiative Decay of Non Radiative Surface Plasmons Excited by Light. *Z. Naturforsch. A* **1968**, *23*, 2135–2136. [CrossRef]
49. Otto, A. Excitation of nonradiative surface plasma waves in silver by the method of frustrated total reflection. *Z. Phys. A Hadron. Nucl.* **1968**, *216*, 398–410. [CrossRef]
50. Liedberg, B.; Nylander, C.; Lundstrom, I. Surface-Plasmon Resonance for Gas-Detection and Biosensing. *Sens. Actuators* **1983**, *4*, 299–304. [CrossRef]
51. Oh, S.H.; Altug, H. Performance metrics and enabling technologies for nanoplasmonic biosensors. *Nat. Commun.* **2018**, *9*, 5263. [CrossRef] [PubMed]
52. Smith, C.L.; Stenger, N.; Kristensen, A.; Mortensen, N.A.; Bozhevolnyi, S.I. Gap and channeled plasmons in tapered grooves: A review. *Nanoscale* **2015**, *7*, 9355–9386. [CrossRef]
53. Maier, S.A. *Plasmonics: Fundamentals and Applications*; Springer: New York, NY, USA, 2007.
54. Srivastava, S.K. Fiber Optic Plasmonic Sensors: Past, Present and Future. *Open Opt. J.* **2013**, *7*, 58–83. [CrossRef]
55. Jiang, N.; Zhuo, X.; Wang, J. Active Plasmonics: Principles, Structures, and Applications. *Chem. Rev.* **2018**, *118*, 3054–3099. [CrossRef]
56. Qi, Z.M.; Wei, M.D.; Matsuda, H.; Honma, I.; Zhou, H.S. Broadband surface plasmon resonance spectroscopy for determination of refractive-index dispersion of dielectric thin films. *Appl. Phys. Lett.* **2007**, *90*, 181112. [CrossRef]
57. Homola, J.; Koudela, I.; Yee, S.S. Surface plasmon resonance sensors based on diffraction gratings and prism couplers: Sensitivity comparison. *Sens. Actuators B Chem.* **1999**, *54*, 16–24. [CrossRef]
58. Piliarik, M.; Homola, J. Surface plasmon resonance (SPR) sensors: Approaching their limits? *Opt. Express* **2009**, *17*, 16505–16517. [CrossRef]
59. Abelès, F. Surface electromagnetic waves ellipsometry. *Surf. Sci.* **1976**, *56*, 237–251. [CrossRef]
60. Kabashin, A.V.; Patskovsky, S.; Grigorenko, A.N. Phase and amplitude sensitivities in surface plasmon resonance bio and chemical sensing. *Opt. Express* **2009**, *17*, 21191–21204. [CrossRef] [PubMed]
61. Shen, S.; Liu, T.; Guo, J. Optical phase-shift detection of surface plasmon resonance. *Appl. Opt.* **1998**, *37*, 1747–1751. [CrossRef]
62. Homola, J.; Yee, S.S. Novel polarization control scheme for spectral surface plasmon resonance sensors. *Sens. Actuators B Chem.* **1998**, *51*, 331–339. [CrossRef]

63. Huang, Y.H.; Ho, H.P.; Kong, S.K.; Kabashin, A.V. Phase-sensitive surface plasmon resonance biosensors: Methodology, instrumentation and applications. *Ann. Phys.* **2012**, *524*, 637–662. [CrossRef]

64. Deng, S.; Wang, P.; Yu, X. Phase-Sensitive Surface Plasmon Resonance Sensors: Recent Progress and Future Prospects. *Sensors* **2017**, *17*, 2819. [CrossRef]

65. Sreekanth, K.V.; Alapan, Y.; ElKabbash, M.; Ilker, E.; Hinczewski, M.; Gurkan, U.A.; De Luca, A.; Strangi, G. Extreme sensitivity biosensing platform based on hyperbolic metamaterials. *Nat. Mater.* **2016**, *15*, 621–627. [CrossRef] [PubMed]

66. Salamon, Z.; Macleod, H.A.; Tollin, G. Coupled plasmon-waveguide resonators: A new spectroscopic tool for probing proteolipid film structure and properties. *Biophys. J.* **1997**, *73*, 2791–2797. [CrossRef]

67. Salamon, Z.; Tollin, G. Optical anisotropy in lipid bilayer membranes: Coupled plasmon-waveguide resonance measurements of molecular orientation, polarizability, and shape. *Biophys. J.* **2001**, *80*, 1557–1567. [CrossRef]

68. Bahrami, F.; Maisonneuve, M.; Meunier, M.; Aitchison, J.S.; Mojahedi, M. An improved refractive index sensor based on genetic optimization of plasmon waveguide resonance. *Opt. Express* **2013**, *21*, 20863–20872. [CrossRef] [PubMed]

69. Byard, C.L.; Han, X.; Mendes, S.B. Angle-Multiplexed Waveguide Resonance of High Sensitivity and Its Application to Nanosecond Dynamics of Molecular Assemblies. *Anal. Chem.* **2012**, *84*, 9762–9767. [CrossRef]

70. Salamon, Z.; Brown, M.I.; Tollin, G. Plasmon resonance spectroscopy: Probing molecular interactions within membranes. *Trends Biochem. Sci.* **1999**, *24*, 213–219. [CrossRef]

71. Linman, M.J.; Abbas, A.; Cheng, Q. Interface design and multiplexed analysis with surface plasmon resonance (SPR) spectroscopy and SPR imaging. *Analyst* **2010**, *135*, 2759–2767. [CrossRef]

72. Liang, Y.Z.; Li, L.X.; Lu, M.D.; Yuan, H.Z.; Long, Z.W.; Peng, W.; Xu, T. Comparative investigation of sensing behaviors between gap and lattice plasmon modes in a metallic nanoring array. *Nanoscale* **2018**, *10*, 548–555. [CrossRef]

73. Liang, Y.; Zhang, H.; Zhu, W.; Agrawal, A.; Lezec, H.; Li, L.; Peng, W.; Zou, Y.; Lu, Y.; Xu, T. Subradiant Dipolar Interactions in Plasmonic Nanoring Resonator Array for Integrated Label-Free Biosensing. *ACS Sens.* **2017**, *2*, 1796–1804. [CrossRef]

74. Dahlin, A.B. Sensing applications based on plasmonic nanopores: The hole story. *Analyst* **2015**, *140*, 4748–4759. [CrossRef]

75. Eftekhari, F.; Escobedo, C.; Ferreira, J.; Duan, X.; Girotto, E.M.; Brolo, A.G.; Gordon, R.; Sinton, D. Nanoholes as nanochannels: Flow-through plasmonic sensing. *Anal. Chem.* **2009**, *81*, 4308–4311. [CrossRef]

76. Al Balushi, A.A.; Zehtabi-Oskuie, A.; Gordon, R. Observing single protein binding by optical transmission through a double nanohole aperture in a metal film. *Biomed. Opt. Express* **2013**, *4*, 1504–1511. [CrossRef]

77. Gupta, N.; Dhawan, A. Bridged-bowtie and cross bridged-bowtie nanohole arrays as SERS substrates with hotspot tunability and multi-wavelength SERS response. *Opt. Express* **2018**, *26*, 17899–17915. [CrossRef]

78. Lu, G.W.; Li, W.Q.; Zhang, T.Y.; Yue, S.; Liu, J.; Hou, L.; Li, Z.; Gong, Q.H. Plasmonic-Enhanced Molecular Fluorescence within Isolated Bowtie Nano-Apertures. *ACS Nano* **2012**, *6*, 1438–1448. [CrossRef]

79. Rindzevicius, T.; Alaverdyan, Y.; Sepulveda, B.; Pakizeh, T.; Kall, M.; Hillenbrand, R.; Aizpurua, J.; de Abajo, F.J.G. Nanohole plasmons in optically thin gold films. *J. Phys. Chem. C* **2007**, *111*, 1207–1212. [CrossRef]

80. Park, T.H.; Mirin, N.; Lassiter, J.B.; Nehl, C.L.; Halas, N.J.; Nordlander, P. Optical properties of a nanosized hole in a thin metallic film. *ACS Nano* **2008**, *2*, 25–32. [CrossRef]

81. Sepulveda, B.; Alaverdyan, Y.; Alegret, J.; Kall, M.; Johansson, P. Shape effects in the localized surface plasmon resonance of single nanoholes in thin metal films. *Opt. Express* **2008**, *16*, 5609–5616. [CrossRef]

82. Regmi, R.; Al Balushi, A.A.; Rigneault, H.; Gordon, R.; Wenger, J. Nanoscale volume confinement and fluorescence enhancement with double nanohole aperture. *Sci. Rep.* **2015**, *5*, 15852. [CrossRef]

83. Qian, Z.; Hastings, S.P.; Li, C.; Edward, B.; McGinn, C.K.; Engheta, N.; Fakhraai, Z.; Park, S.-J. Raspberry-like Metamolecules Exhibiting Strong Magnetic Resonances. *ACS Nano* **2015**, *9*, 1263–1270. [CrossRef] [PubMed]

84. Wu, H.W.; Han, Y.Z.; Chen, H.J.; Zhou, Y.; Li, X.C.; Gao, J.; Sheng, Z.Q. Physical mechanism of order between electric and magnetic dipoles in spoof plasmonic structures. *Opt. Lett.* **2017**, *42*, 4521–4524. [CrossRef] [PubMed]

85. Xi, Z.; Urbach, H.P. Magnetic Dipole Scattering from Metallic Nanowire for Ultrasensitive Deflection Sensing. *Phys. Rev. Lett.* **2017**, *119*, 053902. [CrossRef]

86. Wang, J.Q.; Fan, C.Z.; He, J.N.; Ding, P.; Liang, E.J.; Xue, Q.Z. Double Fano resonances due to interplay of electric and magnetic plasmon modes in planar plasmonic structure with high sensing sensitivity. *Opt. Express* **2013**, *21*, 2236–2244. [CrossRef] [PubMed]

87. Jack, C.; Karimullah, A.S.; Leyman, R.; Tullius, R.; Rotello, V.M.; Cooke, G.; Gadegaard, N.; Barron, L.D.; Kadodwala, M. Biomacromolecular Stereostructure Mediates Mode Hybridization in Chiral Plasmonic Nanostructures. *Nano Lett.* **2016**, *16*, 5806–5814. [CrossRef] [PubMed]

88. Ebbesen, T.W.; Lezec, H.J.; Ghaemi, H.F.; Thio, T.; Wolff, P.A. Extraordinary optical transmission through sub-wavelength hole arrays. *Nature* **1998**, *391*, 667–669. [CrossRef]

89. García de Abajo, F.J. Colloquium: Light scattering by particle and hole arrays. *Rev. Mod. Phys.* **2007**, *79*, 1267–1290. [CrossRef]

90. Koerkamp, K.J.; Enoch, S.; Segerink, F.B.; van Hulst, N.F.; Kuipers, L. Strong influence of hole shape on extraordinary transmission through periodic arrays of subwavelength holes. *Phys. Rev. Lett.* **2004**, *92*, 183901. [CrossRef]

91. Van der Molen, K.L.; Segerink, F.B.; van Hulst, N.F.; Kuipers, L. Influence of hole size on the extraordinary transmission through subwavelength hole arrays. *Appl. Phys. Lett.* **2004**, *85*, 4316–4318. [CrossRef]

92. Baburin, A.S.; Gritchenko, A.S.; Orlikovsky, N.A.; Dobronosova, A.A.; Rodionov, I.A.; Balykin, V.I.; Melentiev, P.N. State-of-the-art plasmonic crystals for molecules fluorescence detection. *Opt. Mater. Express* **2019**, *9*, 1173–1179. [CrossRef]

93. De Abajo, F.J.G.I. Light transmission through a single cylindrical hole in a metallic film. *Opt. Express* **2002**, *10*, 1475–1484. [CrossRef]

94. Sturman, B.; Podivilov, E.; Gorkunov, M. Elementary processes of light transformation for slit structures in real and perfect metals. *Photonics Nanostruct. Fundam. Appl.* **2012**, *10*, 409–415. [CrossRef]

95. Chen, Y.; Kotnala, A.; Yu, L.; Zhang, J.; Gordon, R. Wedge and gap plasmonic resonances in double nanoholes. *Opt. Express* **2015**, *23*, 30227–30236. [CrossRef]

96. Ibrahim, I.A.; Mivelle, M.; Grosjean, T.; Allegre, J.T.; Burr, G.W.; Baida, F.I. Bowtie-shaped nanoaperture: A modal study. *Opt. Lett.* **2010**, *35*, 2448–2450. [CrossRef]

97. Fix, B.; Jaeck, J.; Bouchon, P.; Heron, S.; Vest, B.; Haidar, R. High-quality-factor double Fabry-Perot plasmonic nanoresonator. *Opt. Lett.* **2017**, *42*, 5062–5065. [CrossRef]

98. Yoo, D.; Nguyen, N.C.; Martin-Moreno, L.; Mohr, D.A.; Carretero-Palacios, S.; Shaver, J.; Peraire, J.; Ebbesen, T.W.; Oh, S.H. High-Throughput Fabrication of Resonant Metamaterials with Ultrasmall Coaxial Apertures via Atomic Layer Lithography. *Nano Lett.* **2016**, *16*, 2040–2046. [CrossRef]

99. Yoo, D.; Mohr, D.A.; Vidal-Codina, F.; John-Herpin, A.; Jo, M.; Kim, S.; Matson, J.; Caldwell, J.D.; Jeon, H.; Nguyen, N.-C.; et al. High-Contrast Infrared Absorption Spectroscopy via Mass-Produced Coaxial Zero-Mode Resonators with Sub-10 nm Gaps. *Nano Lett.* **2018**, *18*, 1930–1936. [CrossRef]

100. Chen, J.J.; Gan, F.Y.; Wang, Y.J.; Li, G.Z. Plasmonic Sensing and Modulation Based on Fano Resonances. *Adv. Opt. Mater.* **2018**, *6*, 1701152. [CrossRef]

101. Liang, Y.; Peng, W.; Li, L.; Qian, S.; Wang, Q. Tunable plasmonic resonances based on elliptical annular aperture arrays on conducting substrates for advanced biosensing. *Opt. Lett.* **2015**, *40*, 3909–3912. [CrossRef]

102. Chen, Y.; Chu, J.R.; Xu, X.F. Plasmonic Multibowtie Aperture Antenna with Fano Resonance for Nanoscale Spectral Sorting. *ACS Photonics* **2016**, *3*, 1689–1697. [CrossRef]

103. Lu, H.; Liu, X.; Mao, D.; Wang, G. Plasmonic nanosensor based on Fano resonance in waveguide-coupled resonators. *Opt. Lett.* **2012**, *37*, 3780–3782. [CrossRef]

104. Yesilkoy, F.; Terborg, R.A.; Pello, J.; Belushkin, A.A.; Jahani, Y.; Pruneri, V.; Altug, H. Phase-sensitive plasmonic biosensor using a portable and large field-of-view interferometric microarray imager. *Light-Sci. Appl.* **2018**, *7*, 17152. [CrossRef]

105. McPeak, K.M.; Jayanti, S.V.; Kress, S.J.; Meyer, S.; Iotti, S.; Rossinelli, A.; Norris, D.J. Plasmonic Films Can Easily Be Better: Rules and Recipes. *ACS Photonics* **2015**, *2*, 326–333. [CrossRef]

106. Guay, J.M.; Cala Lesina, A.; Cote, G.; Charron, M.; Poitras, D.; Ramunno, L.; Berini, P.; Weck, A. Laser-induced plasmonic colours on metals. *Nat. Commun.* **2017**, *8*, 16095. [CrossRef]

107. Barchiesi, D.; Grosges, T. Fitting the optical constants of gold, silver, chromium, titanium, and aluminum in the visible bandwidth. *J. Nanophotonics* **2014**, *8*, 083097. [CrossRef]

108. Johnson, P.B.; Christy, R.W. Optical Constants of the Noble Metals. *Phys. Rev. B* **1972**, *6*, 4370–4379. [CrossRef]

109. Yakubovsky, D.I.; Arsenin, A.V.; Stebunov, Y.V.; Fedyanin, D.Y.; Volkov, V.S. Optical constants and structural properties of thin gold films. *Opt. Express* **2017**, *25*, 25574–25587. [CrossRef] [PubMed]

110. Cheng, F.; Yang, X.; Gao, J. Ultrasensitive detection and characterization of molecules with infrared plasmonic metamaterials. *Sci. Rep.* **2015**, *5*, 14327. [CrossRef] [PubMed]

111. Zhong, Y.J.; Malagari, S.D.; Hamilton, T.; Wasserman, D. Review of mid-infrared plasmonic materials. *J. Nanophotonics* **2015**, *9*, 093791. [CrossRef]

112. Gutierrez, Y.; de la Osa, R.A.; Ortiz, D.; Saiz, J.M.; Gonzalez, F.; Moreno, F. Plasmonics in the Ultraviolet with Aluminum, Gallium, Magnesium and Rhodium. *Appl. Sci.* **2018**, *8*, 64. [CrossRef]

113. Dabos, G.; Manolis, A.; Tsiokos, D.; Ketzaki, D.; Chatzianagnostou, E.; Markey, L.; Rusakov, D.; Weeber, J.C.; Dereux, A.; Giesecke, A.L.; et al. Aluminum plasmonic waveguides co-integrated with Si3N4 photonics using CMOS processes. *Sci. Rep.* **2018**, *8*, 13380. [CrossRef]

114. Chen, Q.; Cumming, D.R.S. High transmission and low color cross-talk plasmonic color filters using triangular-lattice hole arrays in aluminum films. *Opt. Express* **2010**, *18*, 14056–14062. [CrossRef]

115. Dai, P.; Wang, Y.; Zhu, X.; Shi, H.; Chen, Y.; Zhang, S.; Yang, W.; Chen, Z.; Xiao, S.; Duan, H. Transmissive structural color filters using vertically coupled aluminum nanohole/nanodisk array with a triangular-lattice. *Nanotechnology* **2018**, *29*, 395202. [CrossRef] [PubMed]

116. Mivelle, M.; van Zanten, T.S.; Neumann, L.; van Hulst, N.F.; Garcia-Parajo, M.F. Ultrabright bowtie nanoaperture antenna probes studied by single molecule fluorescence. *Nano Lett.* **2012**, *12*, 5972–5978. [CrossRef] [PubMed]

117. White, J.S.; Veronis, G.; Yu, Z.F.; Barnard, E.S.; Chandran, A.; Fan, S.H.; Brongersma, M.L. Extraordinary optical absorption through subwavelength slits. *Opt. Lett.* **2009**, *34*, 686–688. [CrossRef] [PubMed]

118. Kunz, J.N.; Voronine, D.V.; Lu, W.; Liege, Z.; Lee, H.W.H.; Zhang, Z.; Scully, M.O. Aluminum plasmonic nanoshielding in ultraviolet inactivation of bacteria. *Sci. Rep.* **2017**, *7*, 9026. [CrossRef] [PubMed]

119. Appusamy, K.; Blair, S.; Nahata, A.; Guruswamy, S. Low-loss magnesium films for plasmonics. *Mater. Sci. Eng. B* **2014**, *181*, 77–85. [CrossRef]

120. Jeong, H.H.; Mark, A.G.; Fischer, P. Magnesium plasmonics for UV applications and chiral sensing. *Chem. Commun.* **2016**, *52*, 12179–12182. [CrossRef]

121. Sterl, F.; Strohfeldt, N.; Walter, R.; Walter, R.; Griessen, R.; Tittl, A.; Giessen, H. Magnesium as Novel Material for Active Plasmonics in the Visible Wavelength Range. *Nano Lett.* **2015**, *15*, 7949–7955. [CrossRef]

122. Wu, P.C.; Losurdo, M.; Kim, T.H.; Giangregorio, M.; Bruno, G.; Everitt, H.O.; Brown, A.S. Plasmonic Gallium Nanoparticles on Polar Semiconductors: Interplay between Nanoparticle Wetting, Localized Surface Plasmon Dynamics, and Interface Charge. *Langmuir* **2009**, *25*, 924–930. [CrossRef]

123. Knight, M.W.; Coenen, T.; Yang, Y.; Brenny, B.J.; Losurdo, M.; Brown, A.S.; Everitt, H.O.; Polman, A. Gallium plasmonics: Deep subwavelength spectroscopic imaging of single and interacting gallium nanoparticles. *ACS Nano* **2015**, *9*, 2049–2060. [CrossRef] [PubMed]

124. Wu, P.C.; Losurdo, M.; Kim, T.-H.; Garcia-Cueto, B.; Moreno, F.; Bruno, G.; Brown, A.S. Ga–Mg Core–Shell Nanosystem for a Novel Full Color Plasmonics. *J. Phys. Chem. C* **2011**, *115*, 13571–13576. [CrossRef]

125. Nguyen, B.H.; Nguyen, V.H. Advances in graphene-based optoelectronics, plasmonics and photonics. *Adv. Nat. Sci.-Nanosci. Nanotechnol.* **2016**, *7*, 013002. [CrossRef]

126. Szunerits, S.; Boukherroub, R. Graphene-based biosensors. *Interface Focus* **2018**, *8*, 20160132. [CrossRef]

127. Rodrigo, D.; Limaj, O.; Janner, D.; Etezadi, D.; de Abajo, F.J.G.; Pruneri, V.; Altug, H. Mid-infrared plasmonic biosensing with graphene. *Science* **2015**, *349*, 165–168. [CrossRef] [PubMed]

128. Hu, H.; Yang, X.; Guo, X.; Khaliji, K.; Biswas, S.R.; Garcia de Abajo, F.J.; Low, T.; Sun, Z.; Dai, Q. Gas identification with graphene plasmons. *Nat. Commun.* **2019**, *10*, 1131. [CrossRef] [PubMed]

129. Hu, H.; Yang, X.; Zhai, F.; Hu, D.; Liu, R.; Liu, K.; Sun, Z.; Dai, Q. Far-field nanoscale infrared spectroscopy of vibrational fingerprints of molecules with graphene plasmons. *Nat. Commun.* **2016**, *7*, 12334. [CrossRef] [PubMed]

130. Losurdo, M.; Yi, C.; Suvorova, A.; Rubanov, S.; Kim, T.H.; Giangregorio, M.M.; Jiao, W.; Bergmair, I.; Bruno, G.; Brown, A.S. Demonstrating the capability of the high-performance plasmonic gallium-graphene couple. *ACS Nano* **2014**, *8*, 3031–3041. [CrossRef]

131. Pau, J.L.; García-Marín, A.; Hernández, M.J.; Lorenzo, E.; Piqueras, J. Optical biosensing platforms based on Ga-graphene plasmonic structures on Cu, quartz and SiO2/Si substrates. *Phys. Status Solidi (b)* **2016**, *253*, 664–670. [CrossRef]

132. Zeng, S.; Sreekanth, K.V.; Shang, J.; Yu, T.; Chen, C.K.; Yin, F.; Baillargeat, D.; Coquet, P.; Ho, H.P.; Kabashin, A.V.; et al. Graphene-Gold Metasurface Architectures for Ultrasensitive Plasmonic Biosensing. *Adv. Mater.* **2015**, *27*, 6163–6169. [CrossRef] [PubMed]

133. Mahigir, A.; Chang, T.W.; Behnam, A.; Liu, G.L.; Gartia, M.R.; Veronis, G. Plasmonic nanohole array for enhancing the SERS signal of a single layer of graphene in water. *Sci. Rep.* **2017**, *7*, 14044. [CrossRef] [PubMed]

134. Barho, F.B.; Gonzalez-Posada, F.; Milla-Rodrigo, M.J.; Bomers, M.; Cerutti, L.; Taliercio, T. All-semiconductor plasmonic gratings for biosensing applications in the mid-infrared spectral range. *Opt. Express* **2016**, *24*, 16175–16190. [CrossRef] [PubMed]

135. Agarwal, A.; Vitiello, M.S.; Viti, L.; Cupolillo, A.; Politano, A. Plasmonics with two-dimensional semiconductors: From basic research to technological applications. *Nanoscale* **2018**, *10*, 8938–8946. [CrossRef] [PubMed]

136. Kim, J.; Choudhury, S.; DeVault, C.; Zhao, Y.; Kildishev, A.V.; Shalaev, V.M.; Alù, A.; Boltasseva, A. Controlling the Polarization State of Light with Plasmonic Metal Oxide Metasurface. *ACS Nano* **2016**, *10*, 9326–9333. [CrossRef] [PubMed]

137. Maccaferri, N.; Inchausti, X.; García-Martín, A.; Cuevas, J.C.; Tripathy, D.; Adeyeye, A.O.; Vavassori, P. Resonant Enhancement of Magneto-Optical Activity Induced by Surface Plasmon Polariton Modes Coupling in 2D Magnetoplasmonic Crystals. *ACS Photonics* **2015**, *2*, 1769–1779. [CrossRef]

138. Caballero, B.; García-Martín, A.; Cuevas, J.C. Hybrid Magnetoplasmonic Crystals Boost the Performance of Nanohole Arrays as Plasmonic Sensors. *ACS Photonics* **2016**, *3*, 203–208. [CrossRef]

139. Naik, G.V.; Shalaev, V.M.; Boltasseva, A. Alternative Plasmonic Materials: Beyond Gold and Silver. *Adv. Mater.* **2013**, *25*, 3264–3294. [CrossRef]

140. Bartholomew, R.; Williams, C.; Khan, A.; Bowman, R.; Wilkinson, T. Plasmonic nanohole electrodes for active color tunable liquid crystal transmissive pixels. *Opt. Lett.* **2017**, *42*, 2810–2813. [CrossRef]

141. Lee, Y.; Kim, S.J.; Yun, J.G.; Kim, C.; Lee, S.Y.; Lee, B. Electrically tunable multifunctional metasurface for integrating phase and amplitude modulation based on hyperbolic metamaterial substrate. *Opt. Express* **2018**, *26*, 32063–32073. [CrossRef]

142. Abbas, A.; Linman, M.J.; Cheng, Q. New trends in instrumental design for surface plasmon resonance-based biosensors. *Biosens. Bioelectron.* **2011**, *26*, 1815–1824. [CrossRef]

143. Tu, L.; Huang, L.; Wang, W. A novel micromachined Fabry-Perot interferometer integrating nano-holes and dielectrophoresis for enhanced biochemical sensing. *Biosens. Bioelectron.* **2019**, *127*, 19–24. [CrossRef] [PubMed]

144. Verschueren, D.V.; Pud, S.; Shi, X.; De Angelis, L.; Kuipers, L.; Dekker, C. Label-Free Optical Detection of DNA Translocations Through Plasmonic Nanopores. *ACS Nano* **2018**, *3*, 61–70. [CrossRef] [PubMed]

145. Wu, D.Y.; Li, J.F.; Ren, B.; Tian, Z.Q. Electrochemical surface-enhanced Raman spectroscopy of nanostructures. *Chem. Soc. Rev.* **2008**, *37*, 1025–1041. [CrossRef] [PubMed]

146. Liao, W.C.; Annie Ho, J.A. Improved activity of immobilized antibody by paratope orientation controller: Probing paratope orientation by electrochemical strategy and surface plasmon resonance spectroscopy. *Biosens. Bioelectron.* **2014**, *55*, 32–38. [CrossRef] [PubMed]

147. Li, N.; Lu, Y.; Li, S.; Zhang, Q.; Wu, J.; Jiang, J.; Liu, G.L.; Liu, Q. Monitoring the electrochemical responses of neurotransmitters through localized surface plasmon resonance using nanohole array. *Biosens. Bioelectron.* **2017**, *93*, 241–249. [CrossRef] [PubMed]

148. Patskovsky, S.; Dallaire, A.-M.; Blanchard-Dionne, A.-P.; Vallée-Bélisle, A.; Meunier, M. Electrochemical structure-switching sensing using nanoplasmonic devices. *Ann. Phys.* **2015**, *527*, 806–813. [CrossRef]

149. Lopez, G.A.; Estevez, M.C.; Soler, M.; Lechuga, L.M. Recent advances in nanoplasmonic biosensors: Applications and lab-on-a-chip integration. *Nanophotonics* **2017**, *6*, 123–136. [CrossRef]

150. Belkin, M.; Chao, S.H.; Jonsson, M.P.; Dekker, C.; Aksimentiev, A. Plasmonic Nanopores for Trapping, Controlling Displacement, and Sequencing of DNA. *ACS Nano* **2015**, *9*, 10598–10611. [CrossRef]

materials

MDPI

Review

Sum-Frequency Generation Spectroscopy of Plasmonic Nanomaterials: A Review

Christophe Humbert [1,*], **Thomas Noblet** [1], **Laetitia Dalstein** [1,†], **Bertrand Busson** [1] and **Grégory Barbillon** [2,*]

1   Univ Paris-Sud, Université Paris-Saclay, Laboratoire de Chimie Physique, CNRS, Batiment 201 P2, 91405 Orsay, France; thomas.noblet@u-psud.fr (T.N.); dalstein@gate.sinica.edu.tw (L.D.); bertrand.busson@u-psud.fr (B.B.)
2   EPF-Ecole d'Ingénieurs, 3 bis rue Lakanal, 92330 Sceaux, France
*   Correspondence: christophe.humbert@u-psud.fr (C.H.); gregory.barbillon@epf.fr (G.B.)
†   Current address: Institute of Physics, Academia Sinica, Taipei 11529, Taiwan.

Received: 21 February 2019; Accepted: 5 March 2019; Published: 12 March 2019

Abstract: We report on the recent scientific research contribution of non-linear optics based on Sum-Frequency Generation (SFG) spectroscopy as a surface probe of the plasmonic properties of materials. In this review, we present a general introduction to the fundamentals of SFG spectroscopy, a well-established optical surface probe used in various domains of physical chemistry, when applied to plasmonic materials. The interest of using SFG spectroscopy as a complementary tool to surface-enhanced Raman spectroscopy in order to probe the surface chemistry of metallic nanoparticles is illustrated by taking advantage of the optical amplification induced by the coupling to the localized surface plasmon resonance. A short review of the first developments of SFG applications in nanomaterials is presented to span the previous emergent literature on the subject. Afterwards, the emphasis is put on the recent developments and applications of the technique over the five last years in order to illustrate that SFG spectroscopy coupled to plasmonic nanomaterials is now mature enough to be considered a promising research field of non-linear plasmonics.

Keywords: nanoparticles; non-linear optics; surface plasmons; sum-frequency generation spectroscopy; interfaces; gold

1. Introduction

Presently, the use of the physico-chemical properties of nanomaterials [1–3] in everyday life has become a reality, whether in electronic instruments or medical applications. Thus, their antiseptic action is, for instance, used in razor blades covered with a coating of silver nanoparticles (AgNPs). They are also introduced in some medical treatments such as in cancer therapy to selectively destroy some tumors. Besides, the society developed an interest for a better understanding of the particle interaction with their environment. More generally, metallic nanoparticles arouse a growing interest by the combination of their unique properties with promising developments in areas as diverse as catalysis [4], biophysics [5] or medicine [6], especially by taking advantage of their optical properties in the visible spectral range as in the case of gold nanoparticles (AuNPs) [7–10]. More technologically advanced applications also appeared, in computer processors for instance, and some of these nanoparticles are incorporated in medical treatments such as cancer therapy wherein they are employed to selectively kill tumors [11]. Nevertheless, at this nanoscale, monitoring the action of nano-objects on their immediate environment at molecular level is a sensitive issue because many of their above-mentioned properties are not well understood, which may pose risks to the living body. To better understand these issues of toxicity and health potential for the society, new research tools adapted to the study of interactions between nano-objects and their environment are absolutely necessary [3]. Among the numerous techniques studying nanoparticles such as X-ray spectroscopies [12] or scanning probe microscopies (Atomic Force Microscopy, Scanning Tunneling Microscopy) [13], non-linear optics based on Sum-Frequency Generation (SFG) spectroscopy [14] is distinguished by its surface specificity which allows addressing various fundamental or applied issues related to the chemical functionalization of materials (metals, semiconductors, insulators) from the microscale to the atomic scale, over both the infrared (IR) and visible spectral ranges. The latter affords complementary information with respect to PM-IRRAS (Polarization Modulation-Infrared Reflection-Adsorption Spectroscopy) [15] and SERS (Surface-Enhanced Raman Spectroscopy) [16–19]. Moreover, SFG is now a competitive tool: its sensitivity has indeed increased over the past few years thanks to technical and scientific developments at the picosecond and femtosecond time scales. The kinetics and dynamics of the vibrational electronic coupling, charge and energy transfers constitute the core of the physico-chemical properties that can be investigated in this way [20–25].

At the nanoscale, SFG was first performed in the 1990s on nanomaterials such as Pd and Pt nanoparticles used as catalysts in a gaseous environment in order to reach higher conversion efficiency of carbon monoxide (CO) in carbon dioxide (CO_2) as a function of pressure or/and temperature with respect to their crystalline counterpart [26,27]. It also includes some very recent work concerning Li-Ion batteries based on silicon nanoparticles [28]. From the beginning, SFG was also used to unveil the adsorption/desorption mechanisms on the catalysts surface as reported in numerous studies [29,30]. The first experiment on nanoparticles at the air/liquid interface was reported in 2000 on a cationic surfactant deposited on AuNPs stabilized in their anionic form and forming a monolayer (not very stable) on a silicon substrate [31]. The presence of the surfactant on the AuNPs was thus demonstrated as well as a potential effect of amplification of the surfactant signal. From this moment, the hypothesis of an amplification of the non-linear optical response of the material by the simultaneous excitation of the AuNP surface plasmon resonance (SPR) has been assumed. Indeed, the classic SFG process is produced by mixing two incident IR and visible laser beams on the sample as developed and explained in the corresponding section. With the visible laser beam, it is likely to probe electronic transitions: metal interband transitions, charge transfers, molecular electronic transitions, and SPRs of nanoparticles.

2. Sum-Frequency Generation Spectroscopy of Metallic Nanoparticles

2.1. Fundamentals of Sum-Frequency Generation Spectroscopy

For molecules, optical processes can be described in terms of energy diagrams. In a practical way, with respect to the surface or bulk properties of materials, non-linear optical spectroscopies up to the order 3 are routinely used, for example through the 4-photon mixing encountered in Coherent Anti-Stokes Raman Scattering (CARS) or Coherent Stokes Raman Scattering (CSRS). The reader interested in a comparison of CARS developments on metallic nanostructures with respect to some SFG results can refer to the review written in 2014 by Lis and co-workers [32]. It is worth noting that in a general way, first order spectroscopies such as IR/UV-Visible absorption and emission allow probing of the linear first order susceptibility $\chi^{(1)}$ of materials, while Second Harmonic Generation (SHG) and Sum/Difference-Frequency Generation (SFG/DFG) spectroscopies give access to the non-linear optical second-order susceptibility $\chi^{(2)}$. To finish, Raman/SERS/CARS spectroscopies probe the non-linear third order susceptibility $\chi^{(3)}$. Figure 1 summarizes in terms of simple molecular energy scheme how non-linear processes of order 2 and 3 occur. It also emphasizes the differences and links that exist between them and the basic techniques of infrared and Raman spectroscopies. The common point of these techniques is obviously the access to specific vibration modes ω_{vg} of the constitutive molecules present in the studied materials. It is worth noting that the Raman and non-linear processes exist even in the absence of real electronic states in the molecule. Conversely, if the visible laser beam is tuned to match the energy of real electronic states, we may significantly enhance the efficiency of these mechanisms and improve the sensitivity detection threshold, as performed in Resonant Raman or

Two-Color Sum-Frequency Generation (2C-SFG) spectroscopies. In the latter case, it becomes possible to highlight vibronic coupling inside molecules, which is called Doubly Resonant Sum-Frequency Generation (DRSFG) process [33]. When molecules are adsorbed on a (nanostructured) interface, 2C-SFG can benefit from the electronic properties of the substrate to amplify, under certain conditions, the molecular response through coupling mechanisms between both constituents: molecular adsorbate and substrate [34–38].

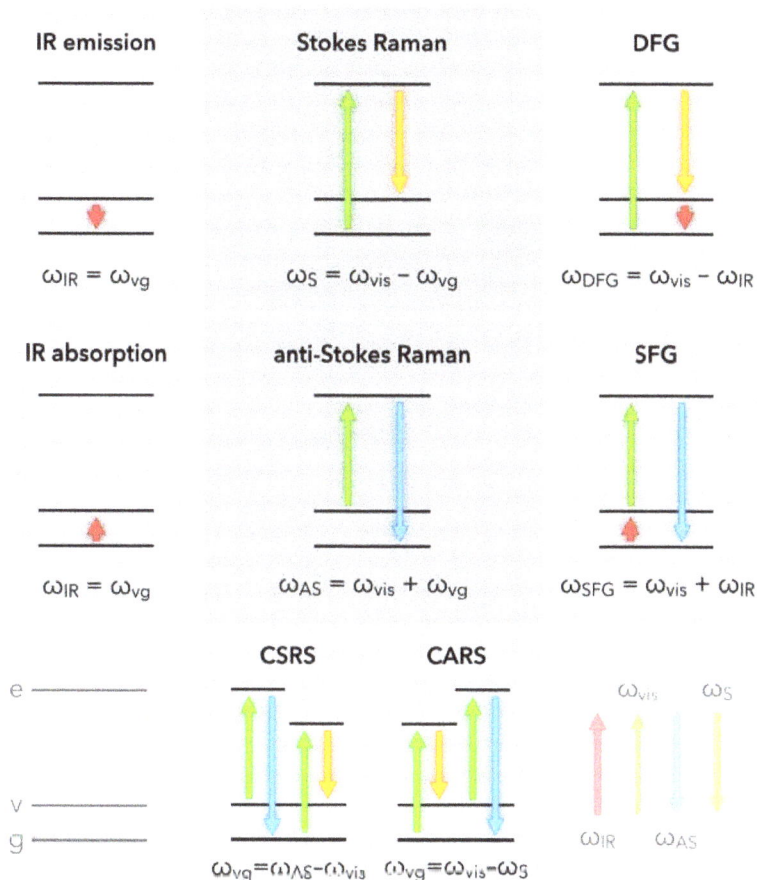

Figure 1. Molecular energy diagrams used for IR, Raman, SFG/DFG and CARS/CSRS spectroscopy. SFG = Sum-Frequency Generation; DFG = Difference-Frequency Generation; CARS = Coherent Anti-Stokes Raman Scattering; CSRS = Coherent Stokes Raman Scattering; S = Stokes; AS = Anti-Stokes; Energy level code: e = virtual electronic state, v = vibrational state, g = ground state.

Fundamentally, SFG spectroscopy is based on a non-linear second-order optical process involving three photons mixed in a medium or at an interface, as illustrated in Figure 2 in the case of functionalized nanoparticles grafted on silicon. To be efficient, it requires intense laser beams to interact coherently at the same point of the probed interface so that the energy ($\omega_{\text{SFG}} = \omega_{\text{VIS}} + \omega_{\text{IR}}$) and the parallel component (*x*-direction) of the momentum ($k_{\text{SFG}} = k_{\text{VIS}} + k_{\text{IR}}$) conservation rules are met. As drawn in this picture, similar relationships are obtained for Difference-Frequency Generation

(DFG) spectroscopy. The interest of this complementary sister spectroscopy will be evidenced later for metallic nanoparticles.

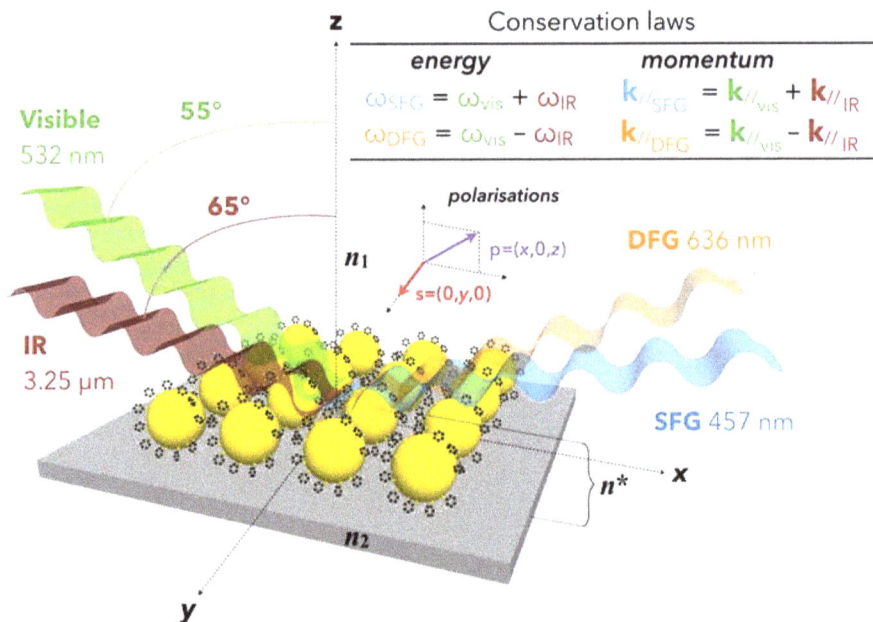

Figure 2. Sketch of SFG/DFG spectroscopy at a nanostructured interface: Si/AuNPs + adsorbed thiophenol/ambient air. The refractive index of each medium constituting the interface is noted n_i (n_1 for air, n_2 for silicon) while that of the interface, noted n^*, must be calculated. The common polarization plane used in SFG spectroscopy depicts the geometry of the linear polarization p (x,0,z) and s (0,y,0) of the three laser beams depending of the chosen experimental scheme.

Due to the complex structure of such an interface, each of the three interfacial components (related to their respective refractive indices n_1 for air, n_2 for silicon and n^* for the AuNPs/thiophenol) drawn in Figure 2 has to be considered in the description of the SFG process. The access to these quantities depends on the choice of the electric field polarization of the IR, Visible, and SFG beams. In our case, as depicted in this figure and following the convention of the literature for the SFG processes, the three beams have always a linear polarization defined either parallel (p) or perpendicular (s) to the incidence plane (xz) of the three involved beams. In other words, p-polarization refers to x- and z-components of the electric fields while s-polarization refers to the y-component of the electric fields. In the framework of the electric dipole approximation, and considering the molecular energy levels depicted in Figure 1, the non-linear SFG/DFG processes simultaneously follow the IR and Raman molecular selection rules. In other words, a vibrational transition is detected in SFG spectroscopy if it is both IR and Raman-active, which generally happens upon adsorption on any substrate. Indeed, the adsorption process soften the spectroscopic exclusion rules between IR and Raman processes, allowing probing of molecules of simple or complex geometry such as carbon monoxide (CO) or Buckminsterfullerenes (C_{60}) in chemical interaction with platinum or silver, respectively. As an interface constitutes the place where the centrosymmetry of the electronic properties in the bulk of two different media is broken, SFG spectroscopy is a choice technique to probe complex surfaces and interfaces because it remains intrinsically blind with respect to the bulk contributions located on both sides of the target area (e.g., molecules in the gas phase or in solution). On the contrary, IR and Raman spectroscopies encompass bulk and interfaces contributions, at the exception of PM-IRRAS and SERS for particular metallic substrates.

The second-order polarizability (hyperpolarizability: β) characterizing molecules in SFG is directly related to a change in the Raman polarizability α and the IR dipole moment μ of the vibrational transitions involved in the process. Moreover, we notice a direct link between β (molecule) and $\chi^{(2)}$ (macroscopic) via a coordinate transformation T making it possible to switch from the molecule reference system to that of the whole sample as illustrated in Figure 3 for the carbon monoxide. Actually, the symmetry rules in the 3-photon mixing of SFG/DFG mean that the quantities β and $\chi^{(2)}$ are third-rank tensors, i.e., they have 27 components measurable by multiple combinations of the possible orientations (polarizations) of the 3 associated electric fields (SFG, Visible, Infrared) in each direction (x, y, z) of our three-dimensional space. Experimentally, we usually measure $\chi^{(2)}$ and we can theoretically retrieve β, taking into account the average over all molecular orientations in the macroscopic interface. Given the complexity of the β formulation illustrated very early by Franken [39], Ward [40] and later by Hirose [41] for transformation T, this is currently done only for simple cases with well-defined geometries and controlled geometries, and when only the molecular signal contributes to the SFG response for instance in catalysts [42]. Nevertheless, this is already partially done for basic structures in order to find the average orientation of molecules deposited on a surface in the case of Self-Assembled Monolayers (SAM) or polymers, and even for biosurfaces [43]. In summary, at least one component of the $\chi^{(2)}$ of any interface should be non-zero in order to detect SFG signal. The researcher interested by a comprehensive and pedagogic description of the microscopic and macroscopic relations between β and $\chi^{(2)}$ on any kind of interface can read the tutorial approach of SFG spectroscopy elaborated by Lambert et al. in reference [44]. Hence provided with the basics of molecular SFG spectroscopy, we can now extend it to nanostructured interfaces in order to highlight the role played by the optical properties of nanoparticles.

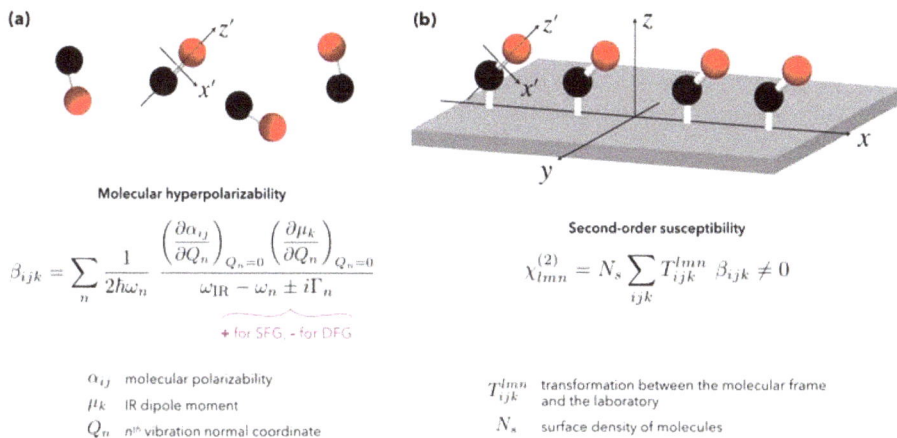

(a)

Molecular hyperpolarizability

$$\beta_{ijk} = \sum_n \frac{1}{2\hbar\omega_n} \frac{\left(\frac{\partial \alpha_{ij}}{\partial Q_n}\right)_{Q_n=0} \left(\frac{\partial \mu_k}{\partial Q_n}\right)_{Q_n=0}}{\omega_{IR} - \omega_n \pm i\Gamma_n}$$

+ for SFG, - for DFG

α_{ij} molecular polarizability
μ_k IR dipole moment
Q_n n^{th} vibration normal coordinate

(b)

Second-order susceptibility

$$\chi^{(2)}_{lmn} = N_s \sum_{ijk} T^{lmn}_{ijk} \beta_{ijk} \neq 0$$

T^{lmn}_{ijk} transformation between the molecular frame and the laboratory
N_s surface density of molecules

Figure 3. Microscopic view of the hyperpolarizability β of a carbon monoxide molecule (**a**) and macroscopic view of the non-linear second-order susceptibility $\chi^{(2)}$ of a carbon monoxide layer adsorbed on a platinum substrate (**b**). β and $\chi^{(2)}$ are third-rank tensors with 27 components that can be probed by SFG spectroscopy as a function of the SFG, Visible and IR laser beams polarization: p or s as defined in Figure 2. These non-linear physical values are related through a coordinate transformation T from the molecular frame to the laboratory (interface).

In Figure 2, we have to take into account all the components of the interface including those of the substrate, by giving a complete description of the effective refractive index n^*. Indeed, for SFG spectroscopy performed at interfaces, the substrate cannot be neglected: for instance, in the case of metal or semiconductor materials, wherein $\chi^{(2)}$ is non-zero, the substrate often gives rise to a more intense SFG signal than the molecules because of the free and/or bound electrons present in the materials. The strong illumination by the incident visible laser beam of these objects translates into the excitation of the electron interband and intraband electronic transitions in metals or the creation of excitons (electron-hole pairs) in semiconductors. In both cases, the substrate generates the so-called non-resonant SFG contribution. At the opposite, the molecular SFG signal is called resonant (with respect to the IR energy) when the incident IR laser beam frequency corresponds to that of the molecular vibration modes. For the sake of clarity, the case of molecules exhibiting electronic transitions in the visible spectral range will not be considered in this report. The interested reader will nevertheless find experimental results showing some adsorbed molecules which are resonant in both IR and visible spectral ranges, evidencing vibro-electronic SFG processes in various kinds of physical [45–47], chemical [33] and biological [48] systems. The description of the SFG processes arising from gold surfaces is a difficult task that has been extensively addressed experimentally and theoretically because it must take into account the non-linear contribution coming from the surface and the bulk of the specific electronic properties of this metal in the visible spectral range. A clear and complete description of the physics at play in its $\chi^{(2)}$ is reported in reference works [49–51], based on SFG measurements performed on molecular monolayers adsorbed on flat gold surfaces. In these works, the molecules act as a reference probe of the electronic properties of gold because the SFG spectral shape of the vibration modes (dips, peak, Fano) is modulated by the color choice of the incident visible laser beam ranging from violet to red. This is based on a three-layer model similar to the image depicted in Figure 2 by considering air/molecules/gold instead of air/AuNPs+thiophenol/Si. It has been shown that the contribution of the free and bound electrons explains the shape modulation of the SFG spectra at this interface [52]. To date, such models have again to be applied to the nanostructured interfaces depicted in Figure 2. Moreover, in the SFG process, the incident visible wavelength is set to 532 nm in order to enhance the AuNPs localized surface plasmon resonance (LSPR) located between 520 and 540 nm corresponding to the case of gold spheres with a diameter of 15 nm. The IR wavelength set at 3.25 µm in the SFG process is chosen to be resonant with the C–H stretching vibration mode of the phenyl ring. This is not related to LSPR properties but to the molecular vibrations of the thiophenol adsorbed on AuNPs. It allows demonstration of the LSPR-SFG coupling, i.e., the enhancement of the C-H vibration mode in the LSPR electric field. In the case of AuNPs, choosing a 632 nm or 785 nm visible wavelength will not allow easy demonstration of a LSPR-SFG coupling because it is located too far from the LSPR maximum at 532 nm (see Figure 1 and Section 2.2).

2.2. Nanoparticles and Sum-Frequency Generation: The Gold Case

The founding SFG spectroscopic measurements on AuNPs were performed as illustrated in Figure 4 by the Davies [31] and Tadjeddine [53] teams in 2000 and 2005, respectively. The nanoparticles were synthesized by the classical Turkevich method [54]. In these conditions, they have a spherical shape of 15 nm diameter and are spread or grafted on silicon or glass substrate, respectively. As for the latter, AuNPs are grafted by silanization of the glass substrate with a monolayer of APTES molecules (3-(aminopropyl)triethoxysilane). The incident visible laser beam is fixed in the green (532 nm) while the incident IR beam is tuned between 2800 cm^{-1} and 3000 cm^{-1}. The molecular probe of the AuNPs surface in Figure 4a is a surfactant co-adsorbed with the AuNPs (dioctadecyldimethylammonium chloride, DODAC) while it is chemically grafted after deposition of the AuNPs in Figure 4b (dodecanethiol, DDT). In both cases, five typical vibration modes are detected by SFG spectroscopy. They come from the DODAC and DDT molecules and are located at IR wavenumbers corresponding to the vibrational fingerprint of the methyl (CH_3) and methylene (CH_2) vibration modes as pointed on both SFG spectra. Nevertheless, we observe that their spectral shapes differ. This effect is characteristic of SFG spectroscopy and, as highlighted in the previous section, finds its origin in the non-resonant contribution of the interface. In other words, the different ways the DODAC and DDT molecules interact with the AuNPs and the silicon/glass substrates after the surface preparation is put in evidence by SFG measurements. In Figure 4, vibration modes appear peak- or dip-shaped. In the case of DODAC on AuNPs/Si, the strong reflectivity of the interface and the high surface density of AuNPs related to the surface preparation favors molecular detection in SFG. The latter is amplified by the LSPR coupled to the vibration modes because of the enhancement of the surrounding electric field. The LSPR of a gold colloidal solution is evidenced by UV-Visible absorption spectroscopy as depicted in Figure 5a.

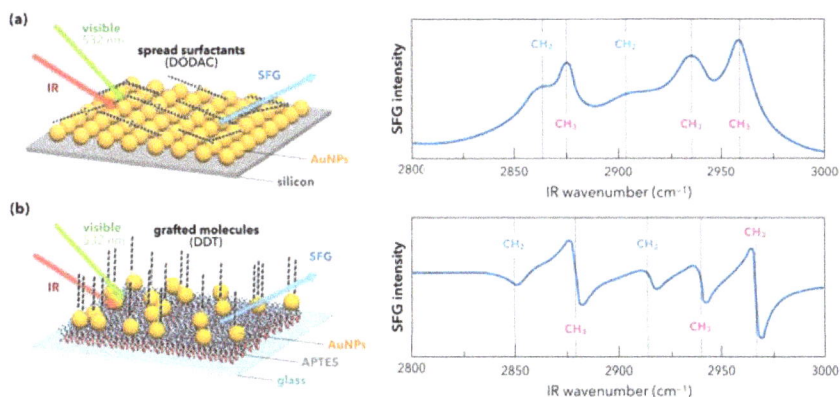

Figure 4. Sketch of the first reference SFG experiments performed on interfaces constituted of 15 nm diameter AuNPs synthesized by Turkevich method. (**a**) Scheme of the DODAC/AuNPs/Si interface (no grafting of AuNPs or DODAC molecules at any step). Each constituent is just deposited or spread, respectively. The corresponding SFG signal displays the five strong methyl and methylene peak-shaped vibration modes of the DODAC molecules (picture adapted from reference [31]). (**b**) Scheme of the DDT/AuNPs/Glass interface (grafting of AuNPs on glass through APTES and chemical functionalization of AuNPs by DDT molecules). The corresponding SFG signal displays the five weak methyl and methylene dip-shaped vibration modes of the DDT molecules (picture adapted from reference [53]).

This enhancement is not clear for DDT adsorbed on AuNPs/glass given the poor reflectivity of glass with respect to silicon and the weaker AuNPs surface density. The broad and strong gold interband $s - d$ electronic transition, coming from the electronic state density of this metal (Figure 5b), is centered around 480 nm. It contributes overwhelmingly to the total SFG signal and interferes destructively with the DDT SFG molecular response, giving an SFG spectrum similar to that encountered on a flat gold surface [50]. Nevertheless, it is worth noting that the gold surface density on this nanostructured interface is 4% with respect to the equivalent gold surface area, which strengthens the fact that SFG spectroscopy is sensitive to the chemistry of functionalized nanoparticles. The LSPR effect is hampered by the interband electronic transition in the SFG process because the SFG wavelength coincides with this latter. We will see further in this review that the tuning of the incident laser beam from blue to red allows playing with the gold LSPR and interband electronic properties in order to favor molecular detection. These interference effects in the SFG process can be easily understood through a simple fitting procedure of the SFG intensity which consists in the square norm of the effective non-linear second-order susceptibility $\chi^{(2)}_{\text{interface}}$ of the interface [53]. Since the optical properties of silicon/glass substrates do not have any significant contribution in the SFG process, they are neglected, which leads to the following expression based on the definitions of mathematical complex values:

$$\left\|\chi_{\text{interface}}^{(2)}\right\|^2 = \left\|\chi_{\text{substrate}}^{(2)} + \chi_{\text{adsorbate}}^{(2)}\right\|^2 = \left\|\chi_{\text{non-resonant}}^{(2)} + \chi_{\text{resonant}}^{(2)}\right\|^2$$

$$= \left\|\chi_{\text{AuNPs}}^{(2)} + \chi_{\text{molecules}}^{(2)}\right\|^2,$$

with:

$$\chi_{\text{AuNPs}}^{(2)} = A\, e^{i\Phi} \quad \text{and} \quad \chi_{\text{molecules}}^{(2)} = \sum_{j=1}^{n} \frac{a_j\, e^{i\varphi_j}}{\omega_{\text{ir}} - \omega_j \pm i\Gamma_j}. \tag{1}$$

This description is intrinsic to SFG/DFG spectroscopy because of the properties of non-linear second-order susceptibilities (third-rank tensors) [55]. In this equation, the molecular non-linear second-order susceptibility is described as a sum of complex Lorentzian functions (+ for SFG, − for DFG in the imaginary part of Equation (1)) corresponding to each vibration mode (i.e., $n = 5$ in the two previous examples of DODAC and DDT). From these considerations, it is clear that an interference factor (phase shift $\Phi - \varphi_j$) between the substrate and the adsorbate will affect more or less strongly the spectral shape as a function of the electronic properties of gold, i.e., its $s - d$ interband transition whether it is nanostructured or not. It is worth noting that the phase shift in the SFG spectra is a unique physical marker of the different components of the interface in the optical range with respect to the information afforded by PM-IRRAS or SERS spectroscopy. .

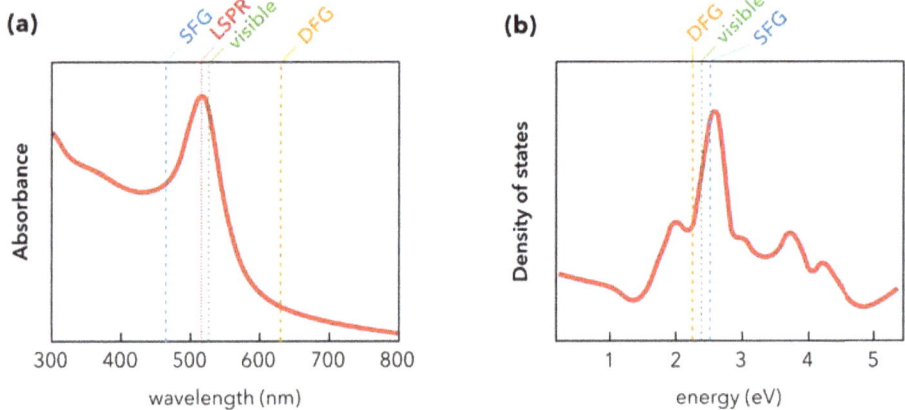

Figure 5. Electronic and optical properties of gold. (**a**) Localized surface plasmon resonance (LSPR) of AuNPs in aqueous solution displayed as a function of the incident wavelength as measured by UV-Visible spectroscopy. (**b**) Scheme of the density of electronic states (DOS) of bulk gold as a function of the incident energy. Pictures (**a**,**b**) are tagged with the SFG and DFG energy for an incident visible beam of 532 nm wavelength and an incident IR beam of 3 μm. In these conditions, it is clear that although the SFG/DFG/visible beams are located near the LSPR maximum in (**a**), the contribution of the gold DOS (main $s - d$ interband electronic transition located at 487 nm) in (**b**) may interfere and drastically hamper a possible LSPR-SFG coupling.

Afterwards in 2006–2007, Benderskii's team used SFG spectroscopy as a sensitive probe of the molecular conformation in a nanoscale geometry, which determines the efficiency of the charge (electron) transfer in applications. They used femtosecond laser beam with an incident visible beam fixed at 800 nm wavelength: this enables to get rid of any LSPR or interband contributions likely to interfere with the DDT SFG signal. Moreover, in order to favor the surface reflectivity on their CaF$_2$ substrate, they performed SFG spectroscopy in ssp-polarization configurations (s: SFG beam, s: Visible beam, p: IR beam) which is more efficient than in ppp-polarization for any glass substrate (see Figure 2 for the definition of the polarization plane used in SFG experiments). Once again, this is related to the tensor nature of the non-linear second-order susceptibility of the interfaces $\chi^{(2)}_{interface}$ which also includes the surface reflectivity in order to quantitatively model the SFG signal intensity as detailed in reference works for the interested reader [14,50]. Benderskii's team deduced information about the average orientation (tilt angle) of the alkane chains of DDT molecules (1.6 nm length) adsorbed on submonolayer films of metallic nanoparticles (DDT-capped gold and silver nanoparticles). It was performed for different nanoparticles diameters ranging from 2 nm (Figure 6a) to 25 nm (Figure 6b) deposited on CaF$_2$ windows [56,57] and the following trend was observed: the relative SFG intensity of the methylene (CH$_2$) vibration modes increased with respect to the methyl (CH$_3$) ones as the AuNP size decreased (Figure 6). Contrary to what is observed on flat gold surfaces where well-ordered DDT molecules build a stable SAM (all-trans conformation) and where only the three methyl vibration modes were detected [50], the presence of methylene vibrations is a well-known indication of the presence of size-dependent gauche conformational defects. In other words, the smaller the AuNP radius, the bigger the surface curvature, which gives rise to gauche defects in the alkane chain due to the greater surface to volume ratio that can be probed by the DDT molecules during adsorption. A comparison with classical IR absorption and Raman measurements was performed on samples with similar preparation protocols but with thicker layers, given the weaker detection limits of those techniques: no systematic/coherent change was detected in the respective intensities of methyl and methylene vibration modes. This experiments perfectly illustrate the high intrinsic sensitivity of SFG to molecular conformation on flat or curved surfaces. Besides, the same trend was observed for similar experiments performed on silver nanoparticles (AgNPs) [57]. Regarding silver, it should also be noted that a 2C-SFG experiment was conducted in 2008 by Traverse and co-workers on a stack of nine layers of AgNPs (1 nm) embedded in Si$_3$N$_4$ layers (9 nm thick) on silicon. In this experiment, a weak SFG signal related to the 2D electronic states of this special structure only appeared when the incident visible beam energy matched that of the AgNP SPR located at 420 nm [58].

A specific experimental improvement was achieved by Tourillon and co-workers from 2007 to 2009 in order to develop highly sensitive platforms based on AuNPs in an original optical scheme where SFG spectroscopy is performed in total internal reflection configuration for the IR, visible and SFG beams (TIR-SFG) as depicted in Figure 7a [59,60]. It was based on a previous experimental demonstration that the metal [61] or molecular [62] SFG yields can be increased by coupling one of the input waves (IR or/and Visible) to a surface polariton at the metal-air interface [62]. Tourillon's idea was to take profit of the building of dense AuNP monolayers or close-packed arrays of AuNPs adsorbed on a quartz prism in order to boost the SFG sensitivity. He could enhance the SFG signal of adsorbed dodecanethiol molecules of at least one magnitude order (with respect to a flat quartz plate where the SFG signal is detected in classical reflection, Figure 7b) [59]. He could also monitor the adsorption of avidin in the biotin-avidin recognition process [60], as the biosensor archetype [63]. It is worth noting that in the TIR-SFG configuration, the symmetry breaking of the interface is also coupled to a supplementary breaking of the local electric fields intensities (exponential decay of the electric field amplitude) involved in the non-linear process around the AuNPs, which could favor even more the enhancement of the SFG signal coupled to the LSPR phenomenon. Accordingly, the TIR-SFG configuration is exploited to address issues related to biological materials as developed in the Section 3.1 entitled "Towards applications of gold nanoparticles".

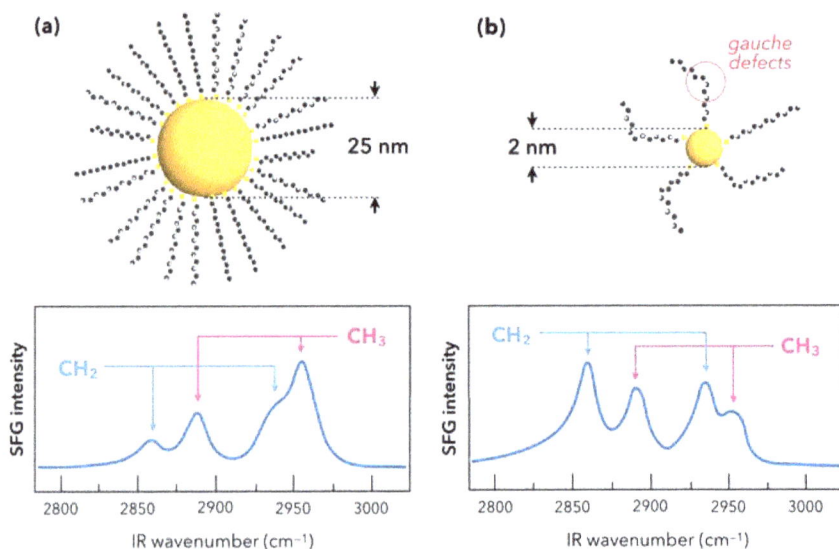

Figure 6. Evolution of the SFG vibrational fingerprint (C-H spectral range) of dodecanethiol molecules (DDT) adsorbed on AuNPs of 2 nm (**a**) and 25 nm (**b**) diameter, respectively. In Figure 6a, the methyl (CH₃) vibration modes intensity dominates the SFG contribution while in Figure 6b, the methylene (CH₂) groups dominate the SFG contribution. As detailed in the text and in references from which the picture is adapted [56,57], this is related to the density of gauche defects in the DDT alkane chains depending on the accessible surface and the ability to move in its surroundings.

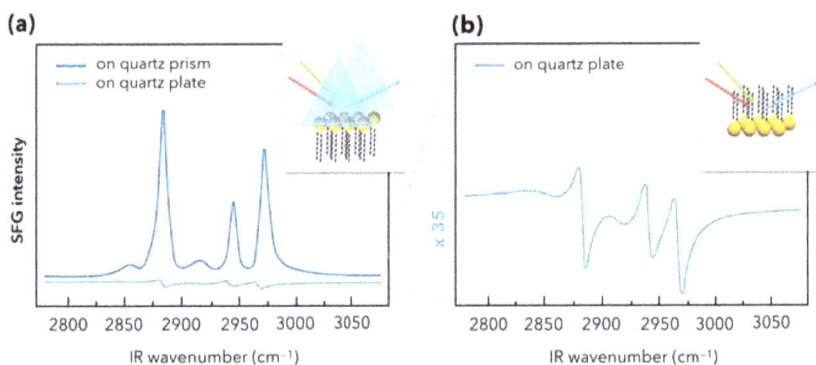

Figure 7. (**a**) SFG spectra of DDT molecules grafted on quartz surface (prism) in a total internal reflection SFG (TIR-SFG) configuration compared with the classical SFG reflection configuration on a quartz surface (plate). (**b**) Zoom of the SFG spectrum of the plate surface, illustrating the great differences in the respective spectral shapes and intensities. It is worth noting that the SFG spectra of the methyl (CH₃) vibration modes pointing out from the probed surface are strongly enhanced in the TIR-SFG configuration (35 times higher). The latter show that DDT molecules adopt a well-organized conformational order inside the self-assembled monolayer (SAM). Picture adapted from reference [59].

All the above-mentioned works do not try to model quantitatively the would-be coupling between AuNPs LSPR and the SFG response of surrounding molecules. A quantitative approach to this problem of combining SFG spectroscopy and plasmonics in AuNPs was first addressed by Pluchery and co-workers in a series of three publications during the 2009–2013 period [64–66]. In these works, the molecular probe was thiophenol (C_6H_5SH), a well-known reference molecule acting as a chemical pollutant to be degraded by (photo)catalysis supported by plasmonics. As a consequence, an abundant literature is related to this chemical organic compound which tackles this subject through linear and non-linear optical spectroscopy. Modelling is elaborated through density functional theory calculations of this molecule adsorbed on gold clusters for instance [67]. In a first step, the idea was to find a reference sample with respect to the nanostructured interface depicted in Figure 2 (Thiophenol/AuNPs/Si) to calculate an enhancement factor directly related to the AuNP LSPR (17 nm diameter) [64]. This naturally led to consider a flat crystalline surface of gold Au(111) covered by an organic SAM adsorbed by dipping the metal substrate in a solution of 1 mM thiophenol. By keeping identical conditions for the SFG measurements on the two samples, it has been shown that the SFG intensity enhancement factor F of the C–H stretching vibration mode (C–H of the phenyl ring: 3058 cm^{-1}) was 21 on the AuNP interface with respect to the Au(111) one, as illustrated in Figure 8a. It is explained by the energy of the incident visible laser beam (at 532 nm) which is located near the LSPR maximum absorbance (520 nm). However, the interband character of gold is always conserved in both samples leading to a dip-shaped spectral feature of the (C–H)-vibration. It is worth noting that the silicon electronic properties in the blue-violet (i.e., matching the SFG beam energy) also contributes to the non-resonant SFG background. We must likewise consider that there is only a filling factor of 10% of gold on the nanostructured sample with respect to Au(111) one when the incident IR and visible beams probe the same area. By comparing the F factor with the enhancement factor T of the local electric field intensity at the surface of a gold sphere in the electrostatic approximation of the Mie theory, the authors also showed that T has a similar value of 20. Besides, a supplementary comparison with the DFG response of the nanostructured sample shows that the dip-shaped feature observed in SFG becomes peak-shaped in DFG because of the sign change in the imaginary part of the Lorentzian function defined in Equation (1). Moreover, as the DFG energy is located at the edge of the interband electronic transition of gold, the non-resonant contribution is weaker than in the SFG case. While this first experimental quantitative illustration of the SFG-LSPR coupling was unambiguous, it was necessary to develop a complete modeling of the SFG intensity of the nanostructured interface related to the fundamentals equations of SFG spectroscopy, especially by calculating the contributions of the surface reflectivity (Figure 8c) to $\chi^{(2)}_{interface}$. It has been addressed and generalized in a second step by considering SFG and DFG measurements performed with the CLIO free electron laser (FEL) in Orsay (France) [65], allowing reaching of a different IR spectral range centered around 10 μm (Figure 8b), i.e., in a vibrational range where the strong Raman-active vibration modes of the thiophenol phenyl ring could be probed, again on samples depicted in Figure 2.

Therefore, the authors use the Maxwell-Garnett formalism in order to calculate the effective dielectric constant $\epsilon^* = (n^*)^2$ of the AuNPs by considering they are embedded in a host air matrix. In this way, it is possible to calculate the weight of the surface reflectivity in the non-linear second-order susceptibility. The knowledge of this optical parameter is crucial because it fixes or not the existence of the LSPR-SFG coupling, beyond a simple absorbance/reflectance effect of the surface. Indeed, it is evidenced that the normal contribution of this reflectivity (i.e., in the z-direction with respect to the sample surface plane) is not sufficient to explain the differences observed in the SFG and DFG spectra intensities of the phenyl ring vibration modes: these measurements and modeling proved that the existence of the coupling of SFG with plasmonics was consistent between the nanoscale (AuNPs) and the microscale (i.e., whatever the description used for the surface area probed by the incident IR and visible laser beams). Besides, at the molecular scale, it was shown that SFG/DFG measurements were consistent with DFT calculations [67] performed on a thiophenol molecule adsorbed on a gold atom, allowing deduction of the nature of the three vibration modes which was a long haul task especially for the 1075 cm^{-1} (v_4 in Figure 8b). The latter appears at this position after undergoing an energy blueshift of 15 cm^{-1} because of the thiophenol adsorption on gold. In summary, it is clear that the energies of the Visible/SFG/DFG processes of the interface of Figure 3 can match more or less efficiently the LSPR energy and favor or not the LSPR-SFG/DFG coupling. The spectral shape of the UV-Visible reflectance conditioned by the silicon reflectivity and the AuNP aggregates as observed in Figure 8c shows that it is possible to tune the incident IR and visible laser beams to fix that point. The best experimental way to answer this question would be to perform 2C-SFG spectroscopy by tuning the incident visible laser beam from blue to red in order to monitor and quantify the LSPR-SFG coupling. This point will be discussed in the section entitled "Recent developments in SFG spectroscopy of plasmonic materials".

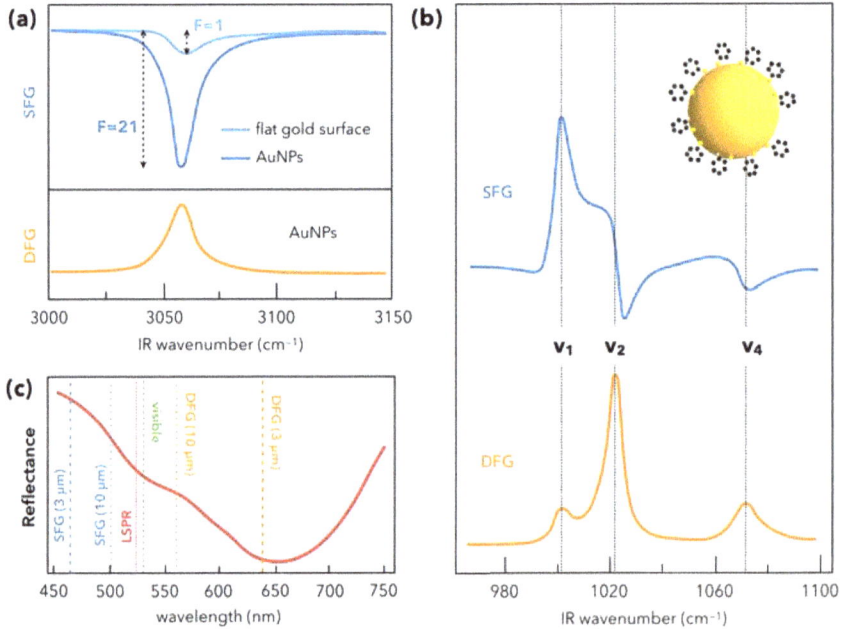

Figure 8. SFG/DFG signals as a function of the UV-Visible optical/electronic properties of nanostructured gold. (**a**) Comparison between thiophenol/Au(111) interface and thiophenol/AuNPs/APTES/Si interface (Visible = 532 nm, IR = 3 µm) in the C–H spectral range (stretching vibration mode of the phenyl ring). In the AuNP case, SFG intensity is enhanced by a factor $F = 21$, corresponding to evidence of LSPR-SFG coupling. The DFG spectrum is also displayed to illustrate the different interference role of the $s - d$ interband electronic transition of gold on the SFG/DFG spectral shapes. (**b**) SFG/DFG spectra of thiophenol/AuNPs/APTES/Si interface (Visible = 532 nm, IR = 10 µm) in the spectral range of C–C vibration modes of the phenyl ring, showing a high SFG and DFG response of this strongly Raman-active modes, also shape-modulated by the interband electronic $s - d$ transition. (**c**) UV-Visible spectrum of the thiophenol/AuNPs/APTES/Si interface. The presence of silicon induces a shape-reversal of the optical response, dominated by the silicon reflectivity. The LSPR maximum is located at 525 nm and couples more or less with the SFG/DFG energy (in addition to the incident visible beam one) as a function of the probed IR spectral range. The LSPR-SFG coupling is favored in the (**b**) case. Figure panel is adapted from references [64,65].

Finally, the monitoring of the AuNPs aggregates depends on their surface coverage on silicon and strongly influences the SFG/DFG spectral shapes as it is discussed exhaustively in the last reference [66] in the series. Considering four different AuNP surface coverages ranging from 1% to 15%, it has been shown that according to the formalism of the effective medium (Maxwell-Garnett or Bruggeman) chosen in the framework of the three-layers model, the deviations to the expected linear law between the vibration mode amplitude (thiophenol C–H streching vibration) and the surface coverage rate by AuNPs could only be explained by the presence of the LSPR. It is worth noting that this linear law is only valid for SFG/DFG if the phase-matching condition between the three beams involved in the non-linear process is satisfied, which is the case in the actual configuration. Otherwise, in the case of dispersed nanoparticles (matrix, solutions), the non-linear phenomena can be scattered in all the directions with respect to the reflected SFG beam coming from a flat surface. In these conditions, the above-mentioned linear relation is not valid [68,69].

3. Recent Developments in SFG Spectroscopy of Plasmonic Materials

3.1. Towards Applications of Gold Nanoparticles

From that moment, thanks to all these fundamental results, it is possible to apply SFG spectroscopy to the study of more complex interfaces related to practical issues. For instance, the increasing role of nanoparticles in consumer goods or medical applications raises questions regarding their possible toxicity for humans. Indeed, as AuNPs are engineered for targeted drug delivery, it is mandatory to understand their physico-chemical interactions with their biological environment, especially in the case of cell damage. SFG spectroscopy/microscopy has already proven to be of great interest in the study of model biointerfaces in the past [70,71]. In fact, the studies of interactions of magnetic nanoparticles [72] or AuNPs [73] with cell membrane models are performed on more or less complex interfaces made of supported lipid bilayers. Indeed, small AuNPs may interact with the cell membrane but the interaction mechanism at the nanoscale is not well understood. To solve these issues, Cecchet's group examined in 2016 how the organization of supported lipid membranes may be affected by the surface charge of small AuNPs (5 nm diameter) and the substrate on which they are adsorbed, all in an aqueous medium [74] as illustrated in Figure 9a. By charging the AuNP surface negatively (Figure 9b) or positively (Figure 9c) depending on their ligands, they showed how the AuNPs induce (or not) structural damages to DPPC (1,2 Dipalmitoyl-*sn*-glycero-3-phosphocholine) membrane models deposited on SiO_2 (negative surface charge) and CaF_2 (positive surface charge) by performing SFG measurements before and after injection of the AuNPs in the water as a function of the exposure time to these latter. It is achieved in the C–H and O–H spectral range to probe the conformational order of the lipid structure and the water molecules in its hydration layer. They observe that all the critical structural changes occur in less than 90 min whether the AuNPs interact (Figure 9c) or not (Figure 9b) with the DPPC bilayer. Moreover, whereas it is not explicitly mentioned in all these works, the LSPR-SFG coupling should also act preponderantly in the high SFG intensity of the vibration modes especially since it is measured in the TIR-SFG configuration with the incident visible beam with its energy matching the SPR one developed by Tourillon as mentioned above [59,60]. They also take advantage of the TIR-SFG configuration in a recent work [75] to extend their investigation to the interaction of cationic AuNPs with supported model membranes carrying negative charges in order to mimic living systems where the outer and inner membrane sides of the cell are negatively charged.

To date, in the research field of AuNPs potentially used as highly sensitive platforms in (bio)sensing for (bio)molecular recognition, there is only one work published in 2015 [34] which treated quantitatively the subject after the emerging work of Tourillon in 2009 [60], but not in TIR-SFG conditions. Contrary to this latter, Dalstein and co-workers choose silicon as a substrate to build a DDT (or thiophenol)/AuNPs/Si interface similar to Figure 2. For such a platform, the idea is to use afterwards silicon as an electronic transducer of the optical SFG response, which requires a good understanding of the optical property's role in the SFG response of the building blocks. In this paper, the focus is put on the deep investigation of the grafting APTES or APTES/MUA (MUA stands for mercaptoundecanoic acid) layer on silicon before and after AuNPs (15 nm diameter) immobilization thanks to this organic layer. The substrate is thus silanized with either amine or mixed aminethiol layer as depicted in Figure 10a. The effect of these small AuNPs on the SFG signal of the grafting layer (6 methyl/methylene stretching vibration modes) is probed by 2C-SFG spectroscopy at different incident visible wavelengths (500 nm, 568 nm, 670 nm) showing a shape-reversal from dips to peaks when tuning the color from blue to red as displayed in Figure 10b. Once again, this interference effect comes from the non-resonant SFG signal of the AuNP interband electronic transition and the silicon optical activity but, this time, it is directly highlighted by tuning the incident visible beam wavelength in the SFG process and not by indirect deduction coming from the SFG and DFG shapes of the vibration modes as evidenced in Figure 8a,c. Thanks to the optical properties of gold, 2C-SFG enables a clear localization of the six weak SFG vibration modes of the disordered APTES layer without taking profit of a TIR-SFG configuration (which is not possible with silicon in the probed optical spectral range).

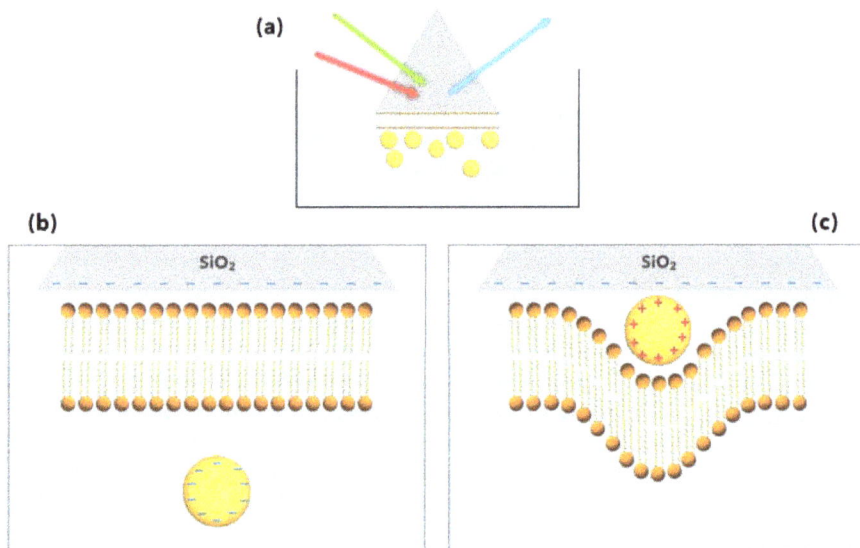

Figure 9. (**a**) Applications of SFG spectroscopy to the study of the interactions of AuNPs with biological material in aqueous medium. (**b**) Effect of a total negative surface charge (SiO$_2$ and AuNPs) on the nanoparticle interaction with a model membrane (lipid bilayer) where no penetration is observed. (**c**) same as (**b**) but for positively charged AuNP surface. This time, the AuNPs crosses and distorts the lipid bilayer to reach the SiO$_2$ surface. Figure panel is reprinted (adapted) with permission from [74], copyright 2016 American Chemical Society.

Despite the presence of APTES or APTES/MUA, it is shown afterwards that the AuNP functionalization by DDT (Vis = 532 nm, Figure 10c) or thiophenol (Vis = 620 nm, Figure 10c) is successful and can be monitored by 2C-SFG spectroscopy even in the presence of a grafting layer. Besides, this works enables to establish a linear relationship between the AuNP absorbance with respect to their surface coverage in UV-Visible spectroscopy, which proves valid up to the aggregation limit: this gives an easy tool to evaluate the AuNP surface coverage of any similar sample from UV-Visible spectroscopy without systematically resorting to Scanning Electron Microscopy (SEM). Moreover, this work shows that the SFG intensities follow a quadratic relation with the UV-Vis absorbance, a direct proof that there exists a close correlation between linear and non-linear optical properties. From the previous work, it is clear that the LSPR-SFG coupling in such interfaces can be probed by 2C-SFG spectroscopy on a wide spectral range of incident visible wavelengths.

(a)

(b)

(c)

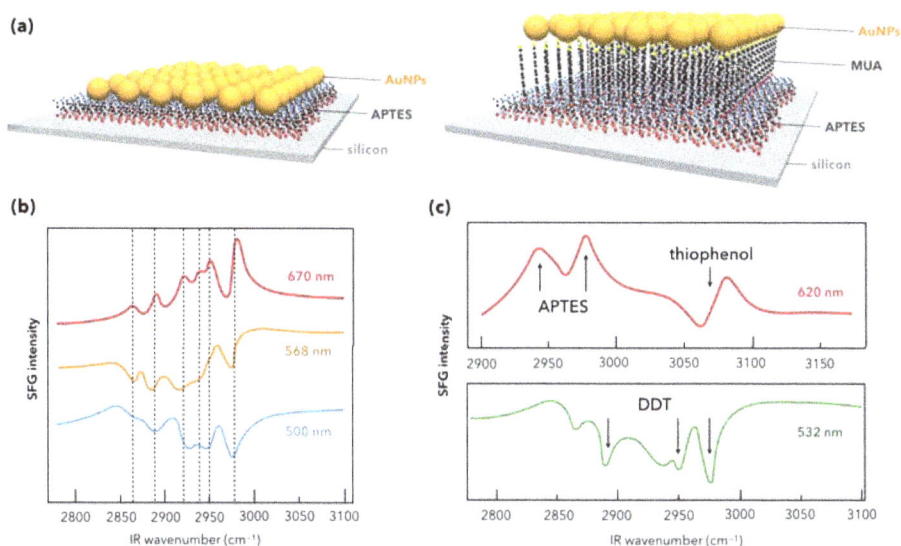

Figure 10. (**a**) AuNPs/APTES/Si (or AuNPs/MUA/APTES/Si) interfaces before functionalization with thiophenol or dodecanethiol (DDT). (**b**) 2C-SFG spectra of the AuNPs/APTES/Si interfaces for three different incident visible wavelengths from blue to red as indicated on the picture. The complete shape-reversal of the APTES vibrational signature between the blue and red curves is well marked. (**c**) SFG spectra for specific incident wavelength revealing the vibrational fingerprint of thiophenol (**top**) or DDT molecules (**bottom**) chemically adsorbed on two different AuNPs/APTES/Si interfaces, respectively. Figure panel is adapted from reference [34].

3.2. Complex Plasmonic Nanostructures

In the previous sections, we have seen that the emphasis has been put on small spherical nanoparticles of gold covered by molecular probes or in interaction with biomaterials. We also observed that different strategies were developed to take advantage of the AuNP LSPR in order to enhance the chemical response in non-linear SFG spectroscopy. While the use of spherical AuNPs is interesting when trying to apply SFG spectroscopy to (bio)chemical or biological applications, the plasmonic contribution is not systematically optimized in such systems. Indeed, as SFG spectroscopy is intrinsically sensitive to the symmetry breaking of the electronic properties of the studied interface, it becomes obvious that spheres (centrosymmetrical materials) are not the best candidate to take full advantage of plasmonics. The point is that as spheres are adsorbed on an interface, the SFG signal appears because of the symmetry breaking related to the different chemistry above and below the AuNPs. In plasmonics, materials can be designed to optimize the SERS signal of highly nanostructured surfaces [76]. This begins to be equally applied in non-linear optics for nanopillars and nanotriangles. In fact, the first 2C-SFG measurements on highly nanostructured platforms were performed on gold nanopillars vertically standing on metallic surfaces (gold or platinum) by Cecchet's group [77]. As displayed in Figure 11a, the nanopillars are electrochemically grown on the substrate with an average height of 106 nm or 100 nm and a mean diameter of 66 nm or 80 nm in the case of gold and platinum, respectively. Within this framework, both samples are functionalized by DDT molecules. Gold nanopillars exhibit two LSPR: a transverse mode (TM) located at 510 nm and a longitudinal mode (LM) located at 710 nm or 660 nm depending on the growth on gold or platinum, respectively. The highlighting of these modes by UV-Visible spectroscopy requires the monitoring of the incident beam polarization (*s* for TM and *p* for LM) as depicted in Figure 11b. In these conditions, it is clear that playing with the polarization scheme in the 2C-SFG process along with the incident visible laser beam wavelength (between 450 and 670 nm) allows fine tuning of the enhancement of each LSPR and therefore optimize the LSPR-SFG coupling as demonstrated in Figure 11c by the enhancement of the molecular SFG signal of the DDT adsorbed on the samples. From these observations, it is shown that the coupling is the most efficient for a *ssp*-polarization scheme with a visible or SFG wavelength in coincidence with the TM LSPR, i.e., 515 nm. It is worth noting that this value of 515 nm is obtained for the SFG wavelength when the incident visible beam is set to 605 nm because the IR probed spectral is centered around 3.45 μm. With respect to the equivalent flat gold or platinum surface covered with DDT, the enhancement factor of the methyl stretching vibration modes in SFG reaches 20 thanks to the nanopillars. A similar (but weaker) behavior is observed when exciting the LM LSPR in *ppp*-polarization scheme but for an incident visible wavelength matching its maximum absorbance at 650 nm. An extended discussion in this paper allows determination that the DDT contribution to the SFG signal mainly comes from the nanopillars due to the local electric field enhancement by the TM and LM LSPR along the lateral surface and the top of the nanopillars, respectively. It is worth noting that the $s - d$ interband electronic transition of gold always interferes with the molecular SFG signal whether there exists an LSPR-SFG coupling or not. Anyway, this work proves that 2C-SFG spectroscopy can probe spatially the chemistry of plasmonic platforms and give complementary information to SERS spectroscopy.

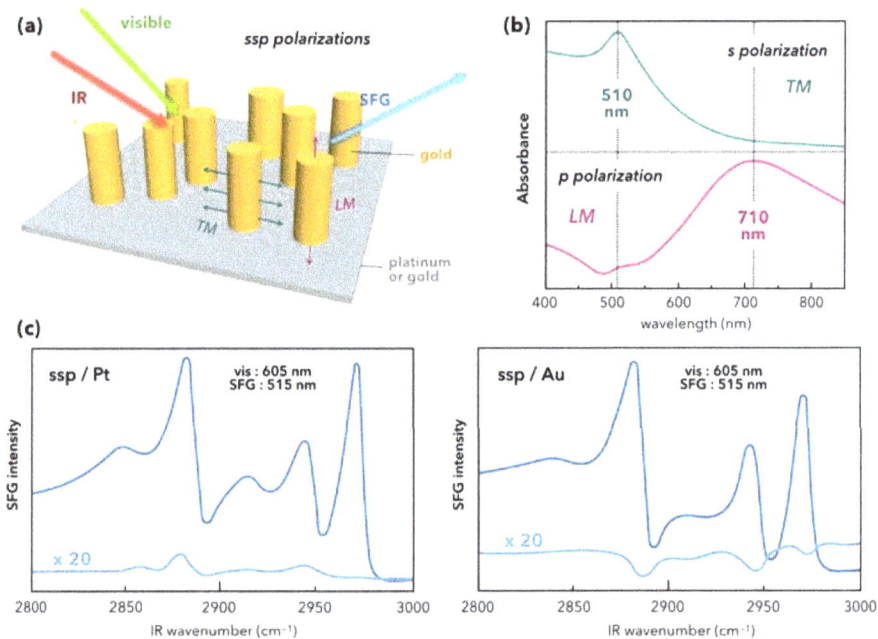

Figure 11. (**a**) Sketch of the 2C-SFG experiment on the gold nanopillars electrochemically grown on a gold or platinum surface in *ssp*-polarization scheme. Samples are functionalized with dodecanethiol (DDT) molecules. The electric yield enhancement of the transverse (TM) and longitudinal (LM) surface plasmon modes is depicted in green and red, respectively. (**b**) UV-Visible absorbance spectra of the sample with a gold substrate. (**c**) Comparison between the SFG spectra of the DDT/Au nanopillars/Pt (**left**) or Au (**right**) interfaces (dark blue) with respect to the DDT/Pt (**left**) or Au (**right**) interfaces (light blue). Figure panel is adapted from reference [77].

To go further, Barbillon and co-workers make the choice in 2018 to use highly ordered nanostructures made of gold nanotriangles (AuNT, height = 60 nm, lateral side = 150 nm) prepared by nanosphere lithography (NSL) technique. The sample consists in a glass substrate covered by a Ti adhesion layer (2 nm) on which a gold film (30 nm) is evaporated as displayed in Figure 12a [78]. The molecular probe is thiophenol, chemically adsorbed on the plasmonic platform. In SERS spectroscopy, its vibrational fingerprint is easily observed in the spectral range of the phenyl ring (10 μm) [19] but remains silent in the C–H spectral range (3 μm). However, it is possible to detect the C–H vibrations modes by SFG spectroscopy, although the highly nanostructured interfaces obtained by NSL do not play in favor of SFG spectroscopy because of its high centrosymmetric character (*p6mm* class symmetry, structured in honeycombs), contrary to the above-mentioned work where nanopillars are randomly dispersed [77].

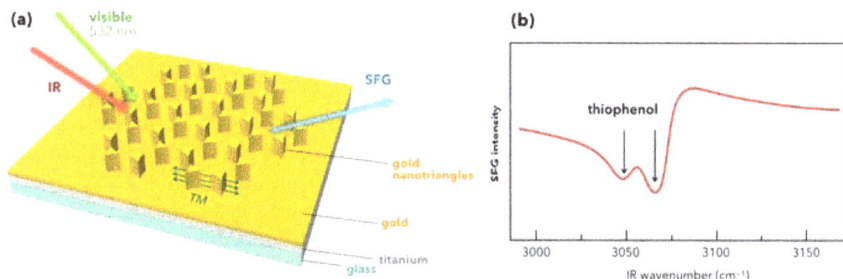

Figure 12. (**a**) Sketch of the SFG experiment performed on gold nanotriangles (AuNT) grafted by nanosphere lithography (NSL) on a gold film (30 nm thickness) deposited on a titanium layer (2 nm thickness) grafted on glass. The AuNT are structured in honeycombs. The AuNT transverse mode (TM) is equally displayed in green. The interface is chemically functionalized by thiophenol molecules. (**b**) SFG spectrum of the sample of (**a**) in *ssp*-polarization, revealing the two C–H (phenyl ring) of the thiophenol. Figure panel is adapted from reference [78].

Nevertheless, as was observed for highly centrosymmetric AuNPs, the authors evidence by SFG in Figure 12b a localized detection of the thiophenol preferentially on the AuNT. It shows up two weak thiophenol ϵ and σ vibration modes [67] at 3050 cm^{-1} and 3071 cm^{-1} (*ssp*-polarization scheme, incident visible wavelength at 532 nm), instead of only one as usually seen [64]. The explanation of the enhancement effect and the LSPR-SFG coupling on AuNT is similar to that encountered for gold nanopillars as explained above because of the TM LSPR mode of AuNT (540 nm). The highlighting of these two vibration modes on a surface is made possible by SFG rather than SERS despite a weak filling factor of AuNT (8%). SERS is mainly useful in solution in order to discriminate between ϵ and σ. The issue is to explain why it works better in SFG than in SERS. It comes from the weak/null enhancement factor (EF) for SERS in the C–H spectral range because of the great spectral redshift (120 nm) between the Raman wavelength (545 nm) and the excitation wavelength (665 nm). In other words, in SERS: $\lambda_{vis} = \lambda_{exc} \neq \lambda_{plasmon}$ with $\lambda_{exc} <<< \lambda_{Raman}$. Conversely, in SFG: $\lambda_{vis} = \lambda_{exc} \approx \lambda_{plasmon}$ (532 nm) with a smaller spectral redshift (70 nm) with respect to λ_{SFG} (460 nm). These simple relations perfectly illustrate the advantages and complementarities of both SERS and SFG spectroscopies. In summary, over the C–H spectral range, (i) for SERS [19]: no infrared excitation in Raman process; very high sensitivity (vs SFG); depends on gold underlayer thickness; no localized detection of molecules in classical configuration; (ii) for SFG: sensitivity depends on surface geometry and is limited in vertical direction due to strong gold $s - d$ interband electronic transitions; infrared and Raman excitations in SFG process; localized detection of molecules in classical configuration. Finally, this work paves the way towards a direct comparison of SERS and SFG vibrational spectroscopies on complex nanostructured metal interfaces, even with a high degree of centrosymmetry, thanks to the presence of distinct LSPR favoring preferential directions as integrated by the non-linear optical response of the interface.

4. Conclusions

Presently, SFG spectroscopy is mostly developed on themes related to chemical and biological issues as for recognition processes leading to a monitored drug delivery for medical issues in the latter case. SFG spectroscopy has become a common optical investigation tool in many research laboratories throughout the world. Restricted to physicists in its early days in the late 1980s, this physical non-linear optical probe of any kind of interface, whatever the nature of its building blocks or its physical state (solid, liquid, gas), has rapidly managed to quickly expand its circle of applications thanks to the involvement of the chemist community. Moreover, it is now possible to find easily commercial femtosecond SFG setups, allowing measurement of dynamic reactions at interfaces immersed or in contact with an aqueous medium. This is also why SFG studies on modified water surface constitute an abundant literature on the subject alongside with specific issues related to biology: protein folding near model cell membranes, flip-flop processes in lipid (hybrid or not) bilayers, as summarized in some references of this review [70,71]. While the study of nanomaterials by SFG has begun 18 years ago, it is today circumscribed to the investigation of interactions in chemical and biological applications. By taking advantage of the LSPR excitation of gold nanoparticles in the SFG process, this latter has the potential to heat the probed interface to facilitate chemical reactions. Optically, when the LSPR energy matches the visible or SFG energy in non-linear optics, it significantly improves the molecular detection threshold. In this review, we have seen that small gold nanoparticles have constituted the core of such SFG studies of nanomaterials so far. It comes from, on the one hand, the presumed nontoxic character of gold for biological applications due to its huge physical and chemical stability and, on the other hand, the fact that its main electronic and optical properties are located in the visible spectral range. The role of its plasmonic and interband electronic properties at nanoscale was the main research field until recently by SFG spectroscopy (picosecond and femtosecond scales) since it is complementary to SERS and PM-IRRAS measurements as non-linear optics provides a different kind of information. Other plasmonic spherical nanoparticles are little or not addressed by SFG due to their lack of chemical stability in air (silver) and their poorer plasmonic properties (copper). The development of 2C-SFG spectroscopy should overcome this limitation within the next years, as it has already been demonstrated on gold for years. Moreover, the use of semiconductor nanoparticles that can be doped and gather free electrons, as those encountered at the surface of metallic nanoparticles, could be relevant if they are combined together in order to manufacture more efficient chemical or biological sensors. Finally, the last part of this review highlights the fact that complementary studies by SERS and SFG spectroscopy of complex nanostructures (nanopillars, nanotriangles) grafted to a substrate open the door to localized non-linear plasmonics. Besides, as the SFG photons are emitted in the visible spectral range, a considerable technical advantage with respect to IR microscopy, it is just a question of time before SFG imaging of nanostructured interfaces be revealed in a close future.

Author Contributions: C.H. wrote all the sections, prepared the original draft, edited the draft; T.N. drew all the Review figures, prepared the original draft, edited the draft; L.D. edited the draft; B.B. edited the draft; G.B. wrote all the sections, prepared the original draft, edited the draft.

Funding: This research received no external funding.

Conflicts of Interest: The authors declare no conflict of interest.

References

1. Daniel, M.-C.; Astruc, D. Gold Nanoparticles: Assembly, Supramolecular Chemistry, Quantum-Size-Related Properties, and Applications toward Biology, Catalysis, and Nanotechnology. *Chem. Rev.* **2004**, *104*, 293–346. [CrossRef]
2. Eustis, S.; El-Sayed, M.A. Why gold nanoparticles are more precious than pretty gold: Noble metal surface plasmon resonance and its enhancement of the radiative and nonradiative properties of nanocrystals of different shapes. *Chem. Soc. Rev.* **2006**, *35*, 209–217. [CrossRef]

3. Das, M.; Shim, K.H.; An, S.S.A.; Yi, D.K. Review on gold nanoparticles and their applications. *Toxicol. Environ. Health Sci.* **2011**, *3*, 193–205. [CrossRef]
4. Miller, J.T.; Kropf, A.J.; Zha, Y.; Regalbuto, J.R.; Delannoy, L.; Louis, C.; Bus, E.; van Bokhoven, J.A. The effect of gold particle size on Au-Au bond length and reactivity toward oxygen in supported catalysts. *J. Catal.* **2006**, *240*, 222–234. [CrossRef]
5. Morel, A.-L.; Boujday, S.; Méthivier, C.; Kraff, J.-M.; Pradier, C.-M. Biosensors elaborated on gold nanoparticles, a PM-IRRAS characterisation of the IgG binding efficiency. *Talanta* **2011**, *85*, 35–42. [CrossRef]
6. Connor, E.E.; Mwamuka, J.; Gole, A.; Murphy, C.J.; Wyatt, M.D. Gold nanoparticles are taken up by human cells but do not cause acute cytotoxicity. *Small* **2005**, *1*, 325–327. [CrossRef]
7. Thompson, D. Michael Faraday's recognition of ruby gold: The birth of modern nanotechnology. *Gold Bull.* **2007**, *40*, 267–269. [CrossRef]
8. Barbillon, G.; Faure, A.-C.; El Kork, N.; Moretti, P.; Roux, S.; Tillement, O.; Ou, M.G.; Descamps, A.; Perriat, P.; Vial, A.; et al. How nanoparticles encapsulating fluorophores allow a double detection of biomolecules by localized surface plasmon resonance and luminescence. *Nanotechnology* **2008**, *19*, 035705. [CrossRef]
9. Barbillon, G.; Bijeon, J.-L.; Lérondel, G.; Plain, J.; Royer, P. Detection of chemical molecules with integrated plasmonic glass nanotips. *Surf. Sci.* **2008**, *602*, L119–L122. [CrossRef]
10. Faure, A.-C.; Barbillon, G.; Ou, M.; Ledoux, G.; Tillement, O.; Roux, S.; Fabregue, D.; Descamps, A.; Bijeon, J.-L.; Marquette, C.A.; et al. Core/shell nanoparticles for multiple biological detection with enhanced sensitivity and kinetics. *Nanotechnology* **2008**, *19*, 485103. [CrossRef]
11. Huang, X.; El-Sayed, M.A. Gold nanoparticles: Optical properties and implementations in cancer diagnosis and photothermal therapy. *J. Adv. Res.* **2010**, *1*, 13–28. [CrossRef]
12. Casaletto, M.P.; Longo, A.; Martorana, A.; Prestianni, A.; Venezia, A.M. XPS study of supported gold catalysts: the role of Au^0 and $Au^{+\delta}$ species as active sites. *Surf. Interface Anal.* **2006**, *38*, 215–218. [CrossRef]
13. Ong, Q.K.; Zhao, S.; Reguera, J.; Biscarini, F.; Stellacci, F. Comparative STM studies of mixed ligand monolayers on gold nanoparticles in air and in 1-phenyloctane. *Chem. Commun.* **2014**, *50*, 10456–10459. [CrossRef]
14. Tian, C.S.; Shen, Y.R. Recent progress on sum-frequency spectroscopy. *Surf. Sci. Rep.* **2014**, *69*, 105–131. [CrossRef]
15. Blaudez, D.; Castano, S.; Desbat, B. PM-IRRAS at liquid interfaces. In *Biointerface Characterization by Advanced IR Spectroscopy*, 1st ed.; Pradier, C.M., Chabal, Y.J., Eds.; Elsevier: Amsterdam, The Netherlands, 2011; pp. 27–55.
16. Stiles, P.L.; Dieringer, J.A.; Shahand, N.C.; Van Duyne, R.P. Surface-Enhanced Raman Spectroscopy. *Annu. Rev. Anal. Chem.* **2008**, *1*, 601–626. [CrossRef]
17. Bryche, J.-F.; Gillibert, R.; Barbillon, G.; Sarkar, M.; Coutrot, A.-L.; Hamouda, F.; Aassime, A.; Moreau, J.; Lamy de la Chapelle, M.; Bartenlian, B.; et al. Density effect of gold nanodisks on the SERS intensity for a highly sensitive detection of chemical molecules. *J. Mater. Sci.* **2015**, *50*, 6601–6607. [CrossRef]
18. Bryche, J.-F.; Gillibert, R.; Barbillon, G.; Gogol, P.; Moreau, J.; Lamy de la Chapelle, M.; Bartenlian, B.; Canva, M. Plasmonic enhancement by a continuous gold underlayer: Application to SERS sensing. *Plasmonics* **2016**, *11*, 601–608. [CrossRef]
19. Bryche, J.-F.; Tsigara, A.; Bélier, B.; Lamy de la Chapelle, M.; Canva, M.; Bartenlian, B.; Barbillon, G. Surface enhanced Raman scattering improvement of gold triangular nanoprisms by a gold reflective underlayer for chemical sensing. *Sens. Actuators B* **2016**, *228*, 31–35. [CrossRef]
20. Arnolds, H.; Bonn, M. Ultrafast surface vibrational dynamics. *Surf. Sci. Rep.* **2010**, *65*, 45–66. [CrossRef]
21. Walter, S.R.; Geiger, F.M. DNA on Stage: Showcasing Oligonucleotides at Surfaces and Interfaces with Second Harmonic and Vibrational Sum Frequency Generation. *J. Phys. Chem. Lett.* **2010**, *1*, 9–15. [CrossRef]
22. Jena, K.C.; Covert, P.A.; Hore, D.K. The Effect of Salt on the Water Structure at a Charged Solid Surface: Differentiating Second- and Third-order Nonlinear Contributions. *J. Phys. Chem. Lett.* **2011**, *2*, 1056–1061. [CrossRef]
23. Kutz, R.B.; Braunschweig, B.; Mukherjee, P.; Dlott, D.D.; Wieckowski, A. Study of Ethanol Electrooxidation in Alkaline Electrolytes with Isotope Labels and Sum-Frequency Generation. *J. Phys. Chem. Lett.* **2011**, *2*, 2236–2240. [CrossRef]
24. Penalber, C.Y.; Baldelli, S. Observation of Charge Inversion of an Ionic Liquid at the Solid Salt–Liquid Interface by Sum Frequency Generation Spectroscopy. *J. Phys. Chem. Lett.* **2012**, *3*, 844–847. [CrossRef]

25. Dalstein, L.; Potapova, E.; Tyrode, E. The elusive silica/water interface: Isolated silanols under water as revealed by vibrational sum frequency spectroscopy. *Phys. Chem. Chem. Phys.* **2017**, *19*, 10343–10349. [CrossRef]

26. Dellwig, T.; Rupprechter, G.; Unterhalt, H.; Freund, H.-J. Bridging the pressure and materials gaps: High pressure sum frequency generation study on supported Pd nanoparticles. *Phys. Rev. Lett.* **2000**, *85*, 776–779. [CrossRef]

27. Rupprechter, G.; Freund, H.-J. Adsorbate-induced restructuring and pressure-dependent adsorption on metal nanoparticles studied by electron microscopy and sum frequency generation spectroscopy. *Top. Catal.* **2000**, *14*, 3–14. [CrossRef]

28. Olson, J.Z.; Johansson, P.K.; Castner, D.G.; Schlenker, C.W. Operando Sum-Frequency Generation Detection of Electrolyte Redox Products at Active Si Nanoparticle Li-Ion Battery Interfaces. *Chem. Mater.* **2018**, *30*, 1239–1248. [CrossRef]

29. Tatsumi, H.; Liu, F.; Han, H.-L.; Carl, L.M.; Sapi, A.; Somorjai, G.A. Alcohol Oxidation at Platinum-Gas and Platinum-Liquid Interfaces: The Effect of Platinum Nanoparticle Size, Water Coadsorption, and Alcohol Concentration. *J. Phys. Chem. C* **2017**, *121*, 7365–7371. [CrossRef]

30. Ouvrard, A.; Ghalgaoui, A.; Michel, C.; Barth, C.; Wang, J.; Carrez, S.; Zheng, W.; Henry, C.R.; Bourguignon, B. CO Chemisorption on Ultrathin MgO-Supported Palladium Nanoparticles. *J. Phys. Chem. C* **2017**, *121*, 5551–5564. [CrossRef]

31. Kawai, T.; Neivandt, D.J.; Davies, P.B. Sum frequency generation on surfactant-coated gold nanoparticles. *J. Am. Chem. Soc.* **2000**, *122*, 12031–12032. [CrossRef]

32. Lis, D.; Cecchet, F. Localized surface plasmon resonances in nanostructures to enhance nonlinear vibrational spectroscopies: Towards an astonishing molecular sensitivity. *Beilstein J. Nanotechnol.* **2014**, *5*, 2275–2292. [CrossRef]

33. Hayashi, M.; Lin, S.H.; Raschke, M.B.; Shen, Y.R. A molecular theory for doubly resonant IR-UV-Vis sum-frequency generation. *J. Phys. Chem. A* **2002**, *106*, 2271–2282. [CrossRef]

34. Dalstein, L.; Ben Haddada, M.; Barbillon, G.; Humbert, C.; Tadjeddine, A.; Boujday, S.; Busson, B. Revealing the interplay between adsorbed molecular layers and gold nanoparticles by linear and nonlinear optical properties. *J. Phys. Chem. C* **2015**, *119*, 17146–17155. [CrossRef]

35. Noblet, T.; Dreesen, L.; Boujday, S.; Méthivier, C.; Busson, B.; Tadjeddine, A.; Humbert, C. Semiconductor quantum dots reveal dipolar coupling from exciton to ligand vibration. *Commun. Chem.* **2018**, *1*, 76. [CrossRef]

36. Frederick, M.T.; Achtyl, J.L.; Knowles, K.E.; Weiss, E.A.; Geiger, F.M. Surface-amplified ligand disorder in CdSe quantum dots determined by electron and coherent vibrational spectroscopies. *J. Am. Chem. Soc.* **2011**, *133*, 7476–7481. [CrossRef]

37. Humbert, C.; Dahi, A.; Dalstein, L.; Busson, B.; Lismont, M.; Colson, P.; Dreesen, L. Linear and nonlinear optical properties of functionalized CdSe quantum dots prepared by plasma sputtering and wet chemistry. *J. Colloid Interface Sci.* **2015**, *445*, 69–75. [CrossRef]

38. Sengupta, S.; Bromley, L.; Velarde, L. Aggregated states of chalcogenorhodamine dyes on nanocrystalline titania revealed by doubly resonant sum frequency spectroscopy. *J. Phys. Chem. C* **2017**, *121*, 3424–3436. [CrossRef]

39. Franken, P.A.; Ward, J.F. Optical harmonics and nonlinear phenomena. *Rev. Mod. Phys.* **1963**, *35*, 23. [CrossRef]

40. Ward, J.F. Calculation of nonlinear optical susceptibilities using diagrammatic perturbation theory. *Rev. Mod. Phys.* **1965**, *37*, 1. [CrossRef]

41. Hirose, C.; Akamatsu, N.; Domen, K. Formulas for the analysis of the surface SFG spectrum and transformation coefficients of cartesian SFG tensor components. *Appl. Spectrosc.* **1992**, *46*, 1051–1072. [CrossRef]

42. Li, X.; Roiaz, M.; Pramhaas, V.; Rameshan, C.; Rupprechter, G. Polarization-Dependent SFG Spectroscopy of Near Ambient Pressure CO Adsorption on Pt(111) and Pd(111) Revisited. *Top. Catal.* **2018**, *61*, 751–762. [CrossRef]

43. Chen, X.; Wang, J.; Boughton, A.P.; Kristalyn, C.B.; Chen, Z. Multiple Orientation of Melittin inside a Single Lipid Bilayer Determined by Combined Vibrational Spectroscopic Studies. *J. Am. Chem. Soc.* **2007**, *129*, 1420–1427. [CrossRef]

44. Lambert, A.G.; Davies, P.B.; Neivandt, D.J. Implementing the Theory of Sum Frequency Generation Vibrational Spectroscopy: A Tutorial Review. *Appl. Spec. Rev.* **2005**, *40*, 103–145. [CrossRef]

45. Caudano, Y.; Silien, C.; Humbert, C.; Dreesen, L.; Mani, A.A.; Peremans, A.; Thiry, P.A. Electron-phonon couplings at C_{60} interfaces: A case study by two-color, infrared-visible sum-frequency generation spectroscopy. *J. Electron Spectrosc. Relat. Phenom.* **2003**, *129*, 139–147. [CrossRef]

46. Kakudji, E.; Silien, C.; Lis, D.; Cecchet, F.; Nouri, A.; Thiry, P.A.; Peremans, A.; Caudano, Y. In situ nonlinear optical spectroscopy of electron–phonon couplings at alkali-doped C_{60}/Ag(111) interfaces. *Phys. Status Solidi B* **2010**, *8*, 1992–1996. [CrossRef]

47. Elsenbeck, D.; Dasa, S.K.; Velarde, L. Substrate influence on the interlayer electron-phonon couplings in fullerene films probed with doubly-resonant SFG spectroscopy. *Phys. Chem. Chem. Phys.* **2017**, *19*, 18519–15528. [CrossRef]

48. Dreesen, L.; Humbert, C.; Sartenaer, Y.; Caudano, Y.; Volcke, C.; Mani, A.A.; Peremans, A.; Thiry, P.A.; Hanique, S.; Frere, J.-M. Electronic and molecular properties of an adsorbed protein monolayer probed by two-colour sum-frequency generation spectroscopy. *Langmuir* **2004**, *20*, 7201–7207. [CrossRef]

49. Covert, P.A.; Hore, D.K. Assessing the Gold Standard: The Complex Vibrational Nonlinear Susceptibility of Metals. *J. Phys. Chem. C* **2015**, *119*, 271–276. [CrossRef]

50. Dalstein, L.; Revel, A.; Humbert, C.; Busson, B. Nonlinear optical response of a gold surface in the visible range: A study by Two-Color Sum-Frequency Generation spectroscopy. I. Experimental determination. *J. Chem. Phys.* **2018**, *148*, 134701. [CrossRef]

51. Mendoza, B.S.; Mochan, W.L.; Maytorena, J.A. Visible-infrared sum and difference frequency generation at adsorbed-covered Au. *Phys. Rev. B* **1999**, *60*, 14334–14340. [CrossRef]

52. Busson, B.; Dalstein, L. Nonlinear optical response of a gold surface in the visible range: A study by two-color sum-frequency generation spectroscopy. III. Simulations of the experimental SFG intensities. *J. Chem. Phys.* **2018**, *149*, 154701. [CrossRef]

53. Humbert, C.; Busson, B.; Abid, J.-P.; Six, C.; Girault, H.H.; Tadjeddine, A. Self-assembled organic monolayers on gold nanoparticles: A study by sum-frequency generation combined with UV-Vis spectroscopy. *Electrochim. Acta* **2005**, *50*, 3101–3110. [CrossRef]

54. Kimling, J.; Maier, M.; Okenve, B.; Kotaidis, V.; Ballot, H.; Plech, A. Turkevich Method for Gold Nanoparticle Synthesis Revisited. *J. Phys. Chem. B* **2006**, *110*, 15700–15707. [CrossRef]

55. Boyd, R.W. *Nonlinear Optics*, 2nd ed.; Academic Press–Elsevier: San Diego, CA, USA, 2003; pp. 1–578.

56. Weeraman, C.; Yatawara, A.K.; Bordenyuk, A.N.; Benderskii, A.V. Effect of Nanoscale Geometry on Molecular Conformation: Vibrational Sum-Frequency Generation of Alkanethiols on Gold Nanoparticles. *J. Am. Chem. Soc.* **2006**, *128*, 14244–14245. [CrossRef]

57. Bordenyuk, A.N.; Weeraman, C.; Yatawara, A.K.; Jayathilake, H.D.; Stiopkin, Y.; Liu, Y.; Benderskii, A.V. Vibrational Sum Frequency Generation Spectroscopy of Dodecanethiol on Metal Nanoparticles. *J. Phys. Chem. C* **2006**, *111*, 8925–8933. [CrossRef]

58. Traverse, A.; Humbert, C.; Six, C.; Gayral, A.; Busson, B. Nonlinear optical properties of Ag nanoparticles embedded in Si_3N_4. *EPL* **2008**, *83*, 64004. [CrossRef]

59. Tourillon, G.; Dreesen, L.; Volcke, C.; Sartenaer, Y.; Thiry, P.A.; Peremans, A. Total internal reflection sum-frequency generation spectroscopy and dense gold nanoparticles monolayer: A route for probing adsorbed molecules. *Nanotechnology* **2007**, *18*, 415301–415308. [CrossRef]

60. Tourillon, G.; Dreesen, L.; Volcke, C.; Sartenaer, Y.; Thiry, P.A.; Peremans, A. Close-packed array of gold nanoparticles and Sum Frequency Generation spectroscopy in total internal reflection: A platform for studying biomolecules and biosensors. *J. Mater. Sci.* **2009**, *44*, 6805–6810. [CrossRef]

61. Fliel, E.R.; van der Ham, E.W.M.; Vrehen, Q.H.F. Enhancing the yield in surface sum-frequency generation by the use of surface polaritons. *Appl. Phys. B* **1999**, *68*, 349–353. [CrossRef]

62. Williams, C.T.; Yan, Y.; Bain, C.D. Total internal reflection sum-frequency spectroscopy: A strategy for studying molecular adsorption on metal surfaces. *Langmuir* **2000**, *16*, 2343–2350. [CrossRef]

63. Dreesen, L.; Sartenaer, Y.; Humbert, C.; Mani, A.A.; Méthivier, C.; Pradier, C.-M.; Thiry, P.A.; Peremans, A. Probing Ligand-Protein Recognition with Sum-Frequency Generation Spectroscopy: The Avidin-Biocytin Case. *ChemPhysChem* **2004**, *5*, 1719–1725. [CrossRef]

64. Pluchery, O.; Humbert, C.; Valamanesh, M.; Lacaze, E.; Busson, B. Enhanced detection of thiophenol adsorbed on Gold Nanoparticles by SFG and DFG nonlinear optical spectroscopy. *Phys. Chem. Chem. Phys.* **2009**, *11*, 7729–7737. [CrossRef]

65. Humbert, C.; Pluchery, O.; Lacaze, E.; Tadjeddine, A.; Busson, B. A Multiscale description of molecular adsorption on gold nanoparticles by nonlinear optical spectroscopy. *Phys. Chem. Chem. Phys.* **2012**, *14*, 280–289. [CrossRef]

66. Humbert, C.; Pluchery, O.; Lacaze, E.; Tadjeddine, A.; Busson, B. Optical spectroscopy of functionalized gold nanoparticles assemblies as a function of the surface coverage. *Gold Bull.* **2013**, *46*, 299–309. [CrossRef]

67. Feugmo, C.G.T.; Liegeois, V. Analyzing the vibrational signatures of thiophenol adsorbed on small gold clusters by DFT calculations. *ChemPhysChem* **2013**, *14*, 1633–1645. [CrossRef]

68. De Aguiar, H.B.; Scheu, R.; Jena, K.C.; de Beer, A.G.F.; Roke, S. Comparison of scattering and reflection SFG: A question of phase matching. *Phys. Chem. Chem. Phys.* **2012**, *14*, 6826–6832. [CrossRef]

69. Roke, S.; Gonella, G. Nonlinear light scattering and spectroscopy of particles and droplets in liquids. *Annu. Rev. Phys. Chem.* **2012**, *63*, 353–378. [CrossRef]

70. Humbert, C.; Busson, B. Sum frequency generation spectroscopy of biointerfaces. In *Biointerface Characterization by Advanced IR Spectroscopy*, 1st ed.; Pradier, C.M., Chabal, Y.J., Eds.; Elsevier: Amsterdam, The Netherlands, 2011; pp. 279–321.

71. Chen, Z. Sum frequency generation vibrational spectroscopy studies on molecular conformation and orientation of biological molecules at interfaces. *Int. J. Mod. Phys. B* **2005**, *19*, 691–713. [CrossRef]

72. Uehara, T.M.; Spolon Marongoni, V.; Pasquale, N.; Miranda, P.B.; Lee, K.-B.; Zucolotto, V.A. Detailed investigation on the interactions between magnetic nanoparticles and cell membrane models. *ACS Appl. Mater. Interfaces* **2013**, *5*, 13063–13068. [CrossRef]

73. Hu, P.; Zhang, X.; Zhang, C.; Chen, Z. Molecular interactions between gold nanoparticles and model cell membranes. *Phys. Chem. Chem. Phys.* **2015**, *17*, 9873–9884. [CrossRef]

74. Toledo-Fuentes, X.; Lis, D.; Cecchet, F. Structural Changes to Lipid Bilayers and Their Surrounding Water upon Interaction with Functionalized Gold Nanoparticles. *J. Phys. Chem. C* **2016**, *120*, 21399–21409. [CrossRef]

75. Toledo-Fuentes, X.; Molinaro, C.; Cecchet, F. Interfacial charges drive the organization of supported lipid membranes and their interaction with nanoparticles. *Colloids Surf. B* **2018**, *172*, 254–261. [CrossRef]

76. Barbillon, G.; Sandana, V.E.; Humbert, C.; Bélier, B.; Rogers, D.J.; Teherani, F.H.; Bove, P.; McClintock, R.; Razeghi, M. Study of Au coated ZnO nanoarrays for surface enhanced Raman scattering chemical sensing. *J. Mater. Chem. C* **2017**, *5*, 3528–3535. [CrossRef]

77. Lis, D.; Caudano, Y.; Henry, M.; Demoustier-Champagne, S.; Ferain, E.; Cecchet, F. Selective Plasmonic Platforms Based on Nanopillars to Enhance Vibrational Sum-Frequency Generation Spectroscopy. *Adv. Opt. Mater.* **2013**, *1*, 244–255. [CrossRef]

78. Barbillon, G.; Noblet, T.; Busson, B.; Tadjeddine, A.; Humbert, C. Localised detection of thiophenol with gold nanotriangles highly structured as honeycombs by nonlinear sum frequency generation spectroscopy. *J. Mater. Sci.* **2018**, *53*, 4554–4562. [CrossRef]

materials

MDPI

Review

Plasmon-Induced Electrocatalysis with Multi-Component Nanostructures

Palaniappan Subramanian [1], Dalila Meziane [2] , Robert Wojcieszak [3] , Franck Dumeignil [3], Rabah Boukherroub [4] and Sabine Szunerits [4,*]

[1] Department of Material Engineering, KU Leuven, Kasteelpark Arenberg 44, P.O. Box 2450, B-3001 Heverlee, Belgium; palan.subramanian@kuleuven.be
[2] Département de Chimie, Faculté des Sciences, Université Mouloud Mammeri, B.P N 17 RP, Tizi Ouzou 15000, Algérie; d_meziane@yahoo.fr
[3] Univ. Lille, CNRS, Centrale Lille, ENSCL, Univ. Artois, UMR 8181-UCCS-Unité de Catalyse et Chimie du Solide, F-59000 Lille, France; robert.wojcieszak@univ-lille.fr (R.W.); franck.dumeignil@univ-lille.fr (F.D.)
[4] Univ. Lille, CNRS, Centrale Lille, ISEN, Univ. Valenciennes, UMR 8520-IEMN, F-59000 Lille, France; rabah.boukherroub@univ-lille.fr
* Correspondence: sabine.szunerits@univ-lille.fr; Tel.: +33-3-6253-1725

Received: 10 December 2018; Accepted: 20 December 2018; Published: 24 December 2018

Abstract: Noble metal nanostructures are exceptional light absorbing systems, in which electron–hole pairs can be formed and used as "hot" charge carriers for catalytic applications. The main goal of the emerging field of plasmon-induced catalysis is to design a novel way of finely tuning the activity and selectivity of heterogeneous catalysts. The designed strategies for the preparation of plasmonic nanomaterials for catalytic systems are highly crucial to achieve improvement in the performance of targeted catalytic reactions and processes. While there is a growing number of composite materials for photochemical processes-mediated by hot charge carriers, the reports on plasmon-enhanced electrochemical catalysis and their investigated reactions are still scarce. This review provides a brief overview of the current understanding of the charge flow within plasmon-enhanced electrochemically active nanostructures and their synthetic methods. It is intended to shed light on the recent progress achieved in the synthesis of multi-component nanostructures, in particular for the plasmon-mediated electrocatalysis of major fuel-forming and fuel cell reactions.

Keywords: plasmonics; catalysis; nanomaterials; electrochemistry; fuel; fuel cells

1. Introduction

Plasmon-accelerated chemical transformation through localized surface plasmon resonance (LSPR) excitation was first experimentally observed during the photocatalytic degradation of formaldehyde (HCHO) and methanol (CH_3OH) on gold nanoparticles (Au NPs) 10 years ago [1]. It was shown that, when irradiated with visible light, Au NPs dispersed on different metal oxide supports (ZrO_2, Fe_2O_3, CeO_2, SiO_2) exhibit significant activity in the oxidation process. Since then, metal-based nanocomposites are widely considered for the construction of photo-redox catalysts. The ability of CdS–Pt nanostructures as photocatalysts for light-driven H_2 production was demonstrated some years later by Wu et al. [2]. Using ultrafast transient absorption spectroscopy, it was demonstrated that the excitons in CdS dissociate by ultrafast electron transfer ($\sim$3.4 ps) to Pt and the charge separated state is long-lived ($\sim$1.2 $\pm$ 0.6 µs) due to hole trapping in CdS. Recently, photocatalytic water splitting using Au NPs was demonstrated using a plasmonic photoelectrode [3], where electron transfer from Au NPs to protons in water generated the photocurrent. As no semiconductor was involved in the catalytic system, no Schottky barrier was formed, and a higher collection efficiency of hot carriers was achieved. These selected examples revealed that harvesting energy from the hot charge carriers of plasmonic structures is promising for energy conversion and photocatalysis alike, paving the way for developments of various catalytic reactions. Most attention has been devoted to plasmon-mediated chemical transformations, and a few reports have investigated plasmon-enhanced electrochemical transformations [4–14]. The progress that has been achieved in the synthesis of materials that are useful for the direct plasmon-accelerated electrocatalysis as well as on metal/semiconductor composites will be outlined here, with the belief that the application of new concepts of electrocatalysis is essential for the advancements of many industrial electrocatalytic processes. These applications might also be of interest for fuel cells, electrochemical sensors, organic electrosynthesis, and so forth. Before describing the synthetic aspects in more detail, we briefly discuss the localized surface plasmon resonance effect and the mechanism of plasmon-enhanced electrocatalysis.

2. Mechanism of Plasmon-Enhanced Electrocatalysis

Metallic nanoparticles that have diameters smaller than the wavelength of light are known to support coherent collective oscillations of delocalized electrons in response to electromagnetic radiation, which are known as localized surface plasmons (LSPs) (Figure 1A). Upon resonant excitation, the collective oscillations of the free electrons give rise to a local electric field enhancement near the surface of the nanoparticles, which strongly concentrates light intensities. The resonances where they occur are named 'localized surface plasmon resonances' (LSPR) to differentiate them from the propagating surface plasmon polaritons of metal surfaces. A tremendous amount of efforts has been devoted to the design of plasmonic structures to adjust the LSPR frequency (λ_{max}), and it is now possible to engineer nanostructures that show LSPR effects from the ultraviolet to the mid-infrared spectral zones (Figure 1B). This is possible as the LSPR properties of plasmonic nanomaterials are strongly dependent on their morphology, size, shape, composition, and even spacing between particle assemblies, allowing the use of several parameters for tuning their λ_{max}. Since the LSPR frequency shifts upon changes of the refractive index of the medium surrounding the particles, embedding metallic nanostructures into other dielectric layers, or forming core–shell plasmonic structures are other means of tuning the LSPR effect.

Following light absorption by nanoparticles and LSPR excitation in the nanoparticles, the plasmons can decay in several competitive pathways (Figure 1C). One is the radiative decay, upon which the plasmon decays into photons, resulting in strong light-scattering effects. This phenomenon is often used for imaging applications and sensing [15,16], and is at the heart of surface-enhanced Raman spectroscopy (SERS) [17,18]. The other route is the non-radiative decay process, which is dominant for small metallic nanoparticles (<40 nm) and leads to the generation of energetic electrons and holes in the plasmonic nanostructures. The excited surface plasmons can then decay by relaxation to generate localized heating effects, which are detrimental for thermal-based applications such as photothermal therapy [19], as well as for plasmon-induced photocatalytic chemical reductions [20]. In addition, excited surface plasmons can transfer "hot" charge carriers to their surroundings, which is primordial for light-driven chemical transformations [20]. Indeed, hot-electron injection is the first mechanism reported for the plasmon-enhanced photoactivity of wide-band gap semiconductors [21,22].

Two situations have to be distinguished in the case of plasmon-mediated electrochemistry: *(1)* pure plasmonic metal nanostructures, and *(2)* metal/semiconductor composites with underlying fundamentally different mechanisms.

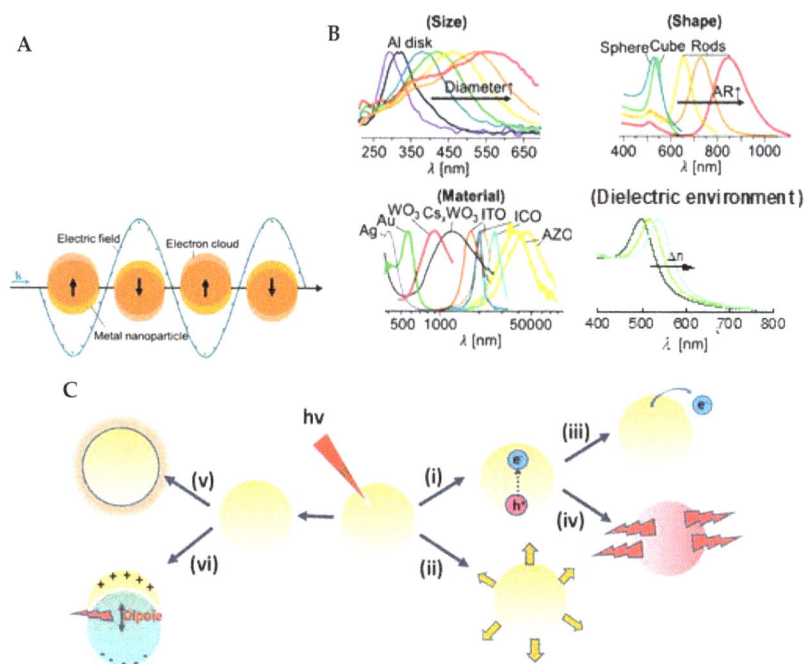

Figure 1. (**A**) Coherent collective oscillations of free electrons of metal nanoparticles in response to light when the diameter of the nanoparticles is smaller than the wavelength of light. (**B**) Plasmonic resonances are engineered by the size [23], shape [24], material composition of nanomaterials [25], and dielectric environment (n = 1.00 to 1.50). (**C**) The decay processes of excited surface plasmon resonance waves: (**i**) non-radiative decay by the excitations of charge carries; (**ii**) radiative decay via scattering, (**iii**) transfer of hot charge carriers to the surrounding; (**iv**) relaxation via heat transfer, (**v**) electromagnetic field enhancement, and (**iv**) dipole resonance energy transfer.

2.1. Indirect Mechanism Using Pure Plasmonic Nanostructures

The plasmon-accelerated electrochemical oxidation of glucose into gluconic was reported by Wang et al. [4] (Figure 2A). An enhanced electrochemical response of glucose oxidation was observed upon the LSPR excitation of Au NPs. Taking into account light intensity, heat effect, and the influence of the LSPR wavelength, the following reaction mechanism was proposed. Upon light absorption and LSPR excitation, the electrons of the Au NPs oscillate collectively and interband excitation occurs, which results in electrons at active states above the Fermi level energy of Au NPs. This excited charge is concentrated on the surface of the Au NPs, and has three possible transfer channels: (*i*) recombination with formed holes, (*ii*) electron transfer, and (*iii*) being removed to the external circuit. In this case, the positive applied potential drives the hot electrons to the external circuit, and the remaining hot holes are driven to the Au NPs surface to accelerate glucose oxidation. This corresponds to the generation and injection of hot charge carriers into adsorbed molecules, which is often identified as an 'indirect' mechanism in the literature.

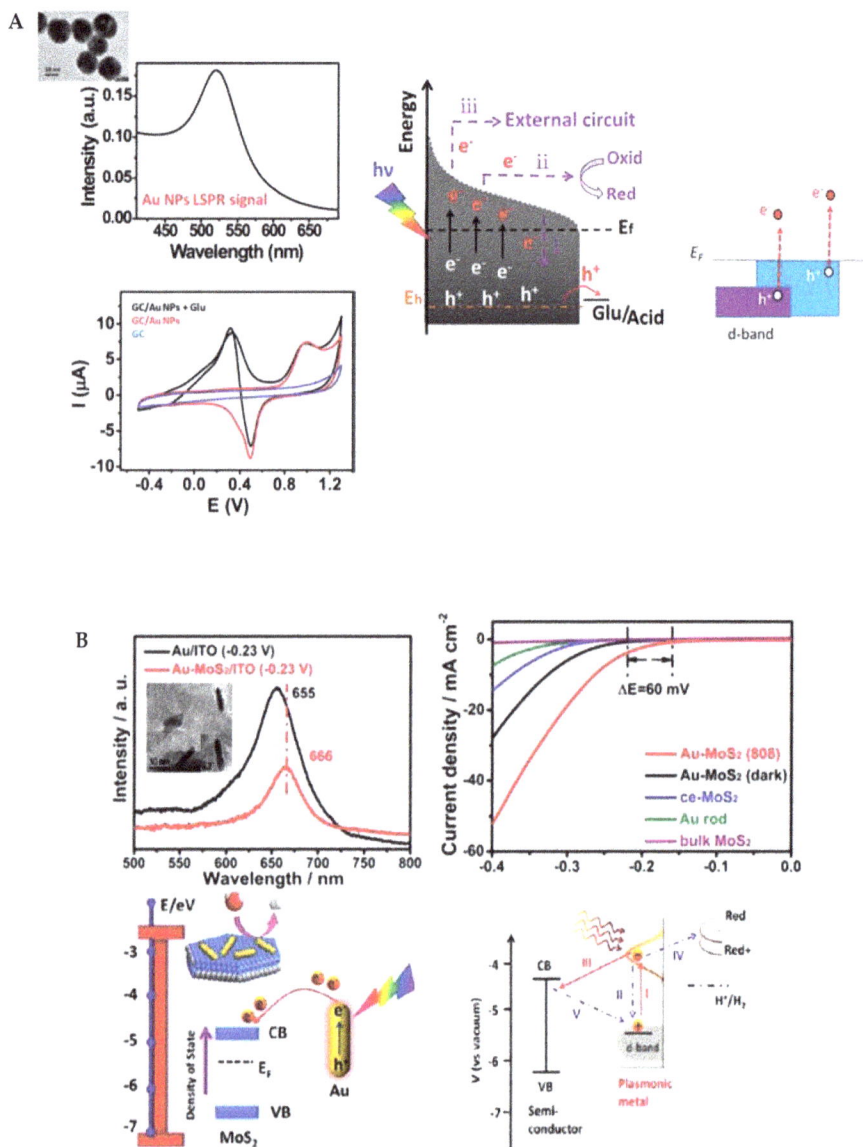

Figure 2. (A) (a) Localized surface plasmon resonance (LSPR) signal of gold nanoparticles (Au NPs) (inset TEM image) and cyclic voltammograms of glassy carbon (GC)/Au NPs in phosphate-buffered saline (PBS) (red) and in glucose (100 mM; black) as well as of GC in PBS (dark blue) together with mechanisms of direct plasmon-accelerated electrochemical reactions using the oxidation of glucose to gluconic acid as an example (reprinted with permission from Ref [4]); **(B)** LSPR spectra of Au rods in Au/Indium Tin Oxide(ITO) and Au−MoS$_2$/ITO (inset: TEM image of Au−MoS$_2$ hybrids and Au NR); polarization curves recorded on Au, MoS$_2$, and Au−MoS$_2$ hybrid (under illumination and dark) and energy level diagram (reprinted with permission from Ref [26]).

2.2. Direct Mechanism by Promotion of an Electron from the Metal to an Empty Molecular Orbital on the Adsorbate

The case of using a non-pure plasmonic substrate but a metal/semiconductor composite is outlined in Figure 2B. The hot electrons are injected from the metal nanoparticles to the conduction band of the semiconductor upon overcoming the Schottky barrier. This process enables the entrapment of hot electrons in the semiconductor particles, and thereby suppresses the electron–hole recombination, promoting redox reactions occurring on the semiconductor nanoparticles. Yi et al. recently demonstrated that the plasmon-excited hot electrons that are generated on Au NRs can be injected to a MoS_2 layer due to the low Schottky barrier between Au NRs and MoS_2 [26]. This system exhibited enhanced electrocatalytic activity toward the hydrogen evolution reaction (HER) due to the increase in charge density on MoS_2 upon the injection of hot electrons. Three probable transfer pathways are described, namely: *(i)* the recombination with holes in Au NRs, *(ii)* injection into the conduction band of the MoS_2, and *(iii)* the direct electrochemical reduction of water on Au NRs by the generated hot electrons (Figure 2B). Semiconductors are used as charge transfer mediators to efficiently collect the excited carriers and thereby promote electrochemical reactions.

3. Synthesis of Plasmonic Electrocatalysts: From Single to Multi-Component Nanostructures

Over the last two decades, a great deal of research has been devoted to the development of nanostructured materials for improved electrocatalysis (Table 1). It is now well established that next to morphology, the size and shape of the nanomaterials are considerably affecting the overall electrocatalytic activity of such systems [5,27]. All of the developed systems focused on electrocatalysis in the dark or under daylight, and a wealth of information on the structure/activity relationship of electrocatalysis on light-induced resonant phenomena have been neglected. Inspired by important advances in using light to enhance photocatalysis [22] as well as photoelectrocatalysis, as recently shown by Thomas et al. using a pure Au NPs plasmonic photo electrode architecture [3], similar effects are expected to be beneficial for plasmon-mediated electrocatalysis.

3.1. Noble Metals-Based Plasmonic Electrocatalysts

Gold (Au), silver (Ag), and copper (Cu)-based nanostructures are the most widely investigated plasmonic nanostructures for electrocatalysis, as they easily allow the tuning of the plasmon resonance from the ultraviolet-visible to the near infrared region, which are the major components of the solar flux. In addition, the advantage of such systems is that the Schottky junction commonly blocking the collection of hot carriers is avoided.

Gold nanoparticles decorating glassy carbon electrodes [4,5] as well as gold nanofiber-based electrodes [11] have been proposed for plasmon-induced electrochemical processes. Wang et al. underlined the importance of hot spots in Au NSs on direct plasmon-enhanced electrochemistry using the oxidation of ascorbic acid as a model [5]. He suggested that an increased number of hot spots, as observed on gold nanostars (Au NSs), results in the best plasmon-enhanced electrochemistry effects.

Silver is another promising material for plasmon-enhanced electrocatalysis due to its high extinction cross-section, but has not been reported so far, which is probably due to the low chemical stability of Ag. Hybrid bimetallic plasmonic nanostructures allow not only a high degree of control over the LSPR decay mechanism [28], but can help to overcome stability issues. Ag–Au nanoparticles were for example used for the plasmon-enhanced electrocatalytic oxidation of glycerol [12]. (Table 1). An Ag–Au catalyst that had been synthesized by a sacrificial support method was proposed recently by Minteer et al. for the enhanced electrocatalytic oxidation of glycerol [12]. The metals were chosen as they are well-known plasmonic materials with LSPR bands in the visible region of the electromagnetic spectrum, and are both able to electrocatalytically oxidize alcohols. When the binary Ag–Au catalyst—immobilized onto carbon-based anode—was illuminated with visible light, a significant increase in current and power output was observed (Figure 3), with an average current density under illumination of 280 $\mu A\ cm^{-2}$ correlating to a power density of 15 $\mu W\ cm^{-2}$. That no heating effect was observed under electrode illumination is in agreement with the concept of hot-electron transfer to glycerol.

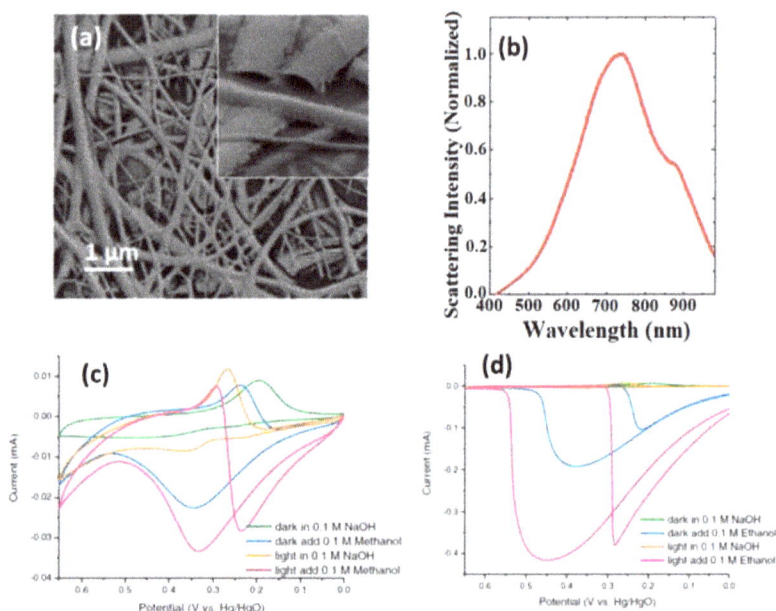

Figure 3. (**a**) SEM image of an Au nanofiber plasmonic electrode; (**b**) UV/Vis spectrum of Au nanofiber electrode; (**c**) cyclic voltammograms of Au nanofiber electrode under light or in the dark recorded in NaOH (0.1 M) in the presence or absence of methanol (0.1 M); (**d**) cyclic voltammograms of Au nanofiber electrode under light or in the dark recorded in NaOH (0.1 M) in the presence or absence of ethanol (0.1 M) (reprinted with the permission of Ref. [11]).

Recently, surface plasmon-enhanced ethylene glycol electrooxidation was conducted by Xu et al. using hollow Pt–Ag nanodendrimers [29]. A 1.7-fold enhancement in catalytic activity under visible light irradiation compared to that under dark conditions was achieved. In addition, 6.2-fold and 7.0-fold enhancements when compared to commercial Pt/C were obtained when the optimized Pt–Ag nanostructures were employed as a photoelectrocatalyst. Indeed, while platinum (Pt) remains one of the most active catalysts for various electrochemical reactions, Pt NPs hardly exhibit an LSPR peak in the visible-light region, limiting their application for plasmon-induced electrochemical applications. The combination of Pt with Ag or Au NPs in the form of bimetallic nanoparticles has thus been investigated in several works [13,29–31].

One approach to increase the performance of a catalytic system is via the incorporation of special metal nanostructures, such as metal-tipped, porous, or needle-like plasmonic structures that have a high electrochemical surface area. Wei et al. recently proposed plasmonic bimetallic structures based on Pt-tipped Au nanorods (Au NRs) for electrochemical water splitting in the visible and the near-infrared region [32]. They demonstrated that these nanostructures outperformed fully Pt-covered nanostructures, which show weak LSPR bands. Pd–Ag hollow nanoflowers have been examined by Du et al. for the electrooxidation of ethylene glycol [33]. Pd atoms were deposited onto the surface of citrate-stabilized Ag seeds during a reducing agent-mediated galvanic replacement process with an electrochemical active surface area, which was determined as 25.8 $m^2 g^{-1}$ for a Pd_1Ag_3–hollow nanostructures, while Pt only exhibits 9.8 $m^2 g^{-1}$.

Materials **2019**, *12*, 43

3.2. Metal–Semiconductor Composites

One of the most widely used semiconductors for photocatalytic applications is titanium, TiO_2, which is low-cost, non-toxic, and has a stable wide band gap (3.2 eV) semiconductor, for which UV light is needed for practical applications. One way to overcome the large energy barrier of TiO_2 is through its hybridization with noble plasmonic nanostructures absorbing in the visible region [34–37]. One of the first reports was proposed by Xu et al. [36], who decorated highly ordered TiO_2 nanotube arrays (TiO_2 NTs) with Au NPs (1.9 at.%) for enhanced ethanol oxidation. To further enhance the catalytic reaction, a bilayer titanium dioxide nanotube (BTNT) decorated periodically with Au NPs was proposed [35]. This heterostructured plasmonic electrode allowed ethanol electrooxidation under visible light illumination with a maximum catalytic current reaching 1.11 mA cm^{-2}, which was 3.6-fold higher than that of conventional Au NPs-decorated monolayer TiO_2 nanotubes. Placing Au NPs into TiO_2 nanocavity arrays resulted in plasmonic electrochemical interfaces with superior oxygen reduction reaction (ORR) activity [34]. It was demonstrated that a 5 nm gold layer deposited on TiO_2 delivered a superior ORR performance with an onset potential of 0.92 V versus reversible hydrogen electrode (RHE), a limiting current density of 5.2 mA cm^{-2}, and an electron transfer number of 3.94. The enhanced reductive activity is attributed to the LSPR effect of isolated Au NPs in TiO_2 nanocavities which suppressed electron recombination. MnO_2 nanosheets decorated with Au NPs were reported by Xu et al. to be an ideal plasmonic electrocatalyst for the oxygen evolution reaction (OER) (Figure 4A) [38]. The confinement of the outer electrons of the Mn cations by plasmonic "hot holes" that were generated on the Au NPs' surface was largely promoted under green light illumination. These hot holes act as efficient electron traps to form active Mn^{n+} species, providing active sites to extract electrons from OH^- and eventually facilitate OER catalysis.

Transition-metal catalysts [8] and metal–organic frameworks (MOFs) [10] are attractive alternatives for the oxygen evolution reaction. Liu et al. demonstrated that when Au NPs are decorated with transition-metal catalysts, such as $Ni(OH)_2$ nanosheets (Figure 4B), they form a plasmonic electrocatalyst, which, upon light illumination, enhances the charge transfer from $Ni(OH)_2$ to Au NPs, and greatly facilitates the oxidation of inactive Ni^{2+} to active $Ni^{3+/4+}$ species, allowing more efficient water oxidation at a lower onset potential [8].

Transition-metal disulfides such as MoS_2, on the other hand, are among the attractive alternatives for catalyzing hydrogen evolution reactions (HERs) [26]. To overcome the limitation of inherently low interparticle conductivity, Shi et al. proposed the incorporation of Au NRs in MoS_2 [26]. The authors found that Au@MoS_2 hybrids drastically improve the HER, with a three-fold increase of the current under excitation of the Au LSPR bands.

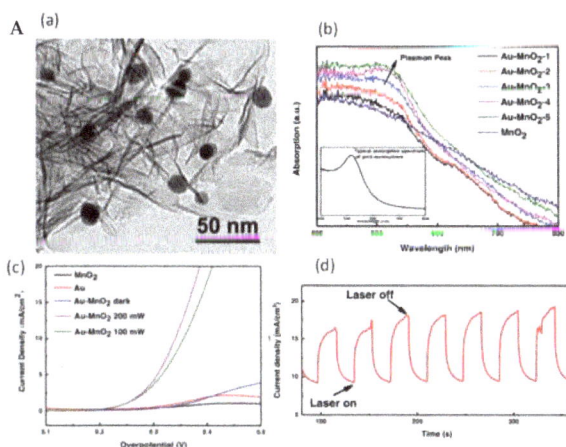

Figure 4. *Cont.*

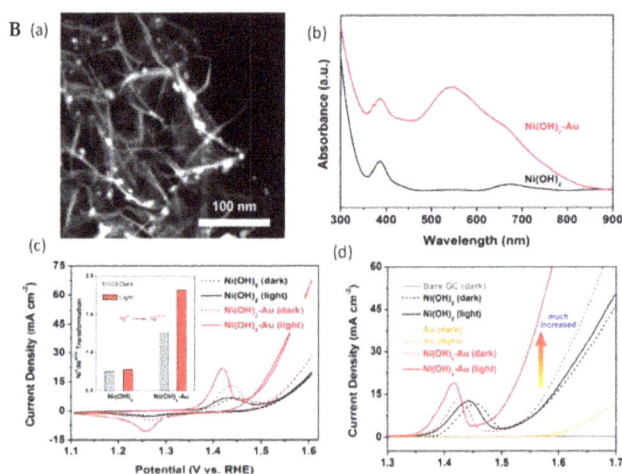

Figure 4. (**A**) (**a**) TEM image of Au–MnO$_2$ nanocomposite, (**b**) UV/Vis absorption spectra of MnO$_2$ nanosheets and Au–MnO$_2$ nanocomposites with various Au loading (inset: LSPR band of gold nanospheres), (**c**) Polarization curve in 0.1 M of KOH with and without 532-nm laser irradiation, (**d**) Chronoamperometric I–t curve of Au@MnO$_2$ nanocomposites with 532-nm laser on and off (reprinted with the permission of Ref. [9]); (**B**) (**a**) HAADF-STEM image of Ni(OH)$_2$–Au hybrid catalyst; (**b**) UV/Vis absorption spectra of Ni(OH)$_2$ nanosheets and Ni(OH)$_2$–Au hybrid catalyst, (**c**) cyclic voltammograms with and without light irradiation of Ni(OH)$_2$ nanosheets and Ni(OH)$_2$–Au hybrid catalyst, and (**d**) oxygen evolution reaction (OER) polarization curves at 10 mV s^{-1} for different electrodes in dark and under light irradiation (532-nm laser, 1.2 W) in 1 M KOH; Ag/AgCl (Saturated KCl) was used as reference electrode (reprinted with the permission of Ref. [8]).

Table 1. Plasmon-mediated electrochemical catalysis.

Plasmonic Catalyst	Electrode	Reaction	Electrolyte	Comments	Ref.
Noble metal-based plasmonic electrocatalysts					
Au NPs	GCE	glucose oxidation	PBS (pH 13.7)	High alkaline conditions to scavenge holes by OH$^-$	[4]
Au nanofiber	GCE	ethanol and methanol oxidation	0.1 M NaOH	Decreased passivation effects	[11]
Ag–Au NPs	GCE	glycerol oxidation	0.1 M NaOH	100% fuel cell power output under visible light	[12]
Au NPs, Au NRs, Au NSs	GCE	ascorbic acid oxidation	PBS (pH 7.4)	Au NPs have weakest effect	[5]
Pt–Ag dendrites	GCE	ethylene glycol oxidation	1.0 M KOH	1.7-fold increase in catalytic activity under light	[29]
Au–Pt NPs	FTO	ethanol oxidation	1.0 M NaOH	2.6 times enhancement	[30]
Ag–Pt nanocages	GCE	ORR	0.1 M KOH	"Hot" electron transfer suppressed formation of peroxide intermediate	[13]
Pt/Fe–Au NRs	GCE	HER	0.5 M H$_2$SO$_4$ 1.0 M KOH	Photothermal effect results in electrocatalysis enhancement	[31]
Pd-tipped Au NRs	GCE	HER	0.5 M H$_2$SO$_4$	High exchange current density of 1.585 mA/cm^2	[32]
PdAg hollow nanoflowers	GCE	Ethylene glycol oxidation	1.0 M KOH	High active surface area of 25.8 m^2 g^{-1} (Pt 9.8 m^2 g^{-1})	[33]

Table 1. *Cont.*

Plasmonic Catalyst	Electrode	Reaction	Electrolyte	Comments	Ref.
Plasmonic metal–semiconductor composites					
Au–TiO$_2$	GCE	ORR	0.1 M NaOH	Activity of 310 mA mg^{-1}	[34]
Au–TiO$_2$ nanotubes	Ti foil	ethanol oxidation	0.5 M H$_2$SO$_4$ 1.0 M KOH	3.6-fold increase with low Au NPs (1.9 at.%)	[35]
Au–MnO$_2$ NPs	GCE	OER	0.1 M KOH	60-mV overpotential	[9]
Ni(OH)$_2$–Au	GCE	OER	1 M KOH	Four-fold enhancement, Tafel slope of 35 mV dec^{-1}	[8]
Au–Co/NiMOF	GCE	OER	1 M KOH	10-fold increase	[10]
Au–CuI NPs	GCE	ethanol oxidation and methylene blue (MB) degradation	1 M KOH	5.6 (ethanol) and 13 times (MB) enhanced activity.	[39]
Au–MoS$_2$	GCE	HER	0.5 M H$_2$SO$_4$	~three-fold increase, turnover of 8.76 s^{-1} at 300 mV	[26]
TiN and doped graphene	GCE	HER	0.5 M H$_2$SO$_4$	Attained an HER current density of 10 mA/cm^2 at a low overpotential of 161 mV.	[14]
Au NP@rGO layer@Pd NS	GCE	Water splitting (OER and HER)	0.1 M KOH	Under visible light irradiation 1.9 and 1.1-fold enhanced HER and OER activity, respectively.	[40]

Au NPs: gold nanoparticles; Au NRs: gold nanorods; Au NSs: gold nanostars, HER: Hydrogen evolution reaction; ORR: Oxygen reduction reaction; OER: Oxygen evolution reaction; GCE: glassy carbon electrode; FTO: Flourine doped tin oxide; MB: Methylene blue; MOF: metal-organic frameworks; PBS: phosphate-buffered saline.

4. Current Trends and Outlook

This short review summarizes the recent developments on nanomaterials for plasmon-enhanced electrochemical reactions with the aim of interesting the communities working in plasmonic and electrochemical processes, providing a common base for jointly progressing in this exciting area of plasmon-mediated electrochemistry. To understand better the design of the nanostructures, the physical fundamental of localized surface plasmon resonance and the various mechanisms for plasmon-enhanced electrochemistry have been provided. Despite the significant advances achieved in the last three years, researchers are facing many challenges in this field. While a large variety of synthetic methods have been developed for the synthesis of these heterostructures, the scale-up of such processes will be an important and imperative aspect for the use of these concepts on a wider scale. The formation of highly reproducible nanostructures with comparable catalytic and plasmonic properties is at the core of the development to envision scale-up, which is an issue that has not been yet resolved. While laser light is often used for the stimulation of the plasmonic effect, developing plasmonic materials that are responsive to sunlight with high catalytic activity represents an important goal in the field of plasmon-mediated chemical/electrochemical reactions. Which guidelines for the future design of plasmonic electrochemical materials can be provided? The formations of bimetallic and metal/semiconductor nanostructures have both shown to be of great promise for plasmon-enhanced electrochemical systems, taking advantage of the catalytic activity and the strong optical effects of these nanostructures. While such simple plasmonic nanostructures have been demonstrated as concentrating light efficiently from the UV to the near-infrared range of the light spectrum and transferring it to adjacent species, thus improving electrochemical transformations, a better understanding of the underlying mechanism of plasmon-enhanced electrochemistry is primordial for optimizing electrochemical-based oxidation and reduction processes. How to separate the generated hot electrons from the holes in an efficient and controlled manner is one of the critical criteria to be investigated, as it is fundamental for enhancing electrochemical reactions. Next to this, the importance of plasmonic heating in plasmon-enhanced electrocatalysis has to be systematically studied. How to distinguish the contribution from the electromagnetic field-enhancement from hot charge carriers-induced enhancement are big experimental and theoretical challenges to be addressed. Further studies focusing on the morphology, composition, heterojunctions, and other nanomaterials-based aspects need to be conducted in order to fully optimize the plasmon-enhanced electrochemical effects. The future of plasmon-mediated electrochemistry might be full of surprises.

Author Contributions: P.S.: writing Sections 2.2 and 3.2; D.M.: writing Section 3.1; R.B.: editing of the draft; F.D.: editing of the draft, catalysis expertise; R.W.: editing of the draft, catalysis expertise; S.S.: writing introduction, original draft preparation.

Funding: This research received no external funding.

Acknowledgments: Financial supports from the Centre National de la Recherche Scientifique (CNRS), the University of Lille, the Hauts-de-France region, Chevreul Institute (FR 2638), FEDER and the CPER "Photonics for Society", are acknowledged.

Conflicts of Interest: The authors declare no conflict of interest.

References

1. Chen, X.; Zhu, H.Y.; Zhao, J.C.; Zheng, Z.F.; Gao, X.P. Visible-light-driven oxidation of organic contaminants in air with gold nanoparticle catalysts on oxide supports. *Angew. Chem. Int. Ed.* **2008**, *47*, 5353–5356. [CrossRef] [PubMed]

2. Wu, K.; Zhu, H.; Liu, Z.; Rodríguez-Córdoba, W.; Lian, T. Ultrafast charge separation and long-lived charge separated state in photocatalytic CdS–Pt nanorod heterostructures. *J. Am. Chem. Soc.* **2012**, *134*, 10337–10340. [CrossRef] [PubMed]

3. Robatjazi, H.; Bahauddin, S.M.; Doiron, C.; Thomann, I. Direct plasmon-driven photoelectrocatalysis. *Nano Lett.* **2015**, *15*, 6155–6161. [CrossRef] [PubMed]

4. Wang, C.; Nie, X.-G.; Shi, Y.; Zhou, Y.; Xu, J.-J.; Xia, X.-H.; Chen, H.-Y. Direct plasmon-accelerated electrochemical reaction on gold nanoparticles. *ACS Nano* **2017**, *11*, 5897–5905. [CrossRef] [PubMed]

5. Wang, C.; Zhao, X.-P.; Xu, Q.-Y.; Nie, X.-G.; Younis, M.R.; Liu, W.-Y.; Xia, X.-H. Importance of Hot Spots in Gold Nanostructures on Direct Plasmon-Enhanced Electrochemistry. *ACS Appl. Nano Mater.* **2018**, *1*, 5805–5811. [CrossRef]

6. Wang, C.; Shi, Y.; Yang, D.-R.; Xia, X.-H. Combining plasmonics and electrochemistry at the nanoscale. *Curr. Opin. Electrochem.* **2017**, *7*, 95–102. [CrossRef]

7. Choi, C.H.; Chung, K.; Nguyen, T.-T.; Kim, D.H. Plasmon-mediated electrocatalysis for sustainable energy: From electrochemical conversion of different feedstocks to fuel cell reactions. *ACS Energy Lett.* **2018**, *3*, 1415–1433. [CrossRef]

8. Liu, G.; Li, P.; Zhao, G.; Wang, X.; Kong, J.; Liu, H.; Zhang, H.; Chang, K.; Meng, X.; Kako, T. Promoting active species generation by plasmon-induced hot-electron excitation for efficient electrocatalytic oxygen evolution. *J. Am. Chem. Soc.* **2016**, *138*, 9128–9136. [CrossRef]

9. Xu, J.; Gu, P.; Birch, D.J.; Chen, Y. Plasmon-Promoted Electrochemical Oxygen Evolution Catalysis from Gold Decorated MnO$_2$ Nanosheets under Green Light. *Adv. Func. Mater.* **2018**, *28*, 1801573. [CrossRef]

10. Wang, M.; Wang, P.; Li, C.; Li, H.; Jin, Y. Boosting Electrocatalytic Oxygen Evolution Performance of Ultrathin Co/Ni-MOF Nanosheets via Plasmon-Induced Hot Carriers. *ACS Appl. Mater. Interfaces* **2018**, *10*, 37095–37102. [CrossRef]

11. Chen, D.; Zhang, R.; Wang, R.; Dal Negro, L.; Minteer, S.D. Gold Nanofiber-Based Electrodes for Plasmon-Enhanced Electrocatalysis. *J. Electrochem. Soc.* **2016**, *163*, H1132–H1135. [CrossRef]

12. Rasmussen, M.; Serov, A.; Artyushkova, K.; Chen, D.; Rose, T.C.; Atanassov, P.; Harris, J.M.; Minteer, S.D. Enhancement of Electrocatalytic Oxidation of Glycerol by Plasmonics. *ChemElectroChem* **2018**. [CrossRef]

13. Lin, S.-C.; Hsu, C.-S.; Chiu, S.-Y.; Liao, T.-Y.; Chen, H.M. Edgeless Ag–Pt Bimetallic Nanocages: In Situ Monitor Plasmon-Induced Suppression of Hydrogen Peroxide Formation. *J. Am. Chem. Soc.* **2017**, *139*, 2224–2233. [CrossRef] [PubMed]

14. Shanker, G.S.; Markad, G.B.; Jagadeeswararao, M.; Bansode, U.; Nag, A. Colloidal Nanocomposite of TiN and N-Doped Few-Layer Graphene for Plasmonics and Electrocatalysis. *ACS Energy Lett.* **2017**, *2*, 2251–2256. [CrossRef]

15. Sil, D.; Gilroy, K.D.; Niaux, A.; Boulesbaa, A.; Neretina, S.; Borguet, E. Seeing is believing: Hot electron based gold nanoplasmonic optical hydrogen sensor. *ACS Nano* **2014**, *8*, 7755–7762. [CrossRef] [PubMed]
16. Szunerits, S.; Boukherroub, R. Sensing using localised surface plasmon resonance sensors. *Chem. Commun.* **2012**, *48*, 8999–9010. [CrossRef] [PubMed]
17. Tabatabaei, M.; McRae, D.; Laguné-Labarthet, F. Recent advances of plasmon-enhanced spectroscopy at bio-Interfaces Frontiers of Plasmon Enhanced Spectroscopy. *ACS Symp. Ser.* **2016**, *2*, 183–207.
18. Demirel, G.; Usta, H.; Yilmaz, M.; Celik, M.; Alidagi, H.A.; Buyukserin, F. Surface-enhanced Raman spectroscopy (SERS): An adventure from plasmonic metals to organic semiconductors as SERS platforms. *J. Mater. Chem. C* **2018**, *6*, 5314–5335. [CrossRef]
19. Huang, X.; El-Sayed, M.A. Plasmonic photo-thermal therapy (PPTT). *Alex. J. Med.* **2011**, *47*, 1–9. [CrossRef]
20. Golubev, A.A.; Khlebtsov, B.N.; Rodriguez, R.D.; Chen, Y. Plasmonic Heting Plays a dominant role in the plasmon-induced photocatalytic reduction of 4-nitrobenzenethiol. *J. Phys. Chem. C* **2018**, *122*, 5657–5663. [CrossRef]
21. Tian, Y.; Tatsuma, T. Plasmon-induced photoelectrochemistry at metal nanoparticles supported on nanoporous TiO_2. *Chem. Commun.* **2004**, 1810–1811. [CrossRef] [PubMed]
22. Zhang, Y.; He, S.; Guo, W.; Hu, Y.; Huang, J.; Mulcahy, J.R.; Wei, W.D. Surface-plasmon-driven hot electron photochemistry. *Chem. Rev.* **2017**, *118*, 2927–2954. [CrossRef] [PubMed]
23. Knight, M.W.; King, N.S.; Liu, L.; Everitt, H.O.; Nordlander, P.; Halas, N.J. Aluminum for plasmonics. *ACS Nano* **2013**, *8*, 834–840. [CrossRef] [PubMed]
24. Chen, H.; Kou, X.; Yang, Z.; Ni, W.; Wang, J. Shape-and size-dependent refractive index sensitivity of gold nanoparticles. *Langmuir* **2008**, *24*, 5233–5237. [CrossRef] [PubMed]
25. Lounis, S.D.; Runnerstrom, E.L.; Llordes, A.; Milliron, D.J. Defect chemistry and plasmon physics of colloidal metal oxide nanocrystals. *J. Phys. Chem. Lett.* **2014**, *5*, 1564–1574. [CrossRef] [PubMed]
26. Shi, Y.; Wang, J.; Wang, C.; Zhai, T.-T.; Bao, W.-J.; Xu, J.-J.; Xia, X.-H.; Chen, H.-Y. Hot electron of Au nanorods activates the electrocatalysis of hydrogen evolution on MoS_2 nanosheets. *J. Am. Chem. Soc.* **2015**, *137*, 7365–7370. [CrossRef]
27. Lv, H.; Li, D.; Strmcnik, D.; Paulikas, A.P.; Markovic, N.M.; Stamenkovic, V.R. Recent advances in the design of tailored nanomaterials for efficient oxygen reduction reaction. *Nano Energy* **2016**, *29*, 149–165. [CrossRef]
28. Aslam, U.; Chavez, S.; Linic, S. Controlling energy flow in multimetallic nanostructures for plasmonic catalysis. *Nat. Nanotechnol.* **2017**, *12*, 1000–1005. [CrossRef]
29. Xu, H.; Song, P.; Fernandez, C.; Wang, J.; Shiraishi, Y.; Wang, C.; Du, Y. Surface plasmon enhanced ethylene glycol electrooxidation based on hollow platinum-silver nanodendrites structures. *J. Taiwan Inst. Chem. Eng.* **2018**, *91*, 316–322. [CrossRef]
30. Yang, H.; He, L.-Q.; Hu, Y.-W.; Lu, X.; Li, G.-R.; Liu, B.; Ren, B.; Tong, Y.; Fang, P.-P. Quantitative Detection of Photothermal and Photoelectrocatalytic Effects Induced by SPR from Au@Pt Nanoparticles. *Angew. Chem. Int. Ed.* **2015**, *54*, 11462–11466. [CrossRef]
31. Guo, X.; Li, X.; Kou, S.; Yang, X.; Hu, X.; Ling, D.; Yang, J. Plasmon-enhanced electrocatalytic hydrogen/oxygen evolution by Pt/Fe–Au nanorods. *J. Mater. Chem A* **2018**, *6*, 7364–7369. [CrossRef]
32. Wei, Y.; Zhao, Z.; Yang, P. Pd-Tipped Au Nanorods for Plasmon-Enhanced Electrocatalytic Hydrogen Evolution with Photoelectric and Photothermal Effects. *ChemElectroChem* **2018**, *5*, 778–784. [CrossRef]
33. Bin, D.; Yang, B.; Zhang, K.; Wang, C.; Wang, J.; Zhong, J.; Feng, Y.; Guo, J.; Du, Y. Design of PdAg hollow nanoflowers through galvanic replacement and their application for ethanol electrooxidation. *Chem. Eur. J.* **2016**, *22*, 16642–16647. [CrossRef]
34. Guo, L.; Liang, K.; Marcus, K.; Li, Z.; Zhou, L.; Mani, P.D.; Chen, H.; Shen, C.; Dong, Y.; Zhai, L.; et al. Enhanced Photoelectrocatalytic Reduction of Oxygen Using Au@TiO_2 Plasmonic Film. *ACS Appl. Mater. Interfaces* **2016**, *8*, 34970–34977. [CrossRef]
35. Jin, Z.; Wang, Q.; Zheng, W.; Cui, X. Highly Ordered Periodic Au/TiO_2 Hetero-Nanostructures for Plasmon-Induced Enhancement of the Activity and Stability for Ethanol Electro-oxidation. *ACS Appl. Mater. Interfaces* **2016**, *8*, 5273–5279. [CrossRef]
36. Xu, Z.; Yu, J.; Liu, G. Enhancement of ethanol electrooxidation on plasmonic Au/TiO_2 nanotube arrays. *Electrochem. Commun.* **2011**, *13*, 1260–1263. [CrossRef]
37. Long, R.; Prezhdo, O.V. Instantaneous generation of charge-seperation state on TiO_2 surface sensitized with plasmonic nanoparticles. *J. Am. Chem. Soc.* **2014**, *136*, 4343–4354. [CrossRef] [PubMed]

38. Kang, Y.; Najmaei, S.; Liu, Z.; Bao, Y.; Wang, Y.; Zhu, X.; Halas, N.J.; Nordlander, P.; Ajayan, P.M.; Lou, J.; et al. Plasmonic Hot Electron Induced Structural Phase Transition in a MoS_2 Monolayer. *Adv. Mater.* **2014**, *26*, 6467–6471. [CrossRef] [PubMed]
39. Sun, M.; Zhai, C.; Hu, J.; Zhu, M.; Pan, J. Plasmon enhanced electrocatalytic oxidation of ethanol and organic contaminants on gold/copper iodide composites under visible light irradiation. *J. Colloid. Interface Sci.* **2018**, *511*, 110–118. [CrossRef] [PubMed]
40. Lee, J.-E.; Marques Mota, F.; Choi, C.H.; Lu, Y.-R.; Boppella, R.; Dong, C.-L.; Liu, R.-S.; Kim, D.H. Plasmon-Enhanced Electrocatalytic Properties of Rationally Designed Hybrid Nanostructures at a Catalytic Interface. *Adv. Mater. Interfaces* **2018**. [CrossRef]

MDPI
St. Alban-Anlage 66
4052 Basel
Switzerland
Tel. +41 61 683 77 34
Fax +41 61 302 89 18
www.mdpi.com

Materials Editorial Office
E-mail: materials@mdpi.com
www.mdpi.com/journal/materials